# FOUNDATIONS OF LASER SPECTROSCOPY

STIG STENHOLM

*Physics Department, KTH, Stockholm, Sweden, and*
*Laboratory of Computational Engineering, HUT, Espoo, Finland*

DOVER PUBLICATIONS, INC.
Mineola, New York

*Copyright*



*Bibliographical Note*

This Dover edition, first published in 2005, is an augmented republication of the work originally published by John Wiley & Sons, Inc., New York, in 1984. The author has provided a new Preface to the Dover Edition, a Bibliography Update, and an Errata list (pages xi-xiii).

*Library of Congress Cataloging-in-Publication Data*

Stenholm, Stig.
Foundations of laser spectroscopy / Stig Stenholm.
p. cm.
"An augmented republication of the work originally published by John Wiley & Sons, Inc., New York, in 1984. The author has provided a new preface to the Dover edition, a bibliography update, and an errata list (pages xi-xiii)"—T.p. verso.
Includes bibliographical references and index.
ISBN 0-486-44498-8 (pbk.)
1. Laser spectroscopy. I. Title.

QC454.L3S75 2005
621.36'6—dc22

2005049733

Manufactured in the United States of America
Dover Publications, Inc., 31 East 2nd Street, Mineola, N.Y. 11501

So what is truth? A mobile army of metaphors, metonyms, anthropomorphisms—in short an aggregate of human relationships which, poetically and rhetorically heightened, become transposed and elaborated, and which, after protracted popular usage, poses as fixed, canonical, obligatory. Truths are illusions whose illusoriness is overlooked.

*Friedrich Nietzsche (1873)*

# Preface to the Dover Edition

This book was written as a complement to more specialized books in the field of laser physics. My feeling was that there had to be a fundamental textbook to show students how to apply quantum mechanics to the problems of laser spectroscopy. Standard texts on quantum mechanics were found to be of limited value only.

The material in this book was based on the developments during the sixties and seventies, when most aspects of laser spectroscopy became understood. It does, however, mirror some of the limitations set by the technology of those days: Most spectral investigations had to be carried out in gas cells where inhomogeneous broadening was an ubiquitous annoyance. Many of the techniques designed to reach the intrinsic line shapes were ingenious, but they have now been partly made obsolete by the use of beam and trap samples. The early laser sources were also rather limited in intensity, which forced many experiments inside an active laser cavity. The ensuing self-consistent response offered complicated problems of interpretation. The methods presented in this book were originally developed to handle such problems. The fact that many of these topics are regarded as common knowledge today derives from the spectacular success of laser technology. In spite of all this, however, much of the theory used in present laser spectroscopy is founded on the material presented here.

The experimental area of laser research has developed and expanded greatly. By choice, the work did not include much material about time-dependent phenomena. Already then, the area of transient spectroscopy comprised many interesting and important applications. In addition, the introduction of femtosecond lasers has offered totally new opportunities to examine time-dependent quantum phenomena.

Just at the time of publication of this book, the concept of a dark state emerged. Unfortunately, this important result came up too late to be included in the book. This phenomenon has later motivated many important experiments. Its applications also include the adiabatic spectroscopic technique called STIRAP and its further developments. It is a pity that the treatment of three-level methods does not include this material. Even other three-level aspects are missing; the shelving and quantum-jump phenomena have offered novel methods of precision spectroscopy.

Transitions in multilevel systems are given an introductory treatment in the book. This was motivated by the emerging high-intensity laser experiments on atomic ionization and molecular dissociation. Today this has expanded into a separate field of research on its own.

Lasers have also opened up the field of optical bistability and its relation to dynamical mappings and chaos. The development of ultrahigh-resolution spectroscopy has offered precisions far beyond what was envisaged twenty years ago. Laser science has suggested even extremely demanding measurements, e.g., those related to detection of gravity waves.

At the time of publishing the book, the precision of laser spectroscopy was mainly limited by the technical noise of the laser sources themselves. The ultimate limits set by quantum fluctuations were still beyond reach. Thus I included a chapter on spectroscopy with fluctuating classical parameters. This material may have become obsolete because of progress in laser stabilization. It still, however, provides an interesting topic, combining laser physics with the methods of stochastic processes.

The recent advances in quantum communications and even esoteric subjects like quantum computing have motivated the need to consider electromagnetic field quantization. Many experiments have investigated both fundamental properties of quantized light including the various aspects of entanglement and quantum noise. Optical physics has become one of the standard tools for fundamental tests of quantum physics. At the time of writing the present book, this was still in the future. At that time, I regarded only an understanding of atomic decay and the spectral shape of spontaneous emission necessary for the spectroscopic scientist. Since then, the area of quantum optics has merged with laser spectroscopy to a united field of broad scope and wide applicability.

At the time of writing, the mechanical manifestations of the light had only been seen as the recoil splitting of spectral lines. The new field of laser cooling and trapping lies entirely outside the scope of the present work. It rests on an interplay between the mechanical effects of quantized light and atomic motion. This research has offered spectacular progress in the last twenty years, and it has made possible the remarkable effect of atomic Bose-Einstein condensation. The theory is termed *semiclassical* when the electromagnetic field remains classical even when acting on quantized matter. It is my contention that the basics of this field are still adequately presented in this book. In spite of the age of the book, it still seems to be used by some students and researchers. Thus it is a great pleasure to find that it can still be made available to the physics community in the form of an enduring Dover edition.

STIG STENHOLM

*Stockholm*
*April 7, 2005*

# Preface

This is a textbook designed to teach the theoretical ideas needed to do calculations in laser spectroscopy. The basic theoretical problems are set up, and their relations to experimental reality are discussed. Some special cases are solved, and various useful calculational techniques are applied. In cases that can be treated algebraically the student can both obtain the results by himself and gain some insight into their physical features. It is my conviction that only such activity can lead to understanding of the complex mental process of relating simplified models to experiments. The book is intended for beginning theorists and experimentalists who lack extensive training in algebraic manipulations. Hence it contains more equations than needed for readers more accustomed to mathematical sophistication. The reader is expected to be able to derive all the results, which presupposes simplified models and approximative solutions. In applications to real laboratory systems, complicated situations must be treated; the actual solution often requires numerical methods.

This book requires the reader to know classical electrodynamics and nonrelativistic quantum theory. It is good if the reader has had a phenomenological course on lasers and their physics. Various laser types are referred to without explanation. The book can be used as a textbook for a course even without previous knowledge of laser physics, but then the lecturer must include descriptions of real systems as illustrations.

The central parts of the book are Chapters 1 and 2. In Chapter 1 I introduce the basic facts from electrodynamics and quantum mechanics. In particular, the density matrix, its equation of motion, and its interpretation are discussed in detail. Chapter 2 derives the response of the medium to strong fields. When this material is mastered the reader can more or less freely choose those parts he wants to learn from the interconnection diagram given on the inside of the cover.

Chapter 3 treats the physical basis of laser operation; it can be read as far as is desired. Questions specific to laser technology are not discussed.

Chapter 4 contains some applications central to laser spectroscopy. These sections are independent and can be read in any order.

Chapter 5 discusses the inclusion of laser fluctuations into the theory. This is a very topical field which has only recently matured, and only the basic understanding can be obtained from this book.

The process of spontaneous emission can be understood only if the field is quantized. In Chapter 6 I present the field quantization and some of its simplest consequences. Those interested in quantum electrodynamics must turn to other books.

This book is intended as a textbook, and I have collected comments and references into separate sections. I have included the historically important references and those that directly add information on the topics treated. In addition, monographs and reviews are mentioned to give the most up-to-date view available. The references are listed in alphabetical order, each with a reference to the sections where it is mentioned.

Most processes belonging to *nonlinear optics* are left out of the work. This includes CARS and related methods as well as parametric generation. Propagation effects and time-dependent phenomena are found in other books. Technical questions of laser operation are not discussed; that includes multimode operation, cavity design, and so forth. Multiphoton processes, which are widely used in laser chemistry and isotope separation, are discussed only briefly, and effects of the continuum are not included.

I wish to thank many colleagues and friends with whom I have worked over the years. I am particularly grateful to have been given the opportunity to work with Prof. W. E. Lamb, Jr. The manuscript has been read in total or part by Prof. J. Eberly, Drs. A. Bambini, R. Salomaa, and J. Javanainen, and Mr. M. Lindberg. I am grateful to them for many valuable corrections and comments.

STIG STENHOLM

*Helsinki, Finland*
*October 1983*

# Bibliography Update

Since the publication of this book, laser-related physics has grown both in scope and sophistication. It is impossible to present a comprehensive view of all these developments. Here I present a selected list of some of the many textbooks. The works omitted here can be found from the references listed.


- M. D. Levenson and S. Kano, *Introduction to Nonlinear Spectroscopy* (Academic Press, New York, 1988)
- S. Mukamel, *Principles of Nonlinear Optical Spectroscopy* (Oxford University Press, Oxford, 1995)
- W. Demtröder, *Laser Spectroscopy* (Springer, Berlin, Heidelberg, 1996)
- S. Svanberg, *Atomic and Molecular Spectroscopy* (Springer, Berlin, Heidelberg, 2001)


These books take the field of laser spectroscopy up to its present state. The latter two go into great detail concerning the technical methods and tools necessary in the laboratory.

A very detailed treatment of the theoretical tools is offered in these volumes:


- B. W. Shore, *The Theory of Coherent Atomic Excitation*: Vol. 1, *Simple Atoms and Fields*; Vol. 2, *Multilevel Atoms and Incoherence* (Wiley, New York, 1990)


The fundamental physics of lasers is presented in many books. Recommended are:


- M. Sargent III, M. O. Scully, and W. E. Lamb, Jr., *Laser Physics* (Addison Wesley, New York, 1974)
- A. E. Siegman, *Lasers* (University Science Books, Sausalito, 1986)
- P. W. Milonni and J. H. Eberly, *Lasers* (Wiley, New York 1988)
- W. T. Silfvast, *Laser Fundamentals* (Cambridge University Press, Cambridge, 1996)
- O. Svelto, *Principles of Lasers* (Plenum, New York, 1998)

The quantized optical field and its applications in quantum optics are presented in:

- C. Cohen-Tannoudji, J. Dupont-Roc, and G. Grynberg, *Photons and Atoms: Introduction to Quantum Electrodynamics* (Wiley, New York, 1989)
- L. Mandel and E. Wolf, *Optical Coherence and Quantum Optics* (Cambridge University Press, Cambridge, 1995)
- M. O. Scully and M. S. Zubairy, *Quantum Optics* (Cambridge University Press, Cambridge, 1997)
- C. C. Gerry and P. L. Knight, *Introductory Quantum Optics*, (Cambridge University Press, Cambridge, 2005)

Unfortunately, most books on field quantization are aimed at high-energy physics and consequently offer little for the quantum optics researcher.

# Errata

Page 34, Eq.(1.124) should read

$$\begin{aligned}\overline{\rho_{nm}(t)} &= e^{-i\omega_{nm}(t-t_0)/T} e^{-(t-t_0)/T} \sum_{N=0}^{\infty} \tfrac{1}{N!} \left(\tfrac{t-t_0}{T}\right)^N \overline{\left(e^{-i\theta}\right)^N} \rho_{nm}(t_0) \\ &= \exp\left[-i\omega_{nm}(t-t_0) - \tfrac{t-t_0}{T}\left(1-\overline{e^{-i\theta}}\right)\right] \rho_{nm}(t_0).\end{aligned}$$

Page 41, The last line of Eq. (1.153) should read

$$\overline{\tilde{\rho}_{12}} = \exp\left[-\frac{D}{2\kappa}\left(t + \frac{1}{2\kappa}\left(e^{-2\kappa t} - 1\right)\right)\right]$$

and Eq. (1.154) should read

$$\overline{\tilde{\rho}_{12}} \propto \exp\left[-\frac{Dt^2}{2}\right].$$

Page 42, Fig.1.15, in the right part the label should be replaced as

$$\frac{D}{\kappa} \to \frac{D}{2\kappa}.$$

Page 49, Eq.(1.187) last line should read

$$\overline{\rho_{nm}(t)} = e^{i\omega_{nm}t} e^{-k^2u^2t^2/4} \rho_{nm}(0).$$

Page 84, Eq.(2.116) should read

$$\chi_n(E_1, E_2, ...) = \chi_n^{(1)} + \sum_m \chi_{nm}^{(2)} E_m + \sum_{ml} \chi_{nml}^{(3)} E_m E_l + ...$$

Page 85, in line 4 of Eq.(2.119)

$$\int \frac{x^2}{x^2+\Gamma_p^2} e^{-x^2/k^2u^2 dx}$$

should read

$$\int \frac{x^2}{x^2+\Gamma_p^2} e^{-x^2/k^2u^2} dx$$

Page 87, line 1, $|\,\Delta\,| \gg \gamma_{13}$ should read $|\,\Delta\,| \gg \gamma_{12}$
Page 135, Eq.(3.120) should read

$$I_{abs} = \left(\frac{\mu_{abs}^2 E^2}{2\hbar^2 \gamma_1^{abs} \gamma_2^{abs}}\right) \frac{\gamma_1^{abs} + \gamma_2^{abs}}{2\gamma_{12}^{abs}} = \xi \eta I,$$

Page 151, Fig. 4.4, The labels $\begin{matrix} E_1 \\ \Omega_1, k_2 \end{matrix}$ should read $\begin{matrix} E_1 \\ \Omega_1, k_1 \end{matrix}$.
Page 155, Eq.(4.46) should read $\tilde{\rho}_{32}^{(2)} =$
Page 202, Eq.(5.46) first line should read

$$(\nu)_{nn'} = \int H_n(\xi) \nu u_{n'}(\xi) d\xi$$

Page 210, Eq.(5.93) second line should read

$$D(n) = \frac{\sqrt{\Gamma D}}{\gamma_2 + n\Gamma} \qquad \text{n odd.}$$

Page 216, In Eq.(5.128), $\Delta\Omega$ should read $\Delta\omega$
Page 219, Eq.(5.145) first line should read

$$\frac{1}{2\pi} \int_{-\pi}^{\pi} \exp\left[-i\lambda \varepsilon_i e^{-i(\varphi+\Delta\Omega_i t)} - i\lambda^* \varepsilon_i e^{i(\varphi+\Delta\Omega_i t)}\right] d\varphi$$

Page 220, Eq.(5.148) first line should read

$$\frac{1}{\pi^2} \int d\lambda' \int d\lambda'' e^{iA\lambda'} e^{iB\lambda''} \exp\left(-\lambda'^2 \overline{|\,\varepsilon\,|^2} - \lambda''^2 \overline{|\,\varepsilon\,|^2}\right)$$

Page 222, Eq.(5.165) last term should read $iL\tilde{\rho}_{21}$

Page 239, Eq.(6.63), the first line should read

$$\Delta E_0 = \sum_{ak\sigma} \frac{\langle b \mid \frac{e}{m}\mathbf{A}\cdot\mathbf{p} \mid a, k\sigma\rangle\langle k\sigma, a \mid \frac{e}{m}\mathbf{A}\cdot\mathbf{p} \mid b\rangle}{-\hbar\omega_k}$$

Page 254, Eq.(6.117) should read

$$\begin{aligned}\Gamma &= \sum_q \lambda(q)^2 \left[\frac{\frac{1}{2}\Gamma}{\nu^2 + (\Gamma/2)^2} + \frac{1}{2}\frac{3\Gamma/4}{(\nu - 2\alpha)^2 + (3\Gamma/4)^2}\right. \\ &\qquad \left. + \frac{1}{2}\frac{3\Gamma/4}{(\nu + 2\alpha)^2 + (3\Gamma/4)^2}\right] \\ &= \pi \sum_q \lambda(q)^2 \left[\delta(\nu) + \frac{1}{2}\delta(\nu - 2\alpha) + \frac{1}{2}\delta(\nu + 2\alpha)\right],\end{aligned}$$

# Contents

# Interconnections between the Parts of Foundations of Laser Spectroscopy

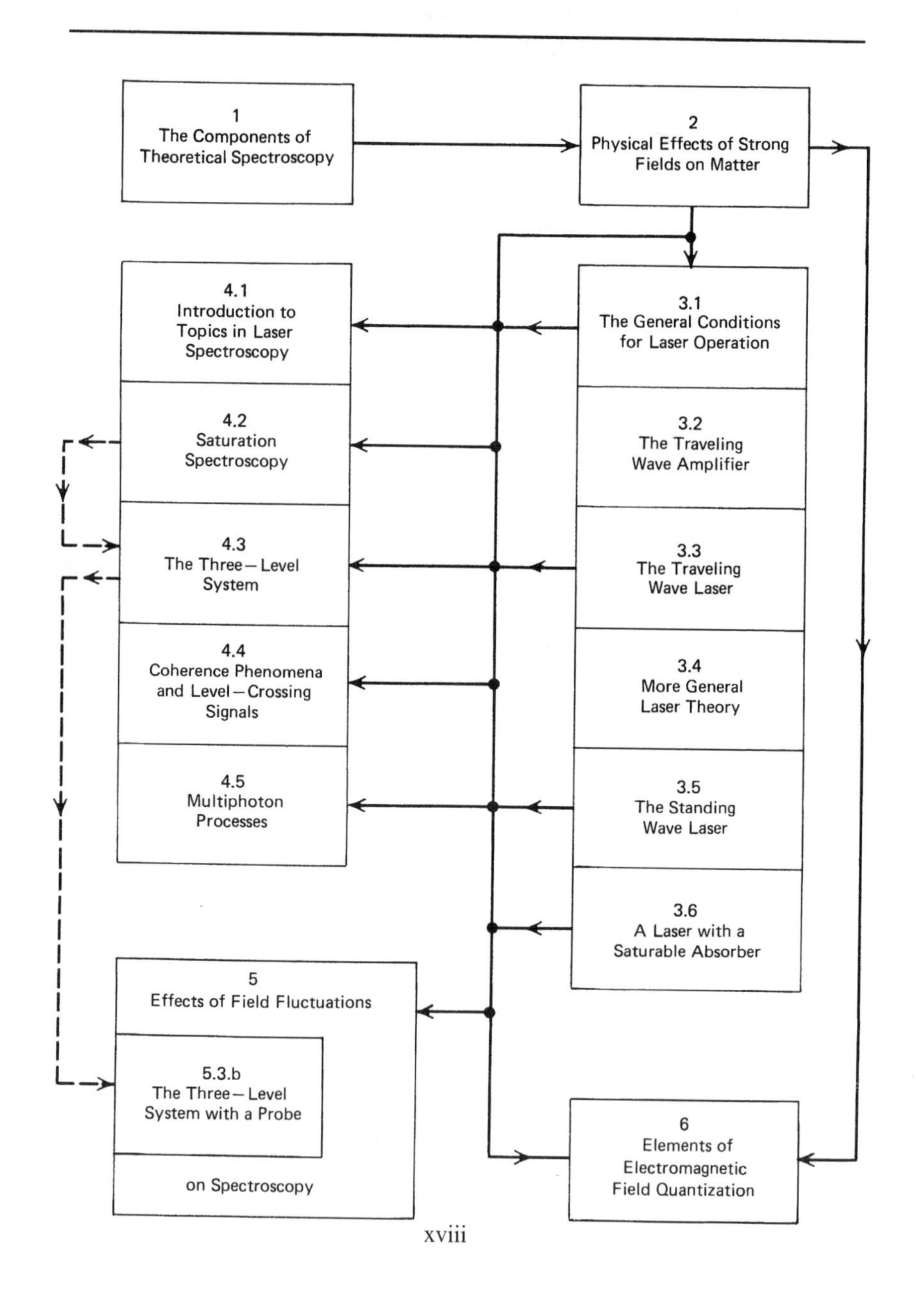

# FOUNDATIONS OF LASER SPECTROSCOPY

CHAPTER 1

# The Components of Theoretical Spectroscopy

## 1.1. INTRODUCTION

The laser is an electromagnetic oscillator that derives its energy directly from the excitation of matter. Often this is found in the internal quantum states of atoms or molecules, and then a quantum description of matter is inevitable. Only recently has it become possible to utilize free electrons for laser operation. The laser provides an optical light source with well-defined phase relationships, and hence it constitutes a straightforward extension of other electromagnetic oscillators. This justifies the use of the term *quantum electronics*. The light itself can, however, be described in a classical way for most applications.

Spectroscopy is one field where much progress is due to the introduction of lasers. Even in linear spectroscopy, the laser is the ideal light source, but its large intensity over a narrow frequency range has allowed new effects; nonlinear spectroscopy becomes important. In this book we mainly discuss those features of laser spectroscopy that derive from its nonlinear properties. Many such features are, however, left out. All propagation problems in a nonlinear medium are neglected. Nonlinear mixing of light and its use to create new signals are not included. Such phenomena are discussed in many books on nonlinear optics. Their omission does lead to the exclusion of many techniques of importance to spectroscopy. We do not discuss Raman spectroscopy and its nonlinear generalizations. Some of these are straightforward extensions of the work in Chapter 4.

To keep our treatment simple, we usually have to restrict the number of levels we choose to introduce. This is most easily justified for atoms; in molecules the density of states is so high that even laser spectroscopy must consider a multitude of levels. The basic theory remains the same, but the treatment becomes so involved that often only numerical work is possible. In solids the situation is even more complex. Thus, in this book, we mostly keep applications to atoms in mind.

In strong fields the bound states of the electrons break up, and atoms and molecules ionize. This is an interesting phenomenon, which has important applications in particle detection and isotope separation. Such transitions between discrete bound and continuum states are left out of this book. They can be included in low-order time-dependent perturbation theory, but to provide a more complete treatment would lead us too far away from the point of view expressed here.

This first chapter of the book reviews the basic knowledge used to formulate problems in laser spectroscopy. This includes the classical description of radiation fields and the quantum theory of matter. We devote most of the space to the microscopic description of bound states, including the effects of various phenomenologically introduced processes. The area of physics needed is vast, and many topics are left incomplete. Many statements cannot easily be justified exactly, but their introduction rests on heuristic or pragmatic arguments. The reader should not be afraid if these arguments are hard to grasp at first. If their use does not later provide the understanding desired, the references in the final section of this chapter may provide some illumination.

The basic applications of the features of this chapter are given in Chapter 2. There the foundations are laid for nonlinear laser physics. After working through that chapter the reader can, more or less at will, read the various application parts independently of each other; to simplify this, a scheme of logical interconnections is provided at the beginning of the book.

## 1.2. CLASSICAL DESCRIPTION OF RADIATION FIELDS

The identity between electromagnetic and optical phenomena was established when Maxwell derived the propagation properties for electromagnetic radiation. Well-defined harmonic vibrations soon became the everyday tool of communication engineering, whereas the source of optical radiation was incandescent bodies. Their light was incoherent and random, and hence optical coherence seemed an elusive and slightly mysterious property. The ordinary radio transmitter, however, emits coherent waves, which is a necessary condition for their reception.

When the laser appeared, the situation was changed. It could be understood as a classical oscillator emitting coherent light with well-defined phase properties. With a laser one can easily carry out optical diffraction experiments that were very hard to perform with thermal light sources. The rise of holography provides the best-known example.

Laser research made it manifestly obvious that physical optics could be based entirely on Maxwell's electrodynamic equations. For large amplitudes of the laser light, a classical description can be used, and a wave equation with well-defined amplitude and phase variables can be applied. The starting point must be Maxwell's equations

$$\nabla \times \mathbf{E} = -\frac{\partial}{\partial t}\mathbf{B} \tag{1.1}$$

$$\nabla \times \mathbf{H} = \mathbf{j} + \frac{\partial}{\partial t}\mathbf{D} \tag{1.2}$$

$$\nabla \cdot \mathbf{D} = \rho \tag{1.3}$$

$$\nabla \cdot \mathbf{B} = 0 \tag{1.4}$$

where $\mathbf{E}, \mathbf{D}$ and $\mathbf{H}, \mathbf{B}$ are the electric and magnetic fields respectively. The density of charges is $\rho$, and their current density is $\mathbf{j}$. If we include only free charges into these, the bound neutrals can be taken to have an electric polarization density $\mathbf{P}$ giving the relation

$$\mathbf{D} = \varepsilon_0 \mathbf{E} + \mathbf{P}, \tag{1.5}$$

whereas the magnetic dipole density $\mathbf{M}$ can be neglected for nonmagnetic media and hence

$$\mathbf{B} = \mu_0 \mathbf{H}. \tag{1.6}$$

In nonlinear spectroscopy we can usually take the field-induced polarization $\mathbf{P}$ to be the main source of the fields in Maxwell's equations, and a major part of the physics is concerned with the calculation of $\mathbf{P}$ for various cases of interest.

It is easily seen that Eqs. (1.1) and (1.4) can be satisfied if we introduce the potentials $\mathbf{A}$ and $\varphi$ so that

$$\mathbf{E} = -\frac{\partial}{\partial t}\mathbf{A} - \nabla\varphi \tag{1.7a}$$

$$\mathbf{B} = \nabla \times \mathbf{A}. \tag{1.7b}$$

The same fields are obtained if the potentials are transformed by an arbitrary gauge function $\chi(\mathbf{r}, t)$ into

$$\mathbf{A}' = \mathbf{A} + \nabla\chi \tag{1.8}$$

$$\varphi' = \varphi - \frac{\partial}{\partial t}\chi. \tag{1.9}$$

In radiation problems one condition on the gauge is set by choosing auxiliary conditions for the potentials. In nonrelativistic calculations it is often advantageous to require that

$$\nabla \cdot \mathbf{A} = 0. \tag{1.10}$$

This is the Coulomb guage.† Its advantage is that the potential $\varphi$ is found to satisfy (1.3) in the form

$$\nabla^2 \varphi = -\frac{\rho}{\varepsilon_0}, \tag{1.11}$$

which solves for the Coulomb interaction between the static charges. This is used to form the bound neutrals, which constitute the object of spectroscopic investigations. To include ions is a minor modification, which we can neglect at present. Quantum chemistry treats the formation of atoms and molecules on the basis of the potentials that can be obtained by solving Eq. (1.11) for a set of point changes with the density $\rho(\mathbf{r}) = \Sigma_i q_i \delta(\mathbf{r} - \mathbf{r}_i)$, where the charge $q_i$ is situated at the position $\mathbf{r}_i$. The solution is the well-known function

$$\varphi(\mathbf{r}) = \frac{1}{4\pi\varepsilon_0} \sum_i \frac{q_i}{|\mathbf{r} - \mathbf{r}_i|}. \tag{1.12}$$

This potential is taken to be entirely static. The remaining part of the equation is then used to describe the radiation fields, which are found to satisfy

$$\nabla \cdot \mathbf{E} = 0, \qquad \nabla \cdot \mathbf{B} = 0; \tag{1.13}$$

they are said to be transverse. These transverse fields contain the optical radiation. In the following discussion we use only the Coulomb gauge.

For spectroscopic purposes it is expedient to separate the formation of neutral bound constituents from the radiation field that induces transitions. The disadvantage is that such a division lacks relativistic invariance and holds only in our chosen frame, the laboratory. Because the binding energies are small, no relativistic energies enter our considerations, and we need to include no relativistic effects anyhow.

If we combine Eqs. (1.1) and (1.2) with (1.5) and (1.6) we obtain

$$\nabla \times (\nabla \times \mathbf{E}) = -\nabla^2 \mathbf{E} = -\mu_0 \frac{\partial^2}{\partial t^2} (\mathbf{P} + \varepsilon_0 \mathbf{E}) \tag{1.14}$$

† In relativistic problems one usually prefers the radiation (or Lorentz) gauge setting

$$\nabla \cdot \mathbf{A} - \frac{\partial}{\partial t} \varphi = 0.$$

where we have used (1.13) too. This gives the equation for an electromagnetic wave propagating with velocity $c = (\varepsilon_0\mu_0)^{-1/2}$ and driven by the oscillating dipole moment **P**. For an oscillating point dipole situated at the origin we have

$$\mathbf{P} = q\mathbf{x}\delta(\mathbf{r}). \tag{1.15}$$

For an observational direction $\hat{\mathbf{n}}$ we can see the radiation driven by the transverse component of **P** only and

$$\begin{aligned}\mathbf{P}_\perp &= q[\mathbf{x} - \hat{\mathbf{n}}(\hat{\mathbf{n}} \cdot \mathbf{x})]\,\delta(\mathbf{r}) \\ &= -q[\hat{\mathbf{n}} \times (\hat{\mathbf{n}} \times \mathbf{x})]\,\delta(\mathbf{r}).\end{aligned} \tag{1.16}$$

At a large distance $R$ we observe at the time $t$ the radiation

$$\mathbf{E}(\mathbf{R}, t) = \frac{q\mu_0}{4\pi}\frac{\hat{\mathbf{n}} \times (\hat{\mathbf{n}} \times \ddot{\mathbf{x}})}{R}, \tag{1.17}$$

where the dipole $\ddot{\mathbf{x}}$ must be evaluated at the retarded time $t - R/c$. Equation (1.17) is the solution of (1.14) with the source (1.15). The magnetic field is given by

$$\mathbf{H} = -\frac{q}{4\pi c}\frac{\hat{\mathbf{n}} \times \ddot{\mathbf{x}}}{R}. \tag{1.18}$$

The energy flux is given by the Poynting vector

$$\begin{aligned}\mathbf{S} &= \mathbf{E} \times \mathbf{H} \\ &= \frac{q^2}{(4\pi)^2\varepsilon_0 c^3 R^2}(\hat{\mathbf{n}} \times \ddot{\mathbf{x}})^2\hat{\mathbf{n}},\end{aligned} \tag{1.19}$$

which shows that the radiation is directed along $\hat{\mathbf{n}}$ and has the intensity $(\hat{\mathbf{n}} \times \ddot{\mathbf{x}})^2 \propto \sin^2\theta$ for $\hat{\mathbf{n}}$ along the polar axis $z$. The total radiated power becomes

$$\begin{aligned}W = \int \mathbf{S} \cdot d\mathbf{a} &= \frac{q^2\mu_0|\ddot{\mathbf{x}}|^2}{(4\pi)^2 cR^2}\int \sin^2\theta R^2 d\varphi\, d\cos\theta \\ &= \frac{1}{4\pi\varepsilon_0}\frac{2q^2}{3c^3}|\ddot{\mathbf{x}}|^2\end{aligned} \tag{1.20}$$

which is the conventional expression for an oscillating dipole.†

† The factor $4\pi\varepsilon_0$ has been separated as this is to be put equal to one to obtain the result in Gaussian units.

If we want to solve (1.14) in a specific geometry, we introduce the eigenfunctions given by

$$\nabla^2 \mathbf{U}_n(\mathbf{r}) + \mathbf{k}_n^2 \mathbf{U}_n(\mathbf{r}) = 0 \tag{1.21}$$

and the boundary conditions appropriate to the system we consider. The functions $\mathbf{U}_n$ are called the cavity eigenfunctions and can be chosen transverse

$$\nabla \cdot \mathbf{U}_n = 0. \tag{1.22}$$

They will form a convenient basis set for all fields in the space we want to include in our considerations. We expand the electric field as

$$\mathbf{E}(\mathbf{r}, t) = \sum_n E_n(t) \mathbf{U}_n(\mathbf{r}), \tag{1.23}$$

where the amplitude $E_n(t)$ is determined by the orthogonality of the eigenfunctions $\mathbf{U}_n$ to be

$$E_n(t) = \frac{\int d^3\mathbf{r}\, \mathbf{E}(\mathbf{r}, t) \cdot \mathbf{U}_n(\mathbf{r})}{\int [\mathbf{U}_n(\mathbf{r})]^2 d^3\mathbf{r}}. \tag{1.24}$$

If the eigenfunctions are normalized to one, the denominator can be omitted.

For the empty cavity, $\mathbf{P} = 0$, Eq. (1.14) gives for $E_n(t)$ the relation

$$\frac{d^2}{dt^2} E_n(t) + \Omega_n^2 E_n(t) = 0, \tag{1.25}$$

where the angular frequency variable is defined by setting

$$\Omega_n = c|\mathbf{k}_n|; \tag{1.26}$$

we usually refer to angular frequencies simply as frequencies. The field $E_n(t)$ oscillates at the frequency $\Omega_n$ and continues to do so eternally if no damping occurs. If the field has a decay time $\tau$ we assume on phenomenological grounds that

$$E_n(t) = E_n(0) e^{\pm i\Omega_n t} e^{-t/2\tau}. \tag{1.27}$$

This leads to the equation

$$\frac{d^2}{dt^2} E_n = -\Omega_n^2 E_n(t) \mp \frac{i\Omega_n}{\tau} E_n(t) + \frac{1}{4\tau^2} E_n(t). \tag{1.28}$$

When the system is damped only slightly, the oscillator quality factor

$$Q_n = \frac{\Omega_n}{1/\tau} = \Omega_n \tau \tag{1.29}$$

is large and (1.28) is equivalent with

$$\frac{d^2}{dt^2} E_n(t) + \frac{1}{\tau} \frac{d}{dt} E_n(t) + \Omega_n^2 E_n(t) = 0. \tag{1.30}$$

As the loss factor is a small modification only, we assume that its effect can be included into (1.14) phenomenologically by a term of the form $(1/\tau)(\partial/\partial t)$ in all cases. For a strongly damped medium this assumption must be reconsidered.

In many cases it is immaterial what type of boundaries one assumes for the space to be considered. It is then convenient to choose plane waves normalized in a box of volume $V$ as the solution to (1.21). They are written in the form

$$\mathbf{U}_n(\mathbf{r}) = \frac{\boldsymbol{\varepsilon}(\lambda_n)}{\sqrt{V}} e^{i\mathbf{k}_n \cdot \mathbf{r}}. \tag{1.31}$$

The vectors $\boldsymbol{\varepsilon}(\lambda_n)$ are chosen as two orthogonal polarization directions satisfying the transversality condition

$$\boldsymbol{\varepsilon}(\lambda_n) \cdot \mathbf{k}_n = 0. \tag{1.32}$$

If we require the box to have periodic boundary conditions, the vectors $\mathbf{k}_n$ become quantized to the values

$$\mathbf{k}_n = \left( n_x \frac{2\pi}{L_x}, n_y \frac{2\pi}{L_y}, n_z \frac{2\pi}{L_z} \right), \tag{1.33}$$

where the box has the dimensions $L_x L_y L_z$, and $n_x$, $n_y$, $n_z$ are three integers.

For microwave radiation the box is often a real metallic cavity. For an ideal metal, the field can have no component along the surface, as this would immediately set up a current. Thus the tangential component must disappear on each surface. If the box is rectangular and placed as in Fig.

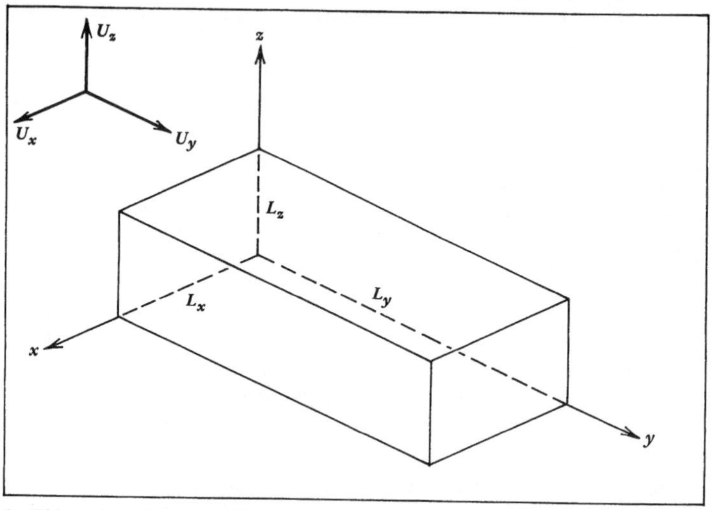


**Fig. 1.1** This picture shows a cavity resonator with metallic walls. The edges are of lengths $L_x$, $L_y$, and $L_z$ respectively, and the eigenmodes consist of vector functions **U(r)**, which are determined by the boundary conditions that the tangential electric field vanishes at the surfaces.

1.1, the solution is easily found to be

$$U_x(x, y, z) = A_x \cos\frac{\pi}{L_x} n_x x \sin\frac{\pi}{L_y} n_y y \sin\frac{\pi}{L_z} n_z z \tag{1.34}$$

$$U_y(x, y, z) = A_y \sin\frac{\pi}{L_x} n_x x \cos\frac{\pi}{L_y} n_y y \sin\frac{\pi}{L_z} n_z z \tag{1.35}$$

$$U_z(x, y, z) = A_z \sin\frac{\pi}{L_x} n_x x \sin\frac{\pi}{L_y} n_y y \cos\frac{\pi}{L_z} n_z z. \tag{1.36}$$

We find directly that the boundary conditions are satisfied;

$$U_x(x, 0, z) = U_x(x, L_y, z) = 0 \tag{1.37}$$

$$U_z(x, 0, z) = U_z(x, L_y, z) = 0, \tag{1.38}$$

and similarly for all other surfaces. The normal components do not disappear because of the choice of solutions. The wave vectors are of the form

$$\mathbf{k}_n = \left( n_x \frac{\pi}{L_x}, n_y \frac{\pi}{L_y}, n_z \frac{\pi}{L_z} \right). \tag{1.39}$$

The transversality condition (1.22) imposes one condition between the components of the vector $\mathbf{A} = (A_x, A_y, A_z)$ in the form

$$\mathbf{A} \cdot \mathbf{k}_n = \pi \left( A_x \frac{n_x}{L_x} + A_y \frac{n_y}{L_y} + A_z \frac{n_z}{L_z} \right) = 0. \tag{1.40}$$

For cavities of more complicated geometries the solution becomes more complicated than (1.34)–(1.36) but the basic idea remains the same.

If one dimension, $L_z$ say, grows to infinity, no boundary condition is needed in this direction. The solution becomes an exponential $\exp(ik_z z)$, and the cavity becomes a wave guide. Such structures are widely used in microwave systems.

The microwave oscillator, the maser, uses metallic cavities as resonance circuits. For optical wavelengths the use of closed metallic cavities is impractical because of the short wavelength. The optical oscillator, the laser, became possible when it was realized that a Fabry–Perot resonator can be used as a cavity.

To achieve resonance in the plane-parallel Fabry–Perot, Fig. 1.2a, the wave must be in phase after traversing the length of the cavity to and fro, which gives the condition $2kL = 2\pi n$, or with $k = 2\pi/\lambda$ the cavity must be an integer number of half-wavelengths

$$L = n\frac{\lambda}{2}, \tag{1.41}$$

and the frequency is given by

$$\Omega_n = \frac{c\pi}{L} n. \tag{1.42}$$

To avoid diffraction losses one must make the plane-parallel resonator of infinite transverse dimensions. By choosing focusing mirrors the energy can be confined to the interior of the Fabry–Perot. The basic mode acquires a Gaussian distribution in the transverse direction and is focused to a beam waist of radius $a$ inside the cavity; see Fig. 1.2b. Here the transverse intensity distribution is given by

$$E(r) \propto e^{-r^2/a^2}. \tag{1.43}$$

There are also higher transverse modes with more complicated dependence on the transverse variables, see examples in Fig. 1.3. For most applications we can assume our discussion to concern the area $r \ll a$, where the transverse variation (1.43) may be neglected. Then we can choose the cavity modes to be determined by (1.42) in the form

$$U_n(z) = \sqrt{\frac{2}{L}} \sin k_n z \tag{1.44}$$

with $k_n = \Omega_n/c$. In addition, two polarization directions are possible. For laser light one polarization is often selected by Brewster windows or polarizers in the apparatus.

A relaxation rate like that in (1.30) can be introduced by a passive absorbing medium homogeneously distributed in space. Even losses in the metallic boundaries can be approximated by a decay time $\tau$. This includes mirror transmission losses for a laser cavity. In addition, the optical cavity has diffraction losses due to the finite radii of the mirrors. This can also be represented by a time $\tau$. The loss rate depends on the mode structure; the

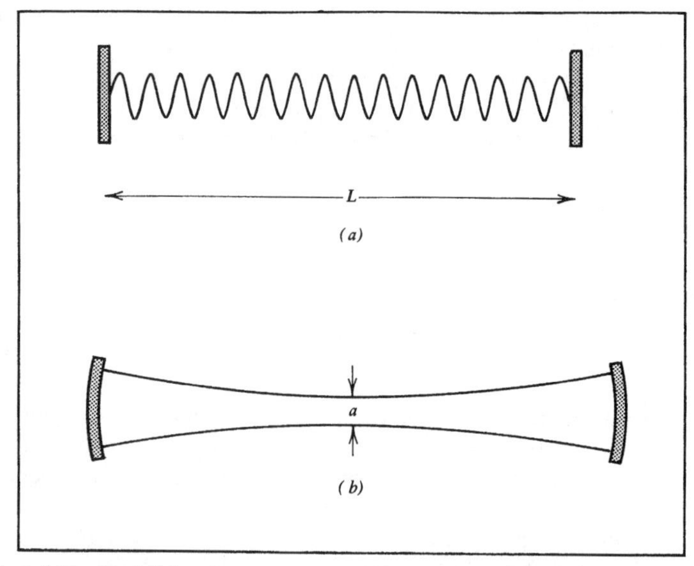


**Fig. 1.2** (*a*) The ideal Fabry–Perot resonator with plane mirrors. The wave pattern repeats itself after one round trip over the distance $2L$. (*b*) A real Fabry–Perot usually employs spherical mirrors with a Gaussian transverse intensity distribution. The beam waist $a$ is defined at the most focused cross section.

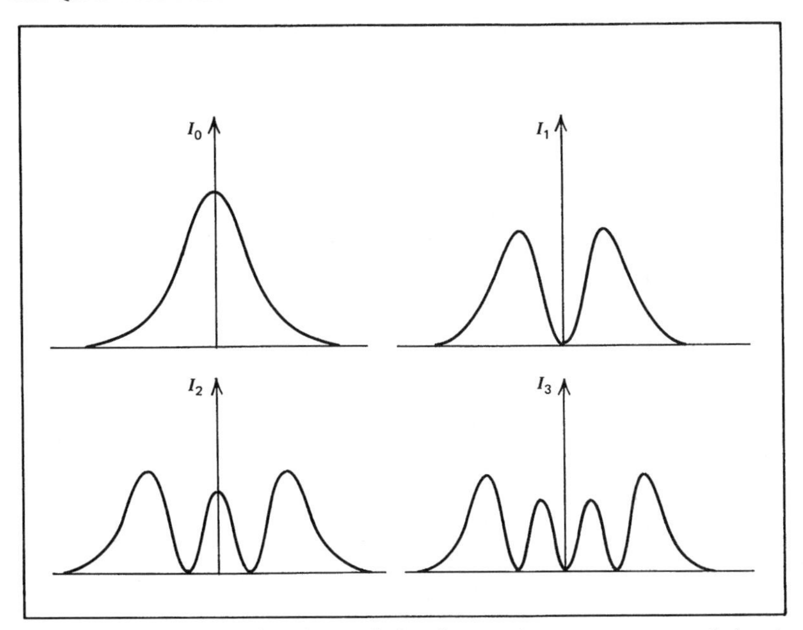


**Fig. 1.3** The transverse modes of an optical cavity contain one, two or more dark points across the transverse section of the beam.

higher transverse modes, especially, have more intensity at the edges and hence experience higher losses. It is often useful to let the loss parameter depend on the mode and set $\tau_n$. For a more detailed discussion of optical resonance modes see the literature listed in the final section of this chapter.

## 1.3. THE QUANTUM DESCRIPTION OF MATTER

The matter subjected to laser radiation consists of atoms and molecules. Their spectrum consists of a set of bound states having discrete energy eigenvalues. These are the solutions of the time-independent Schrödinger equation with the Coulombic potential (1.12). In addition, there is a continuum of free ionic states, which are not discussed in this treatise; see Fig. 1.4. Electronic transitions between the discrete states of ions can be handled in the same way as those of neutral particles.

The atomic Hamiltonian is written

$$H_{\text{at}} = \sum_n |\varphi_n\rangle E_n \langle\varphi_n|. \tag{1.45}$$

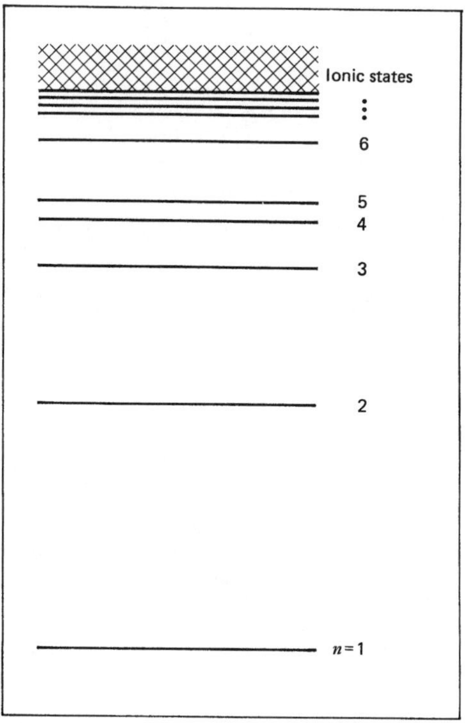


**Fig. 1.4** The bound atomic states become denser toward the ionization limit, above which the states form a continuum.

In laser spectroscopy we subject this to monochromatic radiation at a few frequencies $\{\Omega_i\}$ only. These cause transitions between levels $E_i$ and $E_f$ if the Bohr resonance condition

$$\hbar\Omega_i = E_f - E_i \tag{1.46}$$

is approximately satisfied. As atomic level spacings usually are unequal, we can achieve resonance with only a few transitions; the rest are well out of resonance. Then it is possible to truncate the Hamiltonian and look at only those levels that participate in the resonance interaction. In laser spectroscopy the matter is often assumed to be a two-, three-, or $N$-level system. The transitions into the ionized continuum are neglected in this book.

The time development described by quantum mechanics is unitary, which means that the probability is conserved. If the atomic state is expanded in the energy eigenstates of the Hamiltonian (1.45)

$$|\psi\rangle = \sum_n C_n(t)|\varphi_n\rangle \tag{1.47}$$

unitarity guarantees that

$$\sum_n |C_n(t)|^2 = 1 \tag{1.48}$$

for all times. When we work in a truncated part of the state space only, no unitarity applies. If we consider levels 1 and 2 of Fig. 1.5, they may decay spontaneously or by some other mechanism to levels $a$ and $b$, which remain unobserved. Such a decay is usually exponential and can be described by decay rates added phenomenologically to the Schrödinger equation of the system. From

$$i\hbar\frac{d}{dt}|\psi\rangle = H|\psi\rangle \tag{1.49}$$

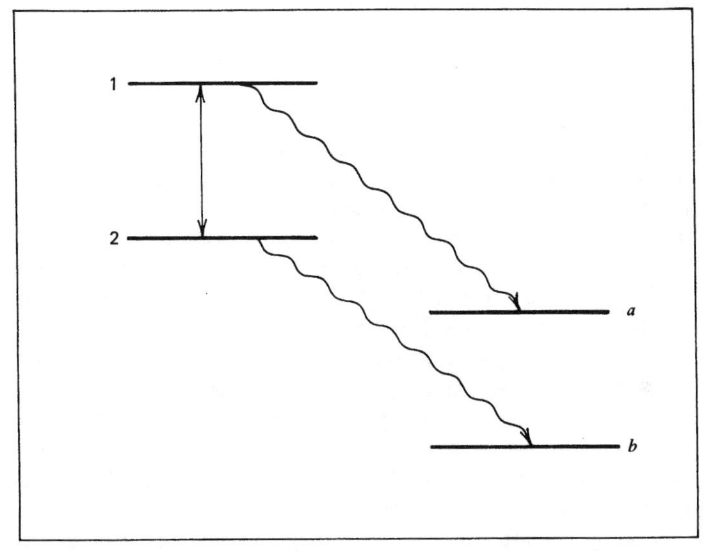


**Fig. 1.5** When states 1 and 2 decay with a constant probability per unit time to unobserved states $a$ and $b$, the decay can be described by an exponential decrease of the probabilities to be in states 1 and 2 with time.

we find that the coefficients $C_n(t)$ satisfy the equations

$$i\hbar\frac{d}{dt}C_n(t) = \sum_m H_{nm}C_m(t) - i\hbar\frac{1}{2}\gamma_n C_n(t), \tag{1.50}$$

where a relaxation term is added. The coefficients $H_{nm}$ are the matrix elements of the total Hamiltonian in the representation given by the energy eigenfunctions of the atomic Hamiltonian (1.45)

$$H_{nm} = \langle\varphi_n|H|\varphi_m\rangle. \tag{1.51}$$

In (1.50) we have added the decay rates in such a way that they give the solution

$$|C_n(t)|^2 = e^{-\gamma_n t}|C_n(0)|^2 \tag{1.52}$$

if no coupling between the states $|\varphi_n\rangle$ occurs. It is obvious that the Eqs. (1.50) do not preserve the probability (1.48). A spontaneous decay between coupled levels, 1 and 2 in Fig. 1.5 say, cannot be described by relaxation terms in the Schrödinger equation.

For an atom the problem is centrally symmetric, and the angular momentum is a good quantum number. In the position representation the eigenfunctions of the energy then become of the form

$$\langle\mathbf{r}|m, l, n\rangle = Y_l^m(\theta, \varphi)R_n(r), \tag{1.53}$$

where $Y_l^m$ is the spherical harmonic with angular momentum $\mathbf{l}^2 = \hbar^2 l(l+1)$ and $l_z = \hbar m$. $R_n(r)$ is the radial wave function, which is known analytically for simple cases like, for example, the hydrogen atom. For this the energy depends only on the principal quantum number, whereas for more complicated atoms the degeneracy with respect to $l$ is lifted, but each level has $(2l+1)$-fold degeneracy due to the quantum number $m$. This is lifted by strong external fields, the Stark and Zeeman effects. The spectrum of a simple model atom is shown in Fig. 1.6.

For real atoms the spectrum is complicated by the occurrence of electronic and nuclear spins, spin-orbit coupling, and other effects. These features make it necessary to specify the levels with care, but do not change the picture of a set of discrete levels acted upon by the radiation field. For molecules the situation becomes even more complicated; vibrational and rotational transitions are to be added.

The coupling of matter to the electromagnetic field must be carried out through the use of the vector potential $\mathbf{A}(\mathbf{r}, t)$. According to the prescription given in quantum mechanics, the Schrödinger equation in a field determined by $(\mathbf{A}, \varphi)$, as in (1.7), is

$$\left[\frac{1}{2m}(\mathbf{p} - q\mathbf{A})^2 + q\varphi\right]\psi(\mathbf{r}, t) = i\hbar\frac{\partial}{\partial t}\psi(\mathbf{r}, t), \tag{1.54}$$

where $\mathbf{p} = -i\hbar\nabla$. Multiplying the equation for the wave function $\psi(\mathbf{r}, t)$ by the quantity $\exp(+iq\chi/\hbar)$, we find the new equation

$$\left[\frac{1}{2m}(\mathbf{p} - q\mathbf{A} - q\nabla\chi)^2 + q\varphi - q\frac{\partial}{\partial t}\chi\right]e^{iq\chi/\hbar}\psi(\mathbf{r}, t) = i\hbar\frac{\partial}{\partial t}e^{iq\chi/\hbar}\psi(\mathbf{r}, t). \tag{1.55}$$

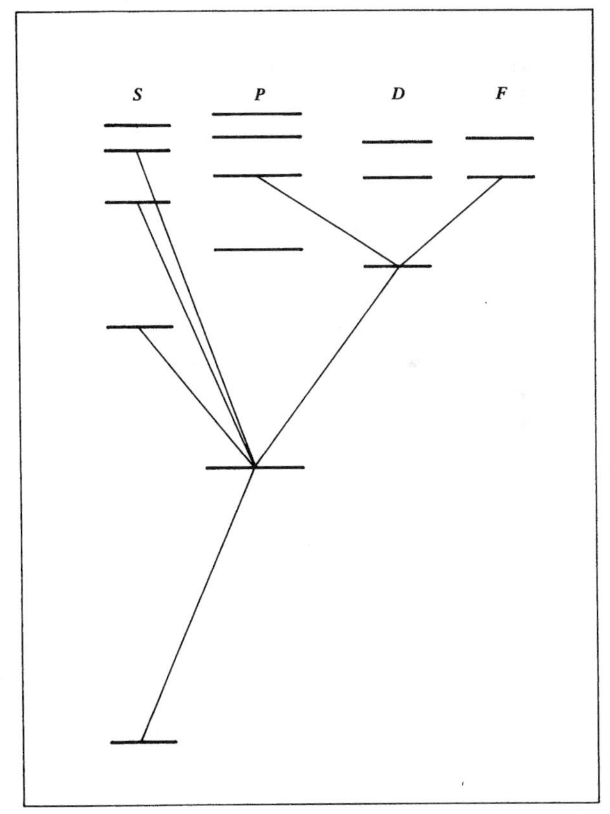


**Fig. 1.6** A schematic energy level diagram of an atom with series of $S$, $P$, $D$, and $F$ levels. Some allowed optical transitions are indicated.

Comparing this equation with (1.8–1.9), we find that the transformation

$$\psi'(\mathbf{r}, t) = \exp\left[+\frac{i\hbar\chi(\mathbf{r}, t)}{\hbar}\right]\psi(\mathbf{r}, t), \tag{1.56}$$

corresponds to a gauge transformation of the potentials.

The interaction in (1.54) leads to a general theory of transitions between the atomic levels. The atomic dimensions $a_0$ are usually much smaller than the wavelength of the light that induces transitions between the levels $\lambda \gg a_0$, and hence the atomic dipole transition is the only one of importance. This amounts to evaluating the radiation field at the position of the atomic center of mass, $\mathbf{R}$ say, and letting the electric field be given by the time derivative $-\dot{\mathbf{A}}$; the magnetic field is taken to induce no transitions.

If, in this situation, we take the gauge function to be

$$\chi(\mathbf{r}, t) = -\mathbf{A}(\mathbf{R}, t) \cdot \mathbf{r}, \tag{1.57}$$

we find that the vector potential disappears

$$\mathbf{A} + \nabla\chi(\mathbf{r}, t) = 0 \tag{1.58}$$

and a new interaction potential appears

$$-q\frac{\partial}{\partial t}\chi = q\mathbf{r} \cdot \dot{\mathbf{A}} = -q\mathbf{r} \cdot \mathbf{E}, \tag{1.59}$$

which is the interaction between the dipole moment $\mu = q\mathbf{r}$ and the electric field. Hence in the dipole approximation we can use the classical interaction (1.59) in the Schrödinger equation. The transformation (1.56) of the wave function is a unitary transformation because $\chi$ is an operator, as it depends on $\mathbf{r}$. We know that the physics is not changed by a unitary transformation. If we calculate correctly, we should obtain the same results before and after the transformation. This is equivalent to the statement that all physical results must be independent of our choice of gauge for the electromagnetic fields.

A unitary transformation remains unitary only if all atomic states are considered. When only a truncated set of states is included, there may be differences between the two ways to calculate observable results. Even if no general argument exists, there are strong indications that the dipole form of the interaction (1.59) is advantageous when one wants to calculate transitions between atomic bound states in the dipole approximation. The situation is, however, complicated, and there is no need to enter into the details of the discussion at this point.

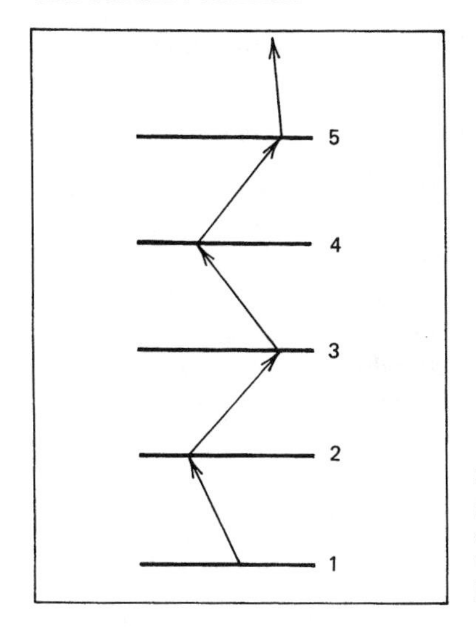


**Fig. 1.7** When light of only one frequency and given polarization impinges on an atom, it is often possible to neglect all the levels except those that are nearly resonantly coupled by the light. They form an excitation ladder.

When angular momentum is a good quantum number, the dipole interaction (1.59) induces transitions only between states such that

$$\langle m, l, n|e\mathbf{r}|m', l', n'\rangle \neq 0, \tag{1.60}$$

which gives the dipole selection rules

$$\Delta l = \pm 1 \tag{1.61}$$

$$\Delta m = 0 \qquad \pi\text{-transition} \tag{1.62}$$

$$\Delta m = \pm 1 \qquad \sigma\text{-transition.} \tag{1.63}$$

For a given experimental geometry $\pi$-transitions can be induced by linearly polarized light, and $\sigma$-transitions by left or right circularly polarized light. For a certain light polarization it is sometimes possible to arrange the coupled energy levels in a sequence $\{E_1, E_2, E_3, \ldots\}$ so that only one state out of each degenerate manifold is coupled to the nearest ones by the fields; see Fig. 1.7. An example is the case of a singlet ground state and $\pi$-transitions, when only the states with $m = 0$ are coupled.

## 1.4. THE DENSITY MATRIX

According to quantum mechanics, the most complete information available about a single system is its state $|\psi\rangle$. To each observable physical quantity belongs a hermitean operator $A$, which has the expectation value

$$\langle A \rangle = \langle \psi | A | \psi \rangle. \tag{1.64}$$

This statement can be verified only on an assembly of systems each prepared with certainty in the state $|\psi\rangle$. For a measurement on a single system nothing can generally be known with certainty. Quantum mechanics, hence, requires an ensemble to verify its propositions.

It is assumed that the measurement of (1.64) can be carried out at a given time $t$. Then the average depends parametrically on time, and the average value $\langle A \rangle$ develops in time. If we use the decomposition of $|\psi\rangle$ given in (1.47), we obtain

$$\langle A(t) \rangle = \sum_{nn'} C_n^*(t) C_{n'}(t) A_{nn'} \tag{1.65}$$

where

$$A_{nn'} = \langle \varphi_n | A | \varphi_{n'} \rangle. \tag{1.66}$$

*The observable quantities must, according to quantum mechanics, be bilinear functions of the state of matter.* This is a necessary feature of a quantum mechanical description, and it cannot be dispensed with. It contains the possibility of interference between various states and wave phenomena of matter, which are in contrast to classical mechanics.

The ideal spectroscopic measurement on an atom or a molecule should be concerned with a single particle. In general, however, one must include several particles to obtain an observable signal, even if recent progress in the sensitivity of laser spectroscopy makes it plausible that future investigations may treat single particles. To observe the ideal atom or molecule, we must use very dilute gases where the particles are undisturbed by mutual interactions. The measured value then constitutes an average over a macroscopic number of particles. It is normally not possible to assemble a set of particles in totally identical states, and hence some properties may vary from one individual to the other. The observation must be regarded as an *ensemble average* over the states of all particles. But this is precisely the situation we need to verify the predictions of quantum mechanics. The observed assembly constituting our sample is taken to be a realization of the statistical ensemble describable by the quantum mechanical averages.

In statistical mechanics one introduces an ensemble because of our incomplete knowledge of the precise state of the object of our investigation. A *density matrix* or *statistical operator* is introduced to describe the state of the ensemble. If known, it determines all the statistical properties of measurements on an assembly of such objects. Atomic spectroscopy investigates the properties of the individual particles but requires a large number of independent particles to obtain an observable effect. Hence we can take the assembly of particles in our real sample to be a realization of the ensemble described by a density matrix. The density matrix thus emerges as a natural way to describe the atomic matter investigated by laser spectroscopy.

If we have $N$ particles, the particle with index $i$ is taken to be in the state $|\psi^{(i)}\rangle$. The quantum mechanical expectation value of the observable $A$ for this state is $\langle\psi^{(i)}|A|\psi^{(i)}\rangle$, and the statistical average over the whole ensemble is given by

$$\overline{\langle A\rangle}=\frac{1}{N}\sum_{i=1}^{N}\langle\psi^{(i)}|A|\psi^{(i)}\rangle. \tag{1.67}$$

Here the average is expressed as a sum over the individual particles making up the ensemble. Using a decomposition of $|\psi^{(i)}\rangle$ according to the basis states $|\varphi_n\rangle$, with the coefficients

$$C_n^{(i)}=\langle\varphi_n|\psi^{(i)}\rangle \tag{1.68}$$

we find for (1.67)

$$\begin{aligned}\overline{\langle A\rangle}&=\frac{1}{N}\sum_{i}\sum_{nn'}C_n^{(i)*}C_{n'}^{(i)}A_{nn'}\\&=\sum_{nn'}\rho_{n'n}A_{nn'}=\operatorname{Tr}\rho A.\end{aligned} \tag{1.69}$$

Here we have defined the new matrix

$$\rho_{n'n}=\frac{1}{N}\sum_{i=1}^{N}C_{n'}^{(i)}C_n^{(i)*}=\overline{C_{n'}C_n^*}. \tag{1.70}$$

This matrix is the *ensemble averaged density matrix* or *statistical matrix*.

We have defined the density matrix in a specific representation, but it is easily seen that it is a matrix representation of an operator that is defined in a representation-independent way. We take the dyadic operator $|\psi^{(i)}\rangle\langle\psi^{(i)}|$, which refers to the particle $i$. We form the ensemble average of this operator as follows

$$\rho=\overline{|\psi\rangle\langle\psi|}=\frac{1}{N}\sum_{i=1}^{N}|\psi^{(i)}\rangle\langle\psi^{(i)}|. \tag{1.71}$$

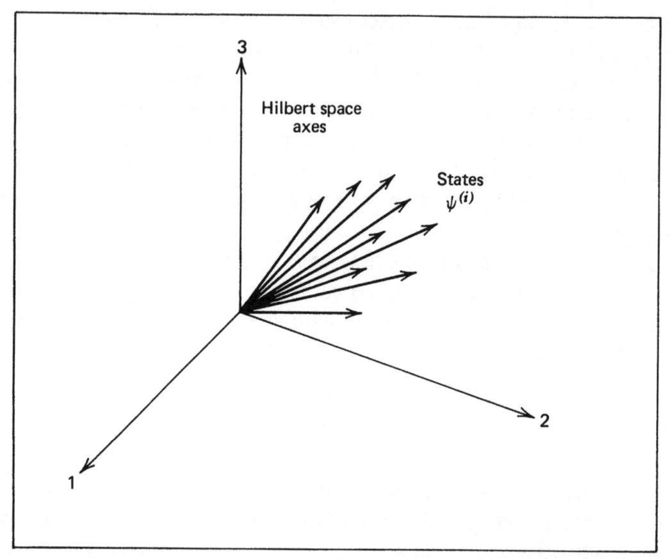


**Fig. 1.8** A quantum mechanical ensemble of independent systems can be represented by a set of unit vectors in the Hilbert space of the system, each one representing one member of the ensemble. This set of state vectors replaces the points in phase space used in classical statistical mechanics. The Hilbert space of a physical system usually has many, often an infinite number of, dimensions; for clarity we show only three in the picture. Another complication is that the state vector in quantum mechanics is a complex quantity, that is, each component in Hilbert space is given by two numbers.

This operator is clearly independent of any choice of basis; see Fig. 1.8. We take its matrix in the $\{\varphi_n\}$ representation and obtain

$$\langle \varphi_{n'}|\rho|\varphi_n\rangle = \frac{1}{N}\sum_{i=1}^{N}\langle \varphi_{n'}|\psi^{(i)}\rangle\langle\psi^{(i)}|\varphi_n\rangle$$

$$= \frac{1}{N}\sum_{i=1}^{N} C_{n'}^{(i)}C_n^{(i)*} = \rho_{n'n} \tag{1.72}$$

by the use of (1.68) and (1.70).

We can write the density matrix in another form by looking at the states $|\psi^{(i)}\rangle$ making up the ensemble of particles. There are no restrictions on these states; in particular, they do not have to be orthogonal or even different. It may be that some of them coincide. Let us denote by $N_\alpha$ the number of times a state $|\psi^{(\alpha)}\rangle$ occurs in the list of states $\{|\psi^{(i)}\rangle\}$. If we sum over all *different states* occurring we must have

$$\sum_{\alpha} N_{\alpha} = N. \tag{1.73}$$

In the sum (1.71) we collect all those states that coincide and sum over the states. Then we find that

$$\begin{aligned}\rho &= \frac{1}{N}\sum_{\alpha} N_{\alpha}|\psi^{(\alpha)}\rangle\langle\psi^{(\alpha)}| \\ &= \sum_{\alpha}|\psi^{(\alpha)}\rangle P(\alpha)\langle\psi^{(\alpha)}|.\end{aligned} \tag{1.74}$$

This is a different way to introduce the density matrix. Instead of a sum over all different individuals as in (1.71) we sum over the *different states* with the appropriate weight for their occurrence

$$P(\alpha) = \frac{N_{\alpha}}{N}. \tag{1.75}$$

Because of (1.73) the weight functions sum to unity;

$$\sum_{\alpha} P(\alpha) = 1.$$

An operator of the form

$$\Pi = |\chi\rangle\langle\chi| \tag{1.76}$$

is called a projection operator because its action on an arbitrary state $|\psi\rangle$ becomes

$$\Pi|\psi\rangle = \langle\chi|\psi\rangle|\chi\rangle, \tag{1.77}$$

which is the component of the state vector $|\psi\rangle$ on the vector $|\chi\rangle$, which is assumed normalized.

We find easily

$$\Pi^2 = |\chi\rangle\langle\chi|\chi\rangle\langle\chi| = \Pi, \tag{1.78}$$

which says that a renewed projection operation only reproduces the previous result. The probability of occurrence of the state $|\chi\rangle$ when a particle is in the state $|\psi\rangle$ is, according to the principles of quantum mechanics, given by

$$\langle\psi|\Pi|\psi\rangle = |\langle\psi|\chi\rangle|^2, \tag{1.79}$$

and hence $\Pi$ is the operator whose expectation value tells the occupation probability of the state $|\chi\rangle$. If we apply the rule (1.69) we find

$$\begin{aligned}\langle\Pi\rangle = \operatorname{Tr}\Pi\rho &= \sum_n \langle\varphi_n|\chi\rangle\langle\chi|\rho|\varphi_n\rangle \\ &= \sum_n \langle\chi|\rho|\varphi_n\rangle\langle\varphi_n|\chi\rangle = \langle\chi|\rho|\chi\rangle.\end{aligned} \tag{1.80}$$

Thus the diagonal element $\langle\chi|\rho|\chi\rangle$ tells the probability of the occurrence of $|\chi\rangle$ in the ensemble described by $\rho$.

If we use the representation (1.74) for the density matrix, we can ask for the occurrence of the state $|\psi^{(\beta)}\rangle$ which is one of the states used to define $\rho$. The result is

$$\langle\psi^{(\beta)}|\rho|\psi^{(\beta)}\rangle = \sum_\alpha P(\alpha)|\langle\psi^{(\alpha)}|\psi^{(\beta)}\rangle|^2. \tag{1.81}$$

If the states $|\psi^{(\alpha)}\rangle$ are nonorthogonal, there is no simple relationship between this probability and $P(\alpha)$. If the states $|\psi^{(\alpha)}\rangle$ are orthogonal we obtain, however,

$$\langle\psi^{(\beta)}|\rho|\psi^{(\beta)}\rangle = P(\beta). \tag{1.82}$$

*Thus* $P(\alpha)$ *can be interpreted as a probability only if the states* $|\psi^{(\alpha)}\rangle$ *are orthogonal*. Otherwise it is only the weight function for the state $|\psi^{(\alpha)}\rangle$ in the ensemble defining the density matrix.

**Example.** A real ensemble must, of course, consist of a tremendously large number of particles. As an exercise, we choose to write out the preceding results explicitly for an ensemble with three particles only; the example is artificial but illuminating. Each particle has two levels. We assume that the particles are in the states

$$|\psi^{(1)}\rangle = \frac{1}{\sqrt{2}}\begin{bmatrix}1\\1\end{bmatrix}; \qquad |\psi^{(2)}\rangle = \frac{1}{\sqrt{2}}\begin{bmatrix}1\\1\end{bmatrix}; \qquad |\psi^{(3)}\rangle = \frac{1}{\sqrt{10}}\begin{bmatrix}1\\-3\end{bmatrix}. \tag{1.83}$$

The density matrix becomes

$$\rho = \frac{1}{3}\left\{\frac{1}{2}\begin{bmatrix}1\\1\end{bmatrix}[1,1] + \frac{1}{2}\begin{bmatrix}1\\1\end{bmatrix}[1,1] + \frac{1}{10}\begin{bmatrix}1\\-3\end{bmatrix}[1,-3]\right\}$$

$$= \frac{1}{30}\begin{bmatrix}11 & 7\\ 7 & 19\end{bmatrix}. \tag{1.84}$$

Here the first two states are identical, and the first two terms of (1.84) can be combined to one with weight 2. The occurrence probability of the state $(1/\sqrt{2})[1,1]$ is, however, not $\frac{2}{3}$ because the third state is not orthogonal but

$$\langle\psi^{(1)}|\rho|\psi^{(1)}\rangle = \frac{1}{\sqrt{2}}[1,1]\frac{1}{30}\begin{bmatrix}11 & 7\\ 7 & 19\end{bmatrix}\frac{1}{\sqrt{2}}\begin{bmatrix}1\\1\end{bmatrix} = \frac{11}{15}. \tag{1.85}$$

The probability of being in the "upper" state $|+\rangle = \begin{bmatrix}1\\0\end{bmatrix}$ can be obtained in two ways. The projection on this state is

$$\Pi = \begin{bmatrix}1\\0\end{bmatrix}[1,0] = \begin{bmatrix}1 & 0\\ 0 & 0\end{bmatrix}, \tag{1.86}$$

and direct calculation gives

$$\langle\Pi\rangle = \operatorname{Tr}\Pi\rho = \tfrac{11}{30} = \langle +|\rho|+\rangle, \tag{1.87}$$

according to the result directly shown in (1.84).

We have introduced the density matrix by forming an ensemble of particles, each in a definite quantum state $|\psi\rangle$, a so-called *pure state*. The ensemble is in a *mixed state*, which directly corresponds to the statistical distribution function of classical statistical mechanics. The Hilbert space of quantum mechanics has only replaced the phase space of classical mechanics, see Fig. 1.8. It can, however, be seen that a density matrix is the general description of a single particle that has interacted with some other physical system ("the rest of the universe"). If this other system can be described by a complete set of states $\{|\xi_\mu\rangle\}$ the general state after the interaction is

$$|\psi\rangle = \sum_{n\mu} C_{n\mu}|\varphi_n\rangle|\xi_\mu\rangle, \tag{1.88}$$

because the states $|\varphi_n\rangle|\xi_\mu\rangle$ form a complete basis in the Hilbert space of the combined systems. The density matrix for such a particle becomes

$$\rho = |\psi\rangle\langle\psi| \tag{1.89}$$

where no average occurs. This is a pure state, which can be verified by noting that $\rho$ is a *projection operator*

$$\rho^2 = \rho; \tag{1.90}$$

this can happen only if $\rho$ contains no statistical mixture.

In many cases in physics we want to consider only a part of a system. We do not want to discuss the probabilities in the system described by $|\xi_\mu\rangle$ but only those described by $|\varphi_n\rangle$. This corresponds to looking at the *marginal probability* distribution in $n$ for the case of classical statistics.

The probability of occurrence of the quantum numbers $n$, $\mu$ is

$$\langle\varphi_n|\langle\xi_\mu|\rho|\xi_\mu\rangle|\varphi_n\rangle = |C_{n\mu}|^2. \tag{1.91}$$

If we do not care about the probability distribution in $\xi$, we sum over its values and obtain

$$\langle\varphi_n|\sum_\mu\langle\xi_\mu|\rho|\xi_\mu\rangle|\varphi_n\rangle = \sum_\mu|C_{n\mu}|^2. \tag{1.92}$$

This gives the probability of occurrence of $\varphi_n$ irrespective of the value of $\xi_\mu$. If we introduce the *reduced density* matrix

$$\tilde{\rho} = \sum_\mu\langle\xi_\mu|\rho|\xi_\mu\rangle = \sum_{nn'}|\varphi_n\rangle\sum_\mu C_{n\mu}C^*_{n'\mu}\langle\varphi_{n'}|, \tag{1.93}$$

this is a density matrix in the Hilbert space of the particle described by the states $|\varphi_n\rangle$. It can, according to our basic understanding of quantum mechanics, be given an interpretation in terms of an *ensemble* of particles: We introduce *formally* the states

$$|\mu\rangle = \frac{\sum_n C_{n\mu}|\varphi_n\rangle}{\sqrt{\sum_n|C_{n\mu}|^2}}. \tag{1.94}$$

We must notice that the denominator is necessary to normalize the states as we have from (1.88)

$$\langle\psi|\psi\rangle = \sum_{n\mu}|C_{n\mu}|^2 = 1. \tag{1.95}$$

The probability of occurrence of a quantum number $\mu$ in $|\psi\rangle$ is

$$P(\mu) = \sum_n|C_{n\mu}|^2. \tag{1.96}$$

We can now rewrite (1.93) as

$$\begin{aligned}\tilde{\rho} &= \sum_{\mu}\sum_{nn'} C_{n\mu}|\varphi_n\rangle\langle\varphi_{n'}|C^*_{n'\mu} \\ &= \sum_{\mu} P(\mu)\frac{\sum_{n} C_{n\mu}|\varphi_n\rangle\sum_{n'}\langle\varphi_{n'}|C^*_{n'\mu}}{\sum_{n}|C_{n\mu}|^2} \\ &= \sum_{\mu}|\mu\rangle P(\mu)\langle\mu|. \end{aligned} \tag{1.97}$$

This shows that the reduced density matrix for an observed particle in an unobserved background can be written in the form of an ensemble averaged density matrix like (1.74). The various members of the ensemble are in the pure states (1.94), which are the *normalized* projections of $|\psi\rangle$ on the states $|\xi_\mu\rangle$. The weight of occurrence of each state is given by $P(\mu)$, which according to (1.96) is the quantum mechanical probability for the occurrence of the quantum number $\mu$ in $|\psi\rangle$. Quantum mechanics thus allows us to use unobserved degrees of freedom as labels for an ensemble of systems. In this way the statistical aspect of quantum mechanics introduces the ensemble point of view in a natural way.

If an observable $A$ operates only on the states of the particle $|\varphi_n\rangle$, we can calculate its expectation value in the state (1.88) as follows:

$$\begin{aligned}\langle A\rangle = \langle\psi|A|\psi\rangle &= \sum_{n'n}\sum_{\mu'\mu} C_{n\mu}C^*_{n'\mu'}\langle\varphi_n|A|\varphi_{n'}\rangle\langle\xi_\mu|\xi_{\mu'}\rangle \\ &= \sum_{nn'} A_{nn'}\sum_{\mu} C^*_{n'\mu}C_{n\mu} = \operatorname{Tr} A\tilde{\rho},\end{aligned}$$

where we have used (1.93). It is thus clear that for observables acting on the states $|\varphi_n\rangle$ only, we can calculate the averages using the reduced density matrix $\tilde{\rho}$, alone. If no observables acting on states $|\xi_\mu\rangle$ are ever investigated, the system can be described entirely by the reduced matrix $\tilde{\rho}$, and the unobserved degrees of freedom remain in the background.

Even if the combined state $|\psi\rangle$ is pure, the density matrix $\tilde{\rho}$ cannot in general be described by any pure states composed entirely of combinations of the states $|\varphi_n\rangle$.

## 1.5. PHYSICAL PROPERTIES OF THE DENSITY MATRIX

In quantum mechanics it is possible to assign the time dependence to the observables if one chooses the Heisenberg picture. Because the density matrix is defined by the state amplitudes, it will be time dependent in the Schrödinger picture. This time dependence can be obtained easily from the Schrödinger equation in the form of (1.50). If we use the definition (1.70) of the density matrix we need the equation

$$i\hbar\frac{d}{dt}C_{n'}C_n^* = i\hbar\frac{d}{dt}C_{n'}C_n^* + i\hbar C_{n'}\frac{d}{dt}C_n^*$$

$$= \sum_m \left(H_{n'm}C_mC_n^* - C_{n'}C_m^*H_{mn}\right) - i\hbar\frac{1}{2}(\gamma_{n'} + \gamma_n)C_{n'}C_n^*, \tag{1.98}$$

where (1.50) has been introduced. The equation (1.98) holds for one particle. If we assume that an ensemble consists of particles with identical Hamiltonians $H$ and decay rates $\gamma$, we can take the ensemble average of (1.98) and obtain directly the density matrix elements. The relaxation matrix $\Gamma$ is defined by the equation

$$\langle\varphi_n|\Gamma|\varphi_m\rangle = \gamma_n\delta_{nm}. \tag{1.99}$$

Then the density matrix equation of motion can be written

$$i\hbar\frac{d}{dt}\rho = [H, \rho] - i\hbar\frac{1}{2}(\Gamma\rho + \rho\Gamma). \tag{1.100}$$

Later, in Sec. 1.6, we discuss more complicated relaxation processes appearing in the density matrix equation.

Next we want to relate the density matrix elements to observable quantities of the system. As the expectation values (1.65) are bilinear in the state amplitudes, they will be linear in the elements of the density matrix as shown in (1.69).

The *diagonal elements* of the density matrix $\rho_{nn}(t)$ in any representation give the probability of occupation of the state $|\varphi_n\rangle$ identified by the quantum number $n$. This was shown in (1.80). We must note that this occupation probability is a function of time; even systems evolving in time have well-defined expectation values because we expect to be able to carry out measurements in arbitrarily short time intervals.

Other operators can be assigned expectation values in the same way. The dipole operator in the interaction (1.59) has the expectation value

$$\langle\mu\rangle = \mathrm{Tr}(e\mathbf{r}\rho) = e\sum_{nn'}\langle n|\mathbf{r}|n'\rangle\rho_{n'n}. \tag{1.101}$$

If the quantum numbers $n$ and $n'$ are related by the dipole selection rules (1.61–1.63), the corresponding matrix elements $\rho_{n'n}$ contribute to the expectation value of the dipole operator $e\langle n|\mathbf{r}|n'\rangle$; the other elements do not. It is often said that these density matrix elements give the dipole moment of the atom. This is an imprecise expression that must be understood in the way explained.

**Example.** We consider a two-level system only, Fig. 1.9, and the dipole matrix element $\langle 1|r|2\rangle$ is nonvanishing. The density matrix is

$$\rho = \begin{bmatrix} \rho_{22} & \rho_{21} \\ \rho_{12} & \rho_{11} \end{bmatrix}. \tag{1.102}$$

The diagonal elements $\rho_{11}$ and $\rho_{22}$ denote the occupation probabilities of states 1 and 2, respectively. Multiplying by the total number of atoms involved, $N$, we obtain the average number of atoms in the ensemble on levels 1 and 2. The expectation value of the dipole matrix element is

$$\begin{aligned} \langle \mu \rangle = \langle er \rangle &= e(r_{12}\rho_{21} + r_{21}\rho_{12}) \\ &= 2e(\mathrm{Re}\, r_{12}\, \mathrm{Re}\, \rho_{21} - \mathrm{Im}\, r_{12} \mathrm{Im}\, \rho_{21}). \end{aligned} \tag{1.103}$$

The phase of $r_{12}$ is determined by the relative phase of the wave functions of the states 1 and 2. These are usually chosen real and hence the dipole matrix element

$$r_{12} = \int \varphi_1(\mathbf{r}) r \varphi_2(\mathbf{r})\, d^3\mathbf{r} \tag{1.104}$$

becomes real. Then (1.103) is

$$\langle \mu \rangle = e r_{12}(\rho_{21} + \rho_{12}). \tag{1.105}$$

In this sense we can say that $\rho_{12}$ is *the dipole moment* of the system.

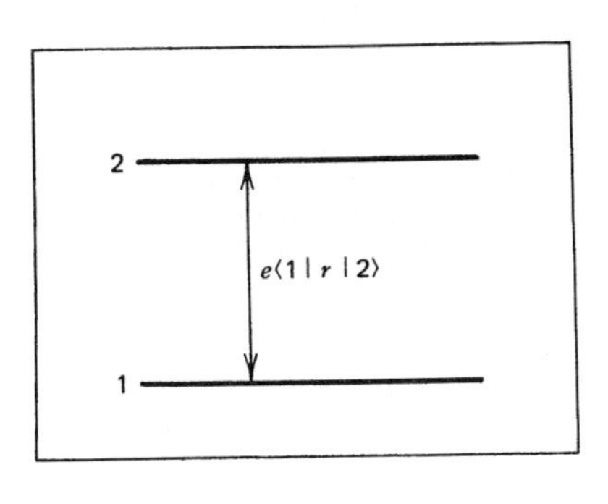


**Fig. 1.9** The "two-level atom," where the states are coupled by the dipole matrix element $e\langle 1|r|2\rangle$.

In theoretical discussions we often choose the matrix elements $r_{nn'}$ to be real for simplicity. Where this assumption needs to be amended, the modifications are straightforward.

If more than two levels are involved there are higher multipole operators. They become tensors and can be expressed in terms of spherical harmonics $Y_l^m$. For given $l$ we have $2l + 1$ components; the case $l = 1$ gives a vector like the dipole moment $\boldsymbol{\mu} = e\mathbf{r}$. The quadrupole operators $Q$ have $l = 2$ and can be represented by $3 \times 3$ hermitean matrices with zero trace. They are easily seen to contain five independent parameters.

For a three-level system with dipole coupling from state 2 to states 1 and 3, we need the element $\rho_{31}$ to express the expectation value of all the components of the quadrupole operator. Hence this matrix element is sometimes referred to as the quadrupole matrix element. This is imprecise, partly because more elements are needed to give all the five components of the quadrupole, and partly because the density matrix enters only the *expectation value* of the quadrupole, not the matrix element of the operator.

For higher multipoles we need further elements of the density matrix. For spectroscopy of atoms and molecules, dipole radiation is dominating, and we need only rarely concern ourselves with higher multipoles.

The formal solution of the equation of motion (1.100) without decay terms is

$$\rho(t) = e^{-iHt/\hbar}\rho(0)e^{iHt/\hbar}, \tag{1.106}$$

and the expectation value of a quantity $A$ is given by

$$\langle A(t)\rangle = \operatorname{Tr} A\rho(t) = \operatorname{Tr} Ae^{-iHt/\hbar}\rho(0)e^{iHt/\hbar}. \tag{1.107}$$

The invariance of the trace under cyclic permutations gives

$$\begin{aligned}\langle A(t)\rangle &= \operatorname{Tr}\left[e^{iHt/\hbar}Ae^{-iHt/\hbar}\rho(0)\right]\\ &= \operatorname{Tr}(A_H(t)\rho(0)),\end{aligned} \tag{1.108}$$

where $A_H(t)$ is the operator $A$ in the Heisenberg picture. Hence we see that the time-dependent observable expectation values can be calculated either in the Heisenberg or Schrödinger picture. In the latter, however, the density matrix remains a constant $\rho(0)$.

In summary we conclude that the measured averages over laboratory samples are taken to constitute a realization of the ensemble averaging procedure implied by quantum mechanics and a distribution of other, unobserved, parameters of the individual particles. The diagonal elements determine the (instantaneous) *occupation probabilities* of the states. The off-diagonal elements give the *expectation values* of *multipole matrix elements*, among which are to be found the main observables of the system.

One advantage of the use of a density matrix is the easy physical interpretation of the results. Another is the possibility to include various physical effects phenomenologically into the equation of motion. This is done in the next section.

## 1.6. RELAXATION TERMS IN THE DENSITY MATRIX EQUATION OF MOTION

One advantage of the use of a density matrix description is that the physical interpretation of its elements allows us to add various processes directly to its equation of motion.

*Decay to Unobserved Levels.* We have already shown in Eq. (1.98) that spontaneous decay to unobserved levels can be represented by a decay term of the form

$$\frac{d}{dt}\rho_{nn'} = -\frac{1}{2}(\gamma_n + \gamma_{n'})\rho_{nn'} + \cdots. \tag{1.109}$$

For the level populations this leads to an exponential decay of their probability of occurrence $\rho_{nn}(t) \propto e^{-\gamma_n t}$. As each spontaneous decay process will emit a photon of radiation, we find from (1.109) that the number of spontaneously emitted photons will be proportional to $\gamma_n \rho_{nn}(t)$, and hence the intensity of spontaneous emission is often taken to be a direct measure of the population on the decaying level.

*Spontaneous Decay between Levels.* If both levels involved in spontaneous decay are of interest, we need to know more than (1.109) about the decay. In the situation of Fig. 1.10 the upper state decays like

$$\frac{d}{dt}\rho_{22} = -\Gamma\rho_{22}, \tag{1.110}$$

where we write out only that term which describes the decay. Each decay process starting at the upper level must end up in the lower level, and hence the rate of growth of probability on level 1 must equal that of loss from level 2. Thus we obtain the term

$$\frac{d}{dt}\rho_{11} = \Gamma\rho_{22} \tag{1.111}$$

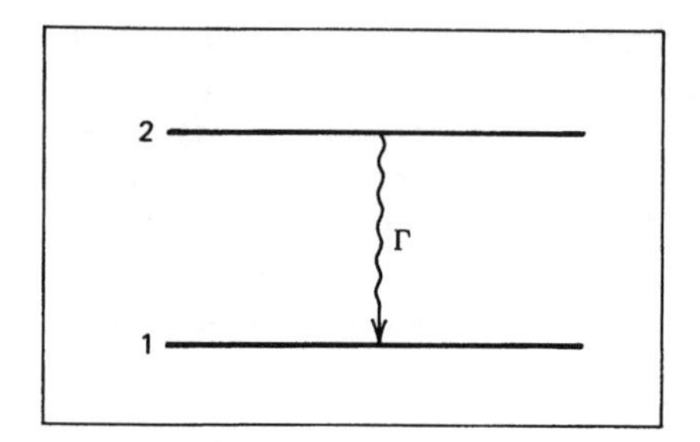


**Fig. 1.10** The "two-level atom" with the upper level decaying spontaneously with the rate Γ to the lower level.

for the lower level. We also need to know the contribution to the off-diagonal element from spontaneous decay. There is no simple way to derive this, only a detailed quantum mechanical analysis can give the decay rate, which is half that of the upper level

$$\frac{d}{dt}\rho_{21} = -\frac{1}{2}\Gamma\rho_{21} + \cdots. \tag{1.112}$$

The spontaneous decay rate is calculated from quantum electrodynamics to be

$$\Gamma = \frac{1}{4\pi\varepsilon_0} \cdot \frac{4}{3} \cdot \frac{\mu_{21}^2 \omega_{21}^3}{\hbar c^3} \tag{1.113}$$

where the energy difference between the levels is $E_2 - E_1 = \hbar\omega_{21}$ and $\mu_{21}$ is the dipole moment matrix element between the levels.

If one level can decay to several final states, we can add the processes directly for the diagonal elements. In the situation of Fig. 1.11, we have

$$\frac{d}{dt}\rho_{11} = -\sum_{k=2}^{5} \Gamma(1 \to k)\rho_{11} + \cdots$$

$$\frac{d}{dt}\rho_{kk} = \Gamma(1 \to k)\rho_{11} + \cdots. \tag{1.114}$$

For the off-diagonal elements a more detailed analysis is needed.

The quantum electrodynamic analysis needed for spontaneous decay is discussed in Chapter 6. For the moment we can consider the effects of spontaneous decay in a phenomenological way according to the treatment presented here.

*Quenching Collisions of Active Levels.* When the spectroscopically active atom meets another particle, the encounter may be violent enough to change

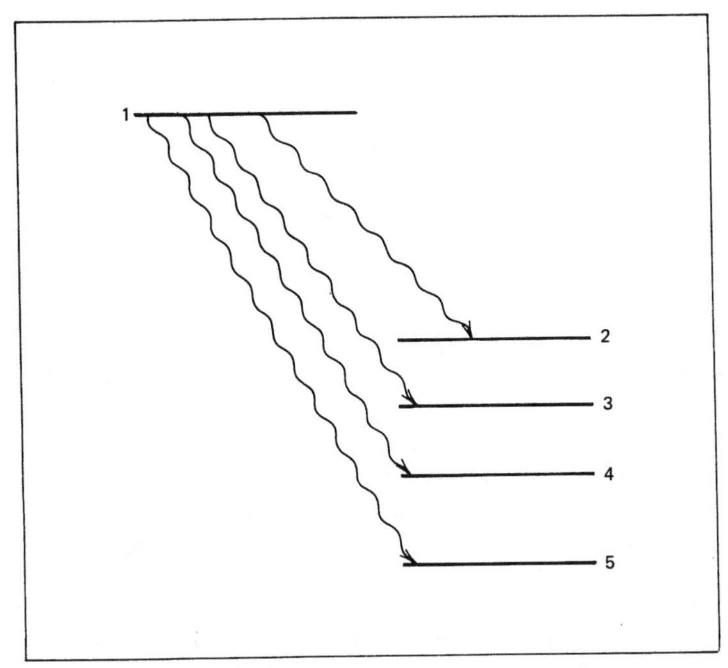


**Fig. 1.11** The total spontaneous decay rate of level $|1\rangle$ is the sum of decays to all possible lower levels $|i = 2,\ldots,5\rangle$. The probability that disappears from $|1\rangle$ is distributed over the levels $|i\rangle$ according to the ratios of the decay rates.

the atom into energy levels not observed in the measurement—or even to ionize it. In this case the interaction with radiation is quenched, and the atom effectively disappears from the measured sample. Such collisions are called *strong*. This can be described as a random process, where the survival probability is an exponential of the form $e^{-\gamma t}$ with $\gamma$ being an encounter rate. Such a process where events occur at totally independent moments is called a Poisson process; the best-known example is radioactive decay. If a density matrix without collisions has the time development $\rho_0(t)$, the probability function for the atom with collisions is

$$\rho(t) = e^{-\gamma t}\rho_0(t). \tag{1.115}$$

In the equation of motion this gives

$$\frac{d}{dt}\rho(t) = -\gamma\rho(t) + e^{-\gamma t}\frac{d}{dt}\rho_0(t). \tag{1.116}$$

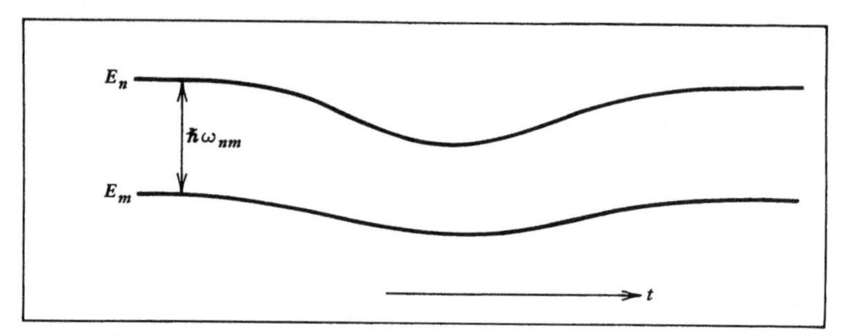


**Fig. 1.12** When two atoms approach each other in a collision, their mutual interaction shifts the energy levels with time. The atomic transition frequency $\omega_{nm}$ then becomes a function of time $t$ during the collision.

The last term gives the ordinary time development as in (1.100), and the quenching collisions add a decay rate $\gamma$ equal for all matrix elements.

*Effects of Weak Collisions.* When an optically active atom collides with another particle, the mutual interaction will displace the energy levels according to Fig. 1.12. The transition frequency $\omega_{nm}$ becomes a function of time. If we can assume the displacement to be so slow that the field sees the instantaneous energy difference only, the level changes are called *adiabatic*.

If we neglect all other causes of time evolution, the off-diagonal density matrix element evolves according to

$$i\hbar\frac{d}{dt}\rho_{nm} = \hbar\omega_{nm}(t)\rho_{nm}. \tag{1.117}$$

The solution of this equation is

$$\begin{aligned}\rho_{nm}(t) &= \exp\left[-i\int_{t_0}^{t}\omega_{nm}(\tau)\,d\tau\right]\rho_{nm}(t_0)\\ &= e^{-i\omega_{nm}(t-t_0)}\exp\left[-i\int_{t_0}^{t}\Delta\omega_{nm}(\tau)\,d\tau\right]\rho_{nm}(t_0),\end{aligned} \tag{1.118}$$

where we have separated the static part of the transition frequency $\omega_{nm}$, valid for infinite separation of the atoms ($t = \pm\infty$), and

$$\Delta\omega_{nm} = \omega_{nm}(t) - \omega_{nm}. \tag{1.119}$$

There are additional parameters needed to specify the collision. The type and state of the colliding particle and the collision geometry are not observed in the process. Only an ensemble average will affect the measurement. We proceed to define a quantity $\overline{e^{-i\theta}}$ which is the collisional phase factor averaged over all possible encounters. Without going into the details of the collision process we write

$$\overline{e^{-i\theta}} = \left\langle \exp\left[ -i \int \Delta\omega_{nm}(\tau)\, d\tau \right] \right\rangle_{\text{one collision}} . \tag{1.120}$$

Let us consider one atomic history. The atom proceeds from $t_0$ to $t_1$ without collision, has one collision at $t_1$ and no further collisions up to the time $t$. The probability of no collisions in the intervals $(t_0, t_1)$ and $(t_1, t)$ are $\exp[-(t_1 - t_0)/T]$ and $\exp[-(t - t_1)/T]$ respectively, where $T$ is the average time between collisions. The probability of one collision in the interval $(t_0, t)$ is $[(t - t_0)/T]$. When we form an ensemble average over all possible collisional histories, we can average over the individual collisions independently. The density matrix solution of (1.117) with a history of one collision becomes then

$$\begin{aligned} \rho_{nm}(t) &= e^{-i\omega_{nm}(t-t_1)} e^{-(t-t_1)/T} \left( \frac{t - t_0}{T} \right) \overline{e^{-i\theta}} \\ &\quad \times e^{-i\omega_{nm}(t_1 - t_0)} e^{-(t_1 - t_0)/T} \rho_{nm}(t_0) \\ &= e^{-i\omega_{nm}(t-t_0)} e^{-(t-t_0)/T} \left( \frac{t - t_0}{T} \right) \overline{e^{-i\theta}} \rho_{nm}(t_0), \end{aligned} \tag{1.121}$$

where the averaged phase factor $\overline{e^{-i\theta}}$ caused by the collision has been introduced. The time of duration of the collision at $t_1$ is assumed brief compared with the other time intervals and the time between collisions $T$.

During the history of an atom we have a series of $N$ independent collisions. They are brief encounters occurring with large intervals, and hence they are Poisson distributed. As the average number expected in our interval $(t_0, t)$ is $(t - t_0)/T$, we find the probability for $N$ collisions

$$P_N = \frac{1}{N!} \left( \frac{t - t_0}{T} \right)^N e^{-(t-t_0)/T}. \tag{1.122}$$

As the collisions are independent, their effects are uncorrelated, and each multiplies on the average the density matrix by the same factor (1.120). The time evolution factors between the collisions combine as in (1.121), and we find the matrix element at time $t$ for a particle experiencing $N$ collisions with each averaged over all possible types. The result is

$$\rho_{nm}(t; N) = e^{-i\omega_{nm}(t-t_0)}\left(\overline{e^{-i\theta}}\right)^N \rho_{nm}(t_0). \tag{1.123}$$

This atomic history occurs with the probability $P_N$, and, forming an averaged density matrix by combining all histories with the appropriate weight, we obtain

$$\begin{aligned}
\overline{\rho_{nm}(t)} &= \sum_{N=0}^{\infty} P_N \rho_{nm}(t; N) \\
&= e^{-i\omega_{nm}(t-t_0)} e^{-(t-t_0)/T} \\
&\quad \times \sum_{N=1}^{\infty} \frac{1}{N!}\left(\frac{t-t_0}{T}\right)^N \left(\overline{e^{-i\theta}}\right)^N \rho_{nm}(t_0) \\
&= \exp\left[-i\omega_{nm}(t-t_0) - \frac{t-t_0}{T}\left(1 - \overline{e^{-i\theta}}\right)\right]\rho_{nm}(t_0).
\end{aligned} \tag{1.124}$$

For this function we have the equation of motion

$$i\frac{d}{dt}\overline{\rho_{nm}} = \left[\omega_{nm} + i\frac{\overline{e^{-i\theta}} - 1}{T}\right]\overline{\rho_{nm}} + \cdots. \tag{1.125}$$

Because the collisional contribution $\overline{e^{-i\theta}}$ usually is a complex number, we find the *collisionally shifted* resonance frequency

$$\omega'_{nm} = \omega_{nm} - \frac{\operatorname{Im} \overline{e^{-i\theta}}}{T}. \tag{1.126}$$

As the average collision rate $T^{-1}$ is proportional to the number of perturbers, the pressure, we call $\omega'_{nm} - \omega_{nm}$ a pressure shift. There also appears a relaxation rate in (1.125) in the form of a *phase perturbation due to collisions*

$$\gamma_{ph} = \frac{\left(1 - \operatorname{Re} \overline{e^{-i\theta}}\right)}{T} \geq 0. \tag{1.127}$$

This is called *pressure broadening*. If the off-diagonal element decays by spontaneous emission according to Eq. (1.109) we find for $\rho_{nm}$ the total decay rate

$$\gamma_{nm} = \tfrac{1}{2}(\gamma_n + \gamma_m) + \gamma_{ph} \geq \tfrac{1}{2}(\gamma_n + \gamma_m). \tag{1.128}$$

Physically speaking, the collisions interrupt the phase of the oscillating dipole moment of the atom at intervals of averaged length $T$; see Fig. 1.13. The spectrum of the radiation is its Fourier transform, and this becomes broadened by an amount of order $T^{-1}$. This is the origin of the phase interruption pressure broadening factor $\gamma_{ph}$ in (1.127).

*Incoherent Pumping.* When atoms are introduced into the atomic levels under consideration by some totally incoherent mechanisms like collisions, light flashes, or chemical processes, we find a constant rate of growth of the probability of occupation of the levels concerned. This can be written as a term

$$\frac{d}{dt}\rho_{nn} = \Lambda_n + \cdots; \tag{1.129}$$

there are no corresponding effects on the off-diagonal elements.

We have considered a large number of phenomenological modifications of the density matrix equation of motion. For spectroscopic applications these are assumed to *occur independently*, and we can *add their effects directly* in the equation of motion. When all this is taken into account, the normalization of the density matrix

$$\operatorname{Tr}\rho = \sum_n \rho_{nn}$$

is usually not conserved.

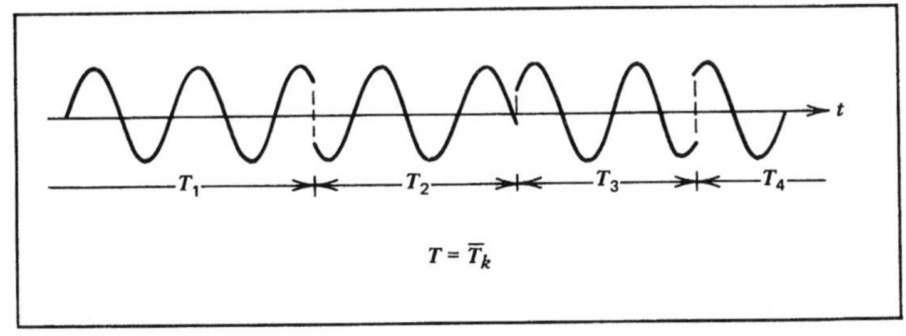


**Fig. 1.13** A periodic oscillation is affected by random jumps in the phase with an average interval of time duration $T$. The spectrum undergoes a broadening of order of magnitude $= T^{-1}$.

## 1.7. COHERENCE AND DEPHASING

When an off-diagonal density matrix element $\rho_{nm}$ differs from zero, it means that some members in the ensemble must be in quantum mechanical states with a linear superposition of $|\varphi_n\rangle$ and $|\varphi_m\rangle$, namely,

$$|\psi\rangle = c_n|\varphi_n\rangle + c_m|\varphi_m\rangle + \cdots. \tag{1.130}$$

The density matrix element is defined as

$$\rho_{nm} = \overline{c_n c_m^*} = \overline{|c_n c_m| \exp[i(\theta_n - \theta_m)]} \tag{1.131}$$

where $\theta_n$ is the phase of $c_n$ and $\theta_m$ that of $c_m$. Even if the coefficients $c_n$, $c_m$ remain finite, the average in (1.131) may go to zero if the phase difference $\Delta\theta = \theta_n - \theta_m$ is distributed over all values; the ensemble contains random phases.

When $\rho_{nm}$ is nonzero, there must be some relationship, correlation, between the phases of the various individuals making up the ensemble. As interference phenomena in quantum mechanics derive from cross-terms between different components of the atomic state as in (1.130), they can occur only if the phases are correlated. Only then can the members of the ensemble cooperate to give the interference. The different individuals must be in coherent states. This coherence can be seen as nonvanishing off-diagonal elements of the density matrix, and consequently these are sometimes referred to as *coherences*. If the states combined in (1.130) are separated by an optical transition, the corresponding matrix element is called *optical coherence*; if they are separated by a radio-frequency transition, they are called *rf-coherences*. Because the latter occur between the magnetic sublevels, they are also called *Zeeman coherences*.

Any physical process that acts differently on the separate individuals of the ensemble will make the relative phases of the components proceed at different rates and a *dephasing* occurs. An example was presented in the preceding section: The collisional phase shifts make the off-diagonal elements in (1.125) decay at an additional rate $\gamma_{ph}$ as compared with the diagonal elements. This is a relaxation process that is not due to any dissipation acting on the individual particles, but it only destroys the coherences between the individuals constituting the ensemble and can be seen only in the averages. In nuclear magnetic resonance (NMR), where these processes were discussed first, the relaxation rate at which energy is dissipated is characterized by a time constant $T_1$, and the relaxation destroying coherence has the time constant $T_2$. Processes acting differently on each particle and thus destroying the coherence are called *inhomogeneous* or $T_2$-*processes*; see Fig. 1.14. Because they do not really dissipate the states but only dephase the coherence, the information can sometimes be recovered. Such an experiment is the *pulse echo*, which can be observed both at radio and optical frequencies.

In the collisional example we met a dephasing process caused by individual brief encounters; in many cases a random perturbation can act continuously to make the phase difference $\Delta\theta$ drift in a diffusionlike manner.

**Example.** As an illustration we take the distribution function $W(\theta)$ of the phase to be governed by the diffusion equation.

$$\frac{\partial}{\partial t} W = D \frac{\partial^2 W}{\partial \theta^2}. \tag{1.132}$$

For a system with a well-defined initial phase we set

$$W(t=0) = \delta(\theta - \theta_0). \tag{1.133}$$

The solution of (1.132–1.133) is known to be

$$W(\theta, t) = \frac{1}{\sqrt{4\pi D t}} e^{-(\theta-\theta_0)^2/4Dt}. \tag{1.134}$$

This describes a diffusion process with

$$\overline{\Delta\theta^2} = 2Dt, \tag{1.135}$$

and the average value of the dephasing factor in off-diagonal density matrix

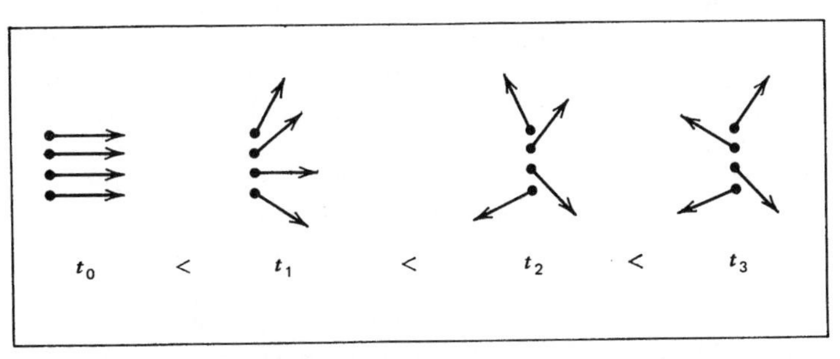


**Fig. 1.14** A set of dipole moments originally in phase at time $t_0$, will dephase if they rotate with different rates. At the time $t_3$ their average value has effectively decayed to zero.

elements like (1.131) becomes

$$\overline{e^{i\Delta\theta}} = \frac{1}{\sqrt{4\pi Dt}} \int_{-\infty}^{+\infty} e^{-\Delta\theta^2/4Dt} e^{i\Delta\theta} d\Delta\theta = e^{-Dt}, \tag{1.136}$$

and an exponential decay appears because of the dephasing. A similar result was obtained in the previous section (1.124). More complicated perturbations can cause other types of decay of the coherences.

We prepare a system in a pure quantum state

$$|\psi\rangle = \sum_n c_n |\varphi_n\rangle, \tag{1.137}$$

where the states $|\varphi_n\rangle$ may be taken to be the energy eigenstates. The density matrix is then given by

$$\rho_{nm} = c_n c_m^*, \tag{1.138}$$

and many nonvanishing coherences exist. Owing to perturbations acting on the system these coherences dephase, and after a time of the order $T_2$ the off-diagonal density matrix elements have decreased appreciably. The system can then no longer be described by a pure state, but can only be represented by the density matrix. This is an example of the reduced density matrix (1.93). When enough time has passed to make the off-diagonal elements vanish, the system can be found in the various states $|\varphi_n\rangle$ with a probability distribution $\rho_{nn} = |c_n|^2$ if no processes but the dephasing have occurred. If we make an assembly of systems to verify these statistical predictions, it is totally consistent with quantum mechanics to assume that each member of this assembly has settled in one of the states $|\varphi_n\rangle$ in such a way that their distribution realizes the probability of occurrence of the various states. Thus the statistical interpretation of quantum mechanics allows us to give a classical probability interpretation to its results after such asymptotically long times for which all coherences have dephased to negligibly small values.

The more complicated a system is, the faster its phase relations become muddled. Thus macroscopic systems have extremely short $T_2$ times and cannot be found in linear superpositions of their quantum states. This appears to explain why they can be described classically.

An interesting dephasing effect occurs when the laser light coupling the atomic levels has a randomly fluctuating phase. This is a common effect for an amplitude-stabilized laser, the phase of which is varying in a stochastic way. We assume the field amplitude to be given by

$$E = \hbar \mathscr{E} e^{i\varphi(t)} \tag{1.139}$$

where $\varphi$ is the random variable. The equation of motion for the two-level density matrix acquires terms of the type

$$i\dot{\rho}_{22} = -\mu\mathscr{E}\left(e^{i\varphi}\rho_{12} - e^{-i\varphi}\rho_{21}\right) + \cdots \tag{1.140}$$

$$i\dot{\rho}_{21} = (\omega_{21} - i\gamma_{21})\rho_{21} + \mu\mathscr{E}e^{i\varphi}(\rho_{11} - \rho_{22}) + \cdots, \tag{1.141}$$

where the uninteresting terms are omitted and only two equations have been written out. These equations define the density matrix elements $\rho_{nn'}$ as nonlinear functions of the random variable $\varphi(t)$, and hence they too become random variables describing a stochastic process.

If we perform the change of variables to the new stochastic process

$$\tilde{\rho}_{21} = e^{-i\varphi}\rho_{21}, \tag{1.142}$$

we find that (1.141) is transformed into

$$i\dot{\tilde{\rho}}_{21} = (\omega_{21} - \dot{\varphi} - i\gamma_{21})\tilde{\rho}_{21} + \mu\mathscr{E}(\rho_{22} - \rho_{11}) + \cdots \tag{1.143}$$

and the only dependence on $\varphi$ is through the derivative $\dot{\varphi}$. Its fluctuations appear as a random modulation of the level spacing $\omega_{21}$, which will blur the phases. The effect of light-phase fluctuations is similar to collisional effects, see Sec. 1.5, but, as $\varphi$ fluctuates symmetrically around zero, we get no frequency shifts in most cases but only a broadening.

If we integrate Eq. (1.143), we find the time evolution

$$\tilde{\rho}_{21}(t) \propto \exp[-i(\omega_{21} - i\gamma_{21})t + i\varphi(t)]. \tag{1.144}$$

When $\varphi$ undergoes its random motion the situation becomes similar to the one already encountered in Eq. (1.131), and a diffusive type assumption for $\varphi(t)$ gives an additional decay due to the field fluctuations exactly as in (1.136). This affects only the off-diagonal elements of the density matrix; the diagonal equations like (1.140) become independent of $\varphi$ after the transformation (1.142). We have thus found that phase diffusions of the laser fields provide an additional dephasing mechanism for the atomic system. These questions are discussed in more detail in Chapter 5.

The spectral line shape derived from an exponential decay is Lorentzian. For large frequencies this goes to zero as $\omega^{-2}$ only; one consequence is that the spectrum lacks a second moment. In a physical system the Lorentzian spectrum must be cut off by some mechanism. To see how this emerges we consider a more detailed Brownian motion model for the phase $\varphi$ than that given by a simple diffusion process such as that in Eq. (1.132). Its equation of motion is taken to be

$$\ddot{\varphi} = -\kappa\dot{\varphi} + F(t), \tag{1.145}$$

where $\kappa$ is a damping rate and $F(t)$ is a random, Langevin force with the correlation property

$$\langle F(t)F(t')\rangle = 2D\delta(t-t'). \tag{1.146}$$

This model has two parameters built into it: One is the correlation time $\kappa^{-1}$ of noise appearing in Eq. (1.145). The theory gives

$$\langle\dot{\varphi}(t)\dot{\varphi}(t')\rangle = \frac{D}{\kappa}e^{-\kappa|t-t'|}; \tag{1.147}$$

see the discussion in Sec. 5.1. The other parameter is $D$, which from Eq. (1.146) is seen to measure the power of the random force. To obtain the diffusion process with its white noise (frequency-independent noise) we must let $\kappa$ go to infinity but keep $(D/\kappa^2)$ constant.

This model satisfies the Fokker-Planck equation

$$\frac{\partial f}{\partial t} = \kappa\frac{\partial}{\partial\dot{\varphi}}[\dot{\varphi}f] + D\frac{\partial^2 f}{\partial\dot{\varphi}^2}, \tag{1.148}$$

where $f(\dot{\varphi}, t)$ is the probability distribution of $\dot{\varphi}$. With the initial condition

$$f(\dot{\varphi}, t=0) = \delta(\dot{\varphi}-\dot{\varphi}_0) \tag{1.149}$$

we obtain the solution†

$$f(\dot{\varphi}, t) = \frac{1}{\sqrt{2\pi}\,\sigma(t)}\exp\left[-\frac{(\dot{\varphi}-m(t))^2}{2\sigma^2(t)}\right] \tag{1.150}$$

where

$$m(t) = \dot{\varphi}_0 e^{-\kappa t} \tag{1.151}$$

$$\sigma^2(t) = \frac{D}{\kappa}[1-e^{-2\kappa t}]. \tag{1.152}$$

† This result is derived in the article by Chandrasekhar in N. Wax, Ed., *Selected Papers on Noise and Stochastic Processes*, Dover, New York, Sec. 1.

If we average the off-diagonal density matrix (1.144), we find

$$\overline{\tilde{\rho}_{21}} \propto \overline{e^{i\varphi(t)}} = \overline{\exp\left(i\int_0^t \dot{\varphi}(t')\,dt'\right)}$$

$$= \lim_{N\to\infty} \prod_{j=0}^{N} \overline{e^{i\dot{\varphi}(t_j)}} = \lim \prod_{j=0}^{N} e^{-\sigma^2(t_j)/2}$$

$$= \exp\left[-\frac{1}{2}\int_0^t \sigma^2(t')\,dt'\right]$$

$$= \exp\left[-\frac{D}{3\kappa}\left(t + \frac{1}{2\kappa}\left(e^{-2\kappa t} - 1\right)\right)\right], \tag{1.153}$$

where we have calculated the average $e^{i\dot{\varphi}}$ using (1.150) and (1.152).

The result in (1.153) gives different behavior in two limits:

1. For short times such that $\kappa t \ll 1$, we obtain

$$\overline{\tilde{\rho}_{21}} \propto \exp[-Dt^2]. \tag{1.154}$$

After a Fourier transform this gives a spectrum of the form

$$\Phi(\omega) = e^{-\omega^2/2D}. \tag{1.155}$$

As this holds for short times, it is the asymptotic behavior for large frequencies $\omega$. Thus the high frequency cutoff is exponential according to (1.155). The actual form derives from our model, but the high-frequency cutoff is always determined by the correlation time $\kappa^{-1}$ of the mechanism causing the line width. This holds both for collisional line widths and line shapes due to light field fluctuations. In the former case the correlation time is of the order of the duration of an individual collision.

2. The long-time behavior for $\kappa t \gg 1$ is given by

$$\overline{\tilde{\rho}_{21}} \propto \exp\left[-\frac{D}{2\kappa}\left(t - \frac{1}{2\kappa}\right)\right] \simeq e^{-Dt/2\kappa}. \tag{1.156}$$

This gives a Lorentzian spectrum of width $(D/2\kappa)$. The two behaviors cross at the time $t_0 = (2\kappa)^{-1}$, and then the height of the spectrum is

$$\Phi_0 = \exp\left[-\frac{2\kappa^2}{D}\right]. \tag{1.157}$$

There are now again two possibilities: (1) Either $D \gg \kappa^2$ and we have

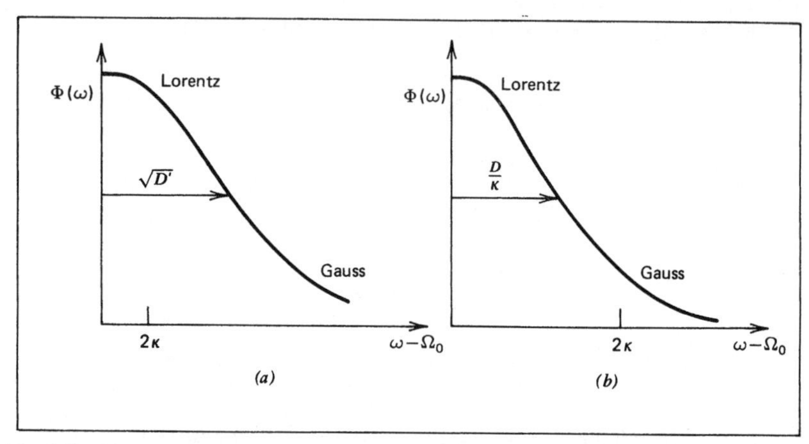


**Fig. 1.15** The crossover between the Gaussian and Lorentzian behavior of a spectral line: (*a*) The Gaussian behavior dominates, and a Lorentzian shape exists only for very low frequencies where its presense cannot be easily verified. (*b*) Most of the line shows a Lorentzian shape, and only the tail of large detunings cuts off faster in a Gaussian way. This corresponds to the usual situation in atomic systems where the cutoff behavior is very difficult to investigate.

essentially a Gaussian line, as shown in Fig. 1.15a, or (2) $D \ll \kappa^2$ and we have essentially a Lorentzian line of width

$$\frac{D}{2\kappa} = \sqrt{\frac{D}{\kappa^2}}\sqrt{D} \ll \sqrt{D}, \tag{1.158}$$

shown in Fig. 1.15b. When $\kappa$ increases, this case becomes dominant and the line width decreases according to (1.158). This is an example of *motional narrowing*: when the noise becomes faster, the line width decreases.

For atomic spectra, the deviations from a Lorentzian behavior usually occur so far in the wings that they are very hard to observe.

An amusing illustration of the role of the dephasing of quantum coherence is offered by Schrödinger's infamous cat. Into a discussion of the basic aspects of quantum mechanics Schrödinger introduced a cat in a closed box. The box also contained a radioactive sample and a counter. At a random time the counter registered emitted radiation and caused a mechanical hammer to break a bottle containing cyanide. This killed the cat, but the outside observer could not verify this fact without opening the box and inspecting its contents.

As the radioactive decay depended on quantum mechanics, there occurred a linear superposition of the two states, an undecayed sample and a live cat and a recorded decay and a dead cat. But what about the cat?

In my opinion, the cat must undoubtedly be regarded as a macroscopic object, which has an extremely short $T_2$ time. After a longer time $t \gg T_2$, the cat is in its two states with the probabilities determined by quantum mechanics, but these are *classical probabilities* verifiable on an ensemble of cats, each one being either dead or alive. Because it takes the observer (and of course the cat) a finite time to verify whether the cat is dead or alive, it is not possible to observe any linear superposition states of such a complicated system. What is the shortest time it takes to verify that something is alive, or that you yourself are alive?

This situation provides an exaggerated example of the measurement process in quantum mechanics. Here the detector providing the permanent record of the decay is the whole system of detector, poison bottle, and the cat. All are fully classical. The paradoxical feature of the situation arises because we choose to introduce a conscious being into the detector. This is, of course, not necessary to obtain a permanent record of the particle detection.

## 1.8. ADIABATIC ELIMINATION PROCEDURES

In physical considerations it is very important to be able to recognize the different time scales occurring. We do this unthinkingly always when we consider an isolated system. In reality every part of the world interacts with something, but the interaction may be weak enough to be seen only over periods of time too long to affect our considerations.

In laser physics it often happens that some part of a system is able to adjust itself to the prevailing situation much more rapidly than the rest of the physical quantities. These fast variables are then able to run to their asymptotic equilibrium values during a time scale over which the other variables remain almost unchanged. Their change can then be regarded as a slow *adiabatic* drift which the fast variables must follow (almost) instantaneously. The fast variables can then be expressed as *algebraic* functions of the slow ones and can be inserted into the equations of motion for these slow variables. The fast variables have then been eliminated, and the adiabatically slow time evolution ensuing concerns only the slow variables. This is called an adiabatic elimination process.

As a simple example let us consider the coupled equations†

$$\dot{x} = -\Omega y - \Gamma x \tag{1.159a}$$

$$\dot{y} = \Omega x - \gamma y. \tag{1.159b}$$

† The equations can be taken to be the Bloch equations in the rotating wave approximation at exact resonance; see Sec. 2.3. In this case we eliminate the dipole moment adiabatically and describe the inversion only.

Eliminating $x$ we find the equation of motion

$$\ddot{y} + (\Omega^2 + \gamma\Gamma)y = -(\Gamma + \gamma)\dot{y}, \tag{1.160}$$

which describes a harmonic oscillator with a linear friction term proportional to the velocity.

Here we want to consider Eqs. (1.159) from a different point of view. We solve for $x(t)$ from (1.159a)

$$x(t) = x(0)e^{-\Gamma t} - \Omega\int_0^t e^{-\Gamma(t-t')}y(t')\,dt'. \tag{1.161}$$

We omit the initial value $x(0)$ and assume $y$ to change very little during times of order $\Gamma^{-1}$. What is the rate of change of $y$? The characteristic time scale for $y$ is given by $\gamma^{-1}$, and we must assume that this is slower than $\Gamma^{-1}$, namely,

$$\Gamma \gg \gamma. \tag{1.162}$$

In addition there is a consistency check to be made at the end of the calculation.

When $\Gamma$ is large enough, the integral in (1.161) contributes only when $t' \approx t$ and picks out the value of $y$ at this point. We find

$$x(t) = -\Omega y(t)\int_0^t e^{-\Gamma(t-t')}\,dt' = -\frac{\Omega}{\Gamma}y(t)(1 - e^{-\Gamma t}). \tag{1.163}$$

When the time interval considered is such that $t \gg \Gamma^{-1}$ we can omit the second term in the result and write

$$x(t) = -\frac{\Omega}{\Gamma}y(t). \tag{1.164}$$

The fast variable $x$ is then found to follow the slow variable $y$ adiabatically.† Its value is taken to be determined by the instantaneous value of $y$, but with the restriction that we look only at variations over times larger than $\Gamma^{-1}$. For a finer time resolution the exact transients of (1.161) must be considered. It is to be noticed that the result (1.164) can be obtained directly from (1.159a) by setting the time derivative $\dot{x}$ equal to zero. The time evolution of $x$ is assumed to run its full course at each moment.

†In the synergetic considerations introduced by H. Haken, the slow variables are said to *slave* the fast ones.

We insert (1.164) into (1.159b) and obtain the adiabatic evolution equation

$$\dot{y} = -\left(\frac{\Omega^2}{\Gamma} + \gamma\right)y. \tag{1.165}$$

This causes a time evolution of $y$ at the rate $(\Omega^2 + \gamma\Gamma)/\Gamma$, which, according to our assumptions, must be much less than $\Gamma$ for the adiabatic condition to prevail. We have

$$\frac{\Omega^2}{\Gamma} + \gamma \ll \Gamma, \tag{1.166}$$

which to the requirement (1.162) adds the necessary condition

$$\Omega \ll \Gamma. \tag{1.167}$$

This is the necessary consistency check on the adiabatic assumption.

If we look at the equations of motion for $y$ and assume it to be overdamped

$$(\Gamma + \gamma)^2 \gg (\Omega^2 + \gamma\Gamma) \tag{1.168}$$

we find that the dissipative term dominates the second-derivative term. The overdamped case can thus be described by the equation of motion

$$\dot{y} = -\frac{(\Omega^2 + \gamma\Gamma)}{\gamma + \Gamma}y = -\left(\frac{\Omega^2}{\Gamma} + \gamma\right)\left(1 - \frac{\gamma}{\Gamma} + \cdots\right)y, \tag{1.169}$$

where correction terms of order $(\gamma/\Gamma)^2$ have been neglected. In the limit $\gamma \ll \Gamma$, Eq. (1.169) agrees with the adiabatic limit of (1.160), but together with (1.162) it also implies (1.165).

In practice often the damping of the slow variable is negligible and

$$\frac{\Omega^2}{\Gamma} \gg \gamma. \tag{1.170}$$

This shows that the term $\Omega^2/\Gamma$ is dominating in (1.169), and we find a simple result. Then we have

$$\Omega\Gamma \gg \Omega^2 \gg \gamma\Gamma \tag{1.171}$$

and

$$\Omega \gg \gamma. \tag{1.172}$$

The oscillator variable $y$ is underdamped. In this case the oscillator equation

(1.160) reduces to

$$\ddot{y} + \Omega^2 y = -\Gamma \dot{y}. \tag{1.173}$$

We must, however, note that from (1.160) it is impossible to tell if $y$ is the slow or the fast variable, because $\gamma$ and $\Gamma$ occur symmetrically in this result. Only from the original equations (1.159) can we see which variable is associated with which decay rate.

In spectroscopy it is often possible to derive simplified equations by an adiabatic elimination procedure. As an example we consider the density matrix equation of motion

$$i\hbar \frac{d}{dt} \rho_{nm} = \sum_l (H_{nl}\rho_{lm} - \rho_{nl}H_{lm}) - i\hbar\gamma_{nm}\rho_{nm}. \tag{1.174}$$

Because of dephasing processes the off-diagonal elements often decay much faster than the diagonal terms. Hence we can assume that the off-diagonal terms go to their equilibrium value fast and they can be eliminated adiabatically. We introduce

$$\begin{aligned} H_{nn} &= \varepsilon_n \\ H_{nm} &= V_{nm} \quad \text{for } n \neq m. \end{aligned} \tag{1.175}$$

To see what happens we solve the off-diagonal elements from (1.174) to lowest order in the coupling $V$. Higher-order terms mainly shift and broaden the resonances. The result is

$$\begin{aligned} \rho_{nm} &= \frac{V_{nm}}{\varepsilon_n - \varepsilon_m - i\hbar\gamma_{nm}} (\rho_{nn} - \rho_{mm}) \\ &= \frac{V_{nm}}{\hbar} \frac{1}{\omega_{nm} - i\gamma_{nm}} (\rho_{nn} - \rho_{mm}). \end{aligned} \tag{1.176}$$

Inserted into the equation for the diagonal elements this gives

$$\begin{aligned} \frac{d}{dt}\rho_{nn} &= -\gamma_{nn}\rho_{nn} - \frac{i}{\hbar}\sum_l (V_{nl}\rho_{ln} - \rho_{nl}V_{ln}) \\ &= -\gamma_{nn}\rho_{nn} - i\sum_l \frac{|V_{nl}|^2}{\hbar^2}\left[\frac{1}{\omega_{ln} - i\gamma_{nl}} - \frac{1}{\omega_{ln} + i\gamma_{nl}}\right](\rho_{ll} - \rho_{nn}) \\ &= -\gamma_{nn}\rho_{nn} + \sum_l W_{ln}(\rho_{ll} - \rho_{nn}) \end{aligned} \tag{1.177}$$

which is a rate equation for the *occupation probabilities* $\rho_{ll}$ of the individual levels. Multiplying by the number $N_0$ of active atoms we find the ordinary rate equations for the *level populations* $N_l = N_0 \rho_{ll}$.

The transfer rate is given by

$$W_{ln} = \frac{2|V_{nl}|^2}{\hbar^2} \frac{\gamma_{nl}}{\omega_{nl}^2 + \gamma_{nl}^2}. \tag{1.178}$$

This result has been derived under the assumption that $\gamma_{nl}$ is large. If we, however, formally let $\gamma$ go to zero we obtain

$$W_{ln} \to 2\pi \frac{|V_{nl}|^2}{\hbar^2} \delta(\omega_{nl}) = \frac{2\pi}{\hbar} |V_{nl}|^2 \delta(E_n - E) \tag{1.179}$$

in agreement with the result of time-dependent perturbation theory. We must, however, remember that the result cannot give the correct time-dependent behavior if $\gamma_{nl} \to 0$; for the steady state result, Eq. (1.177) gives the correct perturbative answer, because then all *time derivatives* are set equal to zero and no differences in time scales are needed to justify the elimination procedure.

Rate equations like (1.177) are often used for the time evolution of laser systems. For qualitative predictions they are valuable, but their quantitative validity must be verified in detail. In addition to a fast decay of off-diagonal elements, they also require that the time evolution that follows from (1.177) be slower than that due to $\gamma_{nm}$. Thus we need a consistency check on the adiabatic assumption, just as in Eq. (1.170).

## 1.9. EFFECTS OF ATOMIC MOTION

When an atom moves toward the observer with velocity $v$, its transition frequency $\omega$ is seen to be displaced by the Doppler shift to the value

$$\omega' = \omega \frac{1 + v/c}{\sqrt{1 - (v/c)^2}} \approx \omega \left[ 1 + \left( \frac{v}{c} \right) + \frac{1}{2} \left( \frac{v}{c} \right)^2 + \cdots \right]. \tag{1.180}$$

Usually the atomic velocities are small compared to the velocity of light, and the higher-order terms can be neglected. For near-resonance radiation the wave vector of the light can be written

$$k = \frac{\Omega}{c} \approx \frac{\omega}{c}, \tag{1.181}$$

and the Doppler-shifted frequency becomes

$$\omega' = \omega + kv. \tag{1.182}$$

The atoms in a gas cell are usually assumed to have a Maxwellian distribution of each component of the velocity

$$W(v) = \frac{1}{\sqrt{\pi}\, u} e^{-v^2/u^2}, \tag{1.183}$$

where the most probable velocity

$$u = \sqrt{\frac{2k_B T}{M}} \tag{1.184}$$

is determined by the temperature $T$ and the particle mass $M$; $k_B$ is Boltzmann's constant. In a gas discharge the velocity distribution may differ from (1.183), but the deviations are expected to cause no appreciable effects.

During the time $t$ the phase of the off-diagonal density matrix element changes by an amount

$$\omega' t = \omega t + kvt, \tag{1.185}$$

where the contribution from the Doppler shift can be written

$$kvt = 2\pi \frac{vt}{\lambda}. \tag{1.186}$$

Because $vt$ is the distance traveled by the atom, we note that the Doppler shift must be taken into account when the atom travels a distance of the order of a wavelength or more during its interaction with the radiation.

For radio-frequency fields the wavelength is such that the Doppler shifts are unimportant; for optical fields the travel of a typical atom usually covers many wavelengths.

The velocity distribution is seen as a source of line broadening for a gaseous sample; because this mechanism refers to a property differing for the individual particles of the sample, not to each atom separately, it is called an *inhomogeneous relaxation* mechanism. Each atom has its own velocity, and in the laboratory that causes a dephasing of the coherence. As a simple illustration consider a sample prepared with an optical coherence $\rho_{nm}(0)$. This oscillates with the frequency $\omega'_{nm}$ which depends on the velocity $v$ of the oscillating atom. For the whole sample of gas we obtain the averaged value

$$\overline{\rho_{nm}(t)} = \int_{-\infty}^{\infty} e^{i\omega'_{nm}t} W(v)\, dv \rho_{nm}(0)$$

$$= e^{i\omega_{nm}t}\rho_{nm}(0)\frac{1}{\sqrt{\pi}\,u}\int e^{ikvt}e^{-v^2/u^2}\, dv$$

$$= e^{i\omega_{nm}t}\rho_{nm}(0)\, e^{-k^2u^2t^2/4}\frac{1}{\sqrt{\pi}\,u}\int e^{-(v-iktu^2/2)^2/u^2}\, dv$$

$$= e^{-i\omega_{nm}t}e^{-k^2u^2t^2/4}\rho_{nm}(0). \tag{1.187}$$

Here we see a Gaussian dephasing different from the exponential one we have met earlier.

The inhomogeneous distribution of atomic velocities plays an essential role in nonlinear spectroscopy of gases. When linear spectroscopy is used in emission or absorption, each atom radiates at its own frequency $\omega'$. The observed response is a superposition of the Lorentzian responses of each atom and because the Doppler width $ku$ usually exceeds the width of the Lorentzian characterizing the individual atom, the *homogeneous width*, we observe a broader response from the gas than that of the individual atoms, see Fig. 1.16.

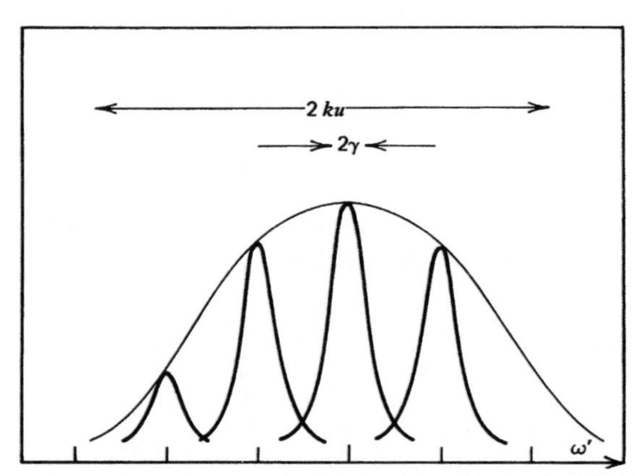


**Fig. 1.16** The individual atom has a Lorentzian response of homogeneous width $\gamma$. Because the atoms move with different velocities the spectral response of the assembly of atoms is given by a Doppler contour of width $ku$.

To introduce the effect of atomic motion into the equation for the density matrix we must collect an ensemble of all those atoms that at time $t$ are situated at $\mathbf{r}$ with the velocity $\mathbf{v}$. We assume that the atoms are labeled by the time $t_0$ and the position $\mathbf{r}_0$ at which they were introduced into the active volume of the sample or the states being considered. This may be due to some pumping process, a collision, or a chemical formation process. The classical trajectory of the particle is written

$$\mathbf{r} = \mathbf{r}_0 + \mathbf{v}(t - t_0). \tag{1.188}$$

If the state of this particle is described by

$$|\psi(t)\rangle = \sum_n C_n(t - t_0)|\varphi_n\rangle, \tag{1.189}$$

we can collect an ensemble density matrix by writing

$$\rho_{nm}(t, \mathbf{r}, \mathbf{v}) = \int d\mathbf{r}_0 \int_{-\infty}^{t} dt_0 \overline{C_n(t - t_0)C_m^*(t - t_0)} \times \delta(\mathbf{r} - \mathbf{r}_0 - \mathbf{v}(t - t_0)). \tag{1.190}$$

This collects all atoms at time $t$ and position $r$ independently of where and when they were introduced. The bar denotes an average over other processes according to our earlier discussions.

The time derivative of (1.190) becomes

$$\frac{\partial}{\partial t}\rho_{nm} = \int d\mathbf{r}_0 \overline{C_n(0)C_m^*(0)}\delta(\mathbf{r} - \mathbf{r}_0) - \mathbf{v}\cdot\frac{\partial}{\partial \mathbf{r}}\rho_{nm} + \int d\mathbf{r}_0 \int dt_0 \left[\frac{d}{dt}\overline{C_n(t - t_0)C_m^*(t - t_0)}\right]\delta(\mathbf{r} - \mathbf{r}_0 - \mathbf{v}(t - t_0)). \tag{1.191}$$

The first term is the pumping rate at the point $\mathbf{r}$, and the last one gives the rate of change because of all other processes as in Eq. (1.100). We write Eq. (1.191) in the form

$$\left[\frac{\partial}{\partial t} + \mathbf{v}\cdot\frac{\partial}{\partial \mathbf{r}}\right]\rho = \Lambda - \frac{i}{\hbar}[H, \rho] - \frac{1}{2}(\Gamma\rho + \rho\Gamma). \tag{1.192}$$

To this we can add all other relaxation rates in accordance with our considerations in Sec. 1.6.

The left-hand side of (1.192) is the total time derivative of a moving particle. The first term on the right-hand side is a pumping matrix, usually only introducing probability on the diagonal elements

$$\Lambda_{nm} = \delta_{nm}\lambda_n W(v), \tag{1.193}$$

with the velocity distribution (1.183). A similar term was discussed in (1.129). It is possible to take different velocity distributions for different states, but normally this is not necessary.

When atoms undergo collisions with perturbing particles, their velocities are changed. Accordingly the atoms are redistributed over the velocity distribution, which we can take into account by adding terms of the form

$$\frac{d}{dt}\rho_{nm}(\mathbf{v}) = \int [K_{nm}(\mathbf{v},\mathbf{v}') - \delta(\mathbf{v}-\mathbf{v}')\gamma_{nm}(\mathbf{v}')] \times \rho_{nm}(\mathbf{v}')\, d\mathbf{v}', \tag{1.194}$$

where the added decay rate is

$$\gamma_{nm}(\mathbf{v}') = \int K_{nm}(\mathbf{v},\mathbf{v}')\, d\mathbf{v}. \tag{1.195}$$

It would take us too long to go into the detailed theory of terms like (1.194), we only refer to the articles mentioned at the end of this part.

In many cases it suffices to add terms like (1.194) to the diagonal elements describing the redistribution of occupation probability. The off-diagonal elements only experience a dephasing but no velocity redistribution.

It is possible to write the equation of motion (1.192) in a different way. We introduce an additional time variable $t'$ and select a subensemble of all particles at time $t'$ but with any position $\mathbf{r}'$ so that only those that reach $\mathbf{r}$ at time $t$ are included. This subensemble is described by the density matrix

$$\bar{\rho}(\mathbf{r},t,t') = \int d\mathbf{r}'\,\delta(\mathbf{r}-\mathbf{r}'-\mathbf{v}(t-t'))\rho(\mathbf{r}',\mathbf{v},t'). \tag{1.196}$$

The equation of motion for this matrix is

$$\begin{aligned}\frac{d}{dt'}\bar{\rho} &= \int d\mathbf{r}'\Big\{-\mathbf{v}\cdot\frac{\partial}{\partial\mathbf{r}'}\delta[\mathbf{r}-\mathbf{r}'-\mathbf{v}(t-t')]\,\rho(\mathbf{r}',\mathbf{v},t') \\ &\qquad +\delta[\mathbf{r}-\mathbf{r}'-\mathbf{v}(t-t')]\frac{\partial}{\partial t'}\rho(\mathbf{r}',\mathbf{v},t')\Big\} \\ &= \int d\mathbf{r}'\,\delta(\mathbf{r}-\mathbf{r}'-\mathbf{v}(t-t'))\left(\frac{\partial}{\partial t'}+\mathbf{v}\cdot\frac{\partial}{\partial\mathbf{r}'}\right)\rho(\mathbf{r}',\mathbf{v},t'),\end{aligned} \tag{1.197}$$

where we have performed a partial integration. Inserting (1.192) and noting that inside the integral

$$H(\mathbf{r}') \equiv H(\mathbf{r} - \mathbf{v}(t - t')) \tag{1.198}$$

we reduce Eq. (1.197) to

$$\frac{d}{dt'}\bar{\rho}(\mathbf{r}, t, t') = \Lambda - \frac{i}{\hbar}[H(\mathbf{r} - \mathbf{v}(t - t')), \bar{\rho}(\mathbf{r}, t, t')] - \frac{1}{2}(\Gamma\bar{\rho} + \bar{\rho}\Gamma). \tag{1.199}$$

Here the time $t$ acts only as a parameter to label the ensemble chosen, but we have to remember that the atom occupies the position $\mathbf{r}$ when we set $t = t'$ by relation (1.196). For practical calculations the same equations emerge from both points of view. The ansatz (1.196) only solves for the flow velocity part of the equation of motion (1.192).

In these considerations we have explicitly used the fact that each atom retains its velocity during its interaction with the fields. When collisions change the atomic velocities, the delta functions can no longer be used. The equation of motion (1.192) remains valid, but the definition (1.196) can be valid only between encounters. It remains, however, possible to add terms like (1.194) to Eq. (1.192).

One way to eliminate Doppler broadening in spectroscopy is to use an atomic or molecular beam in vacuum. If the beam is created by a thermal source, an oven, and collimation of the jet of leaking atoms, it is called a thermal beam; see Fig. 1.17. The transverse velocity distribution is then

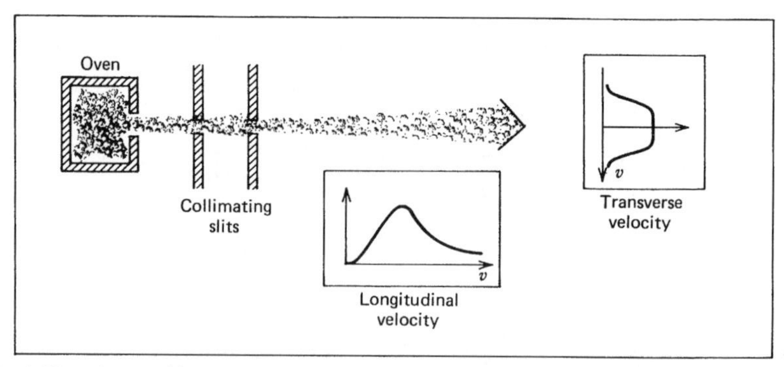


**Fig. 1.17** A beam of hot atoms (or molecules) emerges from an oven. It is collimated by slits, and a beam with a rather well-defined transverse velocity is obtained.

defined by the collimators and the longitudinal one by the velocity distribution times a flux factor proportional to the velocity. The longitudinal flux of particles then becomes

$$W_{\parallel}(v) = Cv^3 e^{-v^2/u^2}. \tag{1.200}$$

The transverse distribution cannot be made very narrow because then the total intensity drops to very low values.

An accelerator produces charged particles with well-defined energy. If such particles are made neutral, a fast beam of monoenergetic particles is obtained. Here the transverse velocity spread is very small, and the longitudinal velocity is nearly Gaussian

$$W_{\parallel}(v) = Ce^{-(v-v_0)^2/u^2}, \tag{1.201}$$

where the average velocity $v_0$ corresponds to the accelerator energy through

$$E = \tfrac{1}{2}Mv_0^2 \tag{1.202}$$

and $u^2$ defines the spread of the accelerator energy. For particles with energies in the gigaelectron volt range relativistic kinematics is necessary. In practice, however, van der Graaf accelerators are used for spectroscopy and the energy is of the order of $10^8$ eV, and a nonrelativistic treatment suffices.

## 1.10. TIME-DEPENDENT AND STEADY STATE SPECTROSCOPY

In laser spectroscopy most experiments contain a time dependence. Only for ions in solids acted upon by a continuous wave does a real steady state ensue. In magnetic resonance spectroscopy this is much more common. For gaseous systems of atoms and molecules the individual particle ordinarily interacts with the field for a finite time only. Each particle ceases to feel the action of the field, but new particles start their history of interaction, and hence the steady state condition is realized on the average. The ensemble-averaged properties of the system can be described by a steady state, which gives the observable properties measured in spectroscopy.

For lasers and many spectroscopic systems, both optically active levels are excited states that can decay spontaneously to lower levels. Thus the interacting particles disappear to unobserved levels which terminates the interaction time. A similar effect is observed when quenching collisions throw the particle out of the active levels, attach it to walls or other solid objects, or effect a chemical reaction that stops the resonant interaction. All these processes can be understood as random processes that can be described by an exponential decay term. These types of processes usually limit the interaction time for atomic spectroscopy with lasers. Only the ground state or some metastable levels live long enough to make other processes come into play.

In applications where a strong laser field is needed the laser source is often pulsed. Then the time dependence of the field determines the interaction time. A similar effect emerges in very long-lived states, especially those in molecular systems, which pass through the light beam without decaying. Then the transit time determines the interaction history. The moving particle traverses the steady state light beam and experiences a time-dependent field amplitude, which is determined by the transverse structure of the laser beam. The distribution of interaction times that emerges is determined by the velocity distribution of the particles. This, in its turn, depends on whether one measures on a particle beam or in a cell. For molecular spectroscopy the transit time often limits the interaction time. This type of limit can, of course, not be well described by exponential decay terms.

We can, however, get a good qualitative understanding of the influence of the interaction time by approximating it by an exponential decay rate $\gamma$. This corresponds to a model where we suddenly inject the atom into the full field. It stays there for a random time and is then suddenly removed. The process is thus Poissonian, and the survival probability decays exponentially. The effective interaction time is taken to be $\gamma^{-1}$.

The rate of interaction between the matter and the field is determined by several parameters: the atomic decay rates $\gamma_{ij}$, the interaction frequencies $\mu_{ij}E/\hbar$, and the detunings $\omega_{ij} - \Omega$. When any one is larger than the inverse of the interaction time, the interaction lasts for a long time. Then we can expect that it can be approximately replaced by an exponential decay. When the interaction time is such that its inverse is of the order of magnitude of the quantities listed we are in the regime of coherent transients, and then no steady state approximation can hold. The treatment must then be genuinely time dependent; see Allen and Eberly (1975).

Even for a long interaction time an exponential approximation cannot be very accurate. We may, however, expect that at the level of accuracy where only one time parameter is used to describe the functional shape of the interaction coupling, the exponential approximation will give a reasonable result. One must, in any case, be aware of the fact that such an approach may be totally misleading when special coherence effects in time are of importance. Its numerical accuracy is, at its best, only qualitative. As a first estimate it may, however, often be adequate.

## 1.11. COMMENTS AND REFERENCES

The text assumes knowledge of the basic structure of quantum mechanics and classical electrodynamics. There are many excellent textbooks on quantum physics, and the reader can choose a favorite. In electrodynamics the textbook by Jackson (1975) is particularly useful. Its Chapter 8 contains the theory of wave guides and Chapter 9, radiation theory. The adaptation to optical resonators and wave guides is presented in a simple way by Yariv (1976). This book also discusses optical detection and other technical questions of laser physics.

Our treatment of the density matrix is rather self-contained. It was first introduced by Landau (1927) and von Neumann (1927), and a beautiful exposition can be found in the book by Tolman (1938). Very useful review articles are those by Fano (1957) and ter Haar (1961). The former article is formally more elaborate, but the latter is closer to physical applications of the density matrix. Its use in magnetic resonance is discussed well by Slichter (1978), Chapter 5, and the origin and manifestations of relaxation processes in NMR are presented by Abragam (1961).

The extensive use of density matrix methods in resonance spectroscopy made it a natural tool for laser spectroscopy too. The inclusion of atomic motion, necessary for optical phenomena, was discussed early by Lamb and his collaborators (Lamb and Sanders 1960, Wilcox and Lamb 1960, and Lamb 1964). The phase-interruption effects of collisions on optical spectra were first discussed by Weisskopf (1932 and 1933); further foundations were provided by Anderson (1949). Later a large body of work has been devoted to this field. I cannot give proper credit to all these works, and I must hence refer to the literature. Especially, the collisional effects on laser spectroscopy have been reviewed briefly by Berman (1975) and in much greater theoretical detail in Berman (1978).

The theory of random noise and its effects on physical systems has been discussed thoroughly during all of this century. The standard reference, which provides all information we need, is Wax (1954). Many modern works can easily be found.

There exists a rapidly increasing literature on quantum measurement theory. Unfortunately it is of little use for practical applications. The example of Schrödinger's cat was contained in a series of articles on the basic formulation of quantum mechanics (Schrödinger 1935), which he wrote to summarize its achievements and bring out those features that were less satisfactory. The articles still provide interesting reading today.

Subsystems that evolve on different time scales can be separated into entities that form hierarchical structures. Such systems also have a wide range of applications in chemistry, biology, and even in sociology. A rate equation for the occupation probabilities of quantum states was first derived by Pauli (1928). This field has later developed into a formal method of wide applicability (Jancel 1969).

The Doppler shift was introduced into laser physics in a systematic way by Lamb (1964) and has been applied by many authors to laser problems, see, for example, Sargent et al. (1974) and Letokhov and Chebotaev (1977). The applications of lasers to nonlinear spectroscopy of atoms and molecules are well reviewed by Levenson (1982). This book provides an excellent complement to the present one. The special questions related to time-dependent phenomena are discussed by Allen and Eberly (1975).

CHAPTER 2

# Physical Effects of Strong Fields on Matter

## 2.1. THE BASIC CONCEPTS OF LIGHT-INDUCED EFFECTS ON ATOMIC MATTER

In this chapter we consider the modifications of atomic properties caused by the interaction with a strong light field. In the dipole approximation there will be a coupling to the atomic dipole moment operator only, and consequently the main observable modification will be the appearance of an induced dipole moment. Because this will be proportional to the expectation value

$$\boldsymbol{\mu} = \langle a | e\mathbf{r} | b \rangle$$

$$= e \int \varphi_a(\mathbf{r}) \mathbf{r} \varphi_b(\mathbf{r}) \, d^3\mathbf{r}, \tag{2.1}$$

only states of *different parity* can give a nonvanishing dipole moment. For $a = b$ the integral vanishes because $|\varphi_a(\mathbf{r})|^2$ is symmetric for states of given parity and $\mathbf{r}$ is antisymmetric. Atoms ordinarily have no dipole moments, and one appears only due to an interaction with the field.

For molecules there may occur an electric dipole moment even in the stationary eigenstates of the Hamiltonian. In addition, states can have mixed parity, and the parity selection rule on dipole matrix elements loses its significance. These two complications, however, require only minor modifications in the notation to fit into the theory we are developing.

For an isotropic medium like a gas of atoms or molecules, the induced polarization will be in the direction of the polarization vector of the driving field, and it goes to zero together with the field. We can, hence, write for the polarization

$$\mathbf{P} = \chi \mathbf{E}, \tag{2.2}$$

where $\chi$ is called the susceptibility of the medium. For weak fields we are in a linear regime, and the susceptibility $\chi$ is a constant; for stronger fields it becomes a function of $|E|$. Owing to isotropy, $\chi$ does not depend on the direction of the field. The modification of $\chi$ from its value at low fields is called *saturation*. Physically this comes about because the original distribution of atoms over their energy levels becomes modified by the field. Then transitions start to appear that were absent originally because the field found no atoms in the corresponding initial levels. In particular, for a two-level system, the downward transitions start to compete with the upward ones.

Because of the proportionality between **P** and **E** in gases we need not worry too much about the vector properties of our field quantities. Only when some special effect requires the inclusion of light polarization and propagation direction vectors do we write out vector indices explicitly. Especially in a crystal environment it is possible that **P** has a direction different from **E**; in this case the susceptibility $\chi$ becomes a tensor. Then $\chi$ can change in quite complicated ways when the direction of **E** varies; the crystal lattice structure is responsible for this.

The induced dipole moment beats with the field and absorbs or emits energy. Which one occurs is determined by the relative phase $\varphi$ between the field oscillation and the dipole. If the field is

$$E(t) = E\cos\Omega t, \tag{2.3}$$

the polarization must follow with the same frequency, but a phase shift is possible

$$P(t) = P\cos(\Omega t + \varphi) = P\cos\varphi\cos\Omega t - P\sin\varphi\sin\Omega t. \tag{2.4}$$

This shows that the polarization can be resolved into one component in phase with the field and one 90° out of phase. Because the rate of energy exchange with the field is given by the time average

$$-\overline{E\frac{dP}{dt}} = \Omega EP\sin\varphi\,\overline{\cos^2\Omega t} = \frac{1}{2}\Omega EP\sin\varphi \tag{2.5}$$

only the out-of-phase component proportional to $\sin\varphi$ contributes to the energy exchange.

When the field interacts with the dipole moment the coupling energy is $\mu E$. According to simple quantum mechanics, this makes the probability of occupancy flip back and forth between the levels with the frequency

$$\omega_R = \frac{2\mu E}{\hbar}, \tag{2.6}$$

which is called the *Rabi flipping frequency*. This parameter determines the time $2\pi/\omega_R$ it takes to transfer the population from one level to the other. If the interaction lasts for a time $T$, we have a simple measure of the degree of saturation in the dimensionless number $\omega_R T$. When this is small very little population transfer has had time to occur.

In steady state operation the length of interaction time is determined by the time it takes the atom to decay out of the optically interacting levels. If we consider the two-level system $|2\rangle \leftrightarrow |1\rangle$, it turns out to be expedient to define the averaged interaction time as

$$T = \frac{1}{2\sqrt{\gamma_1 \gamma_2}}, \tag{2.7}$$

where the $\gamma$'s are the relaxation rates of the levels as introduced in Sec. 1.6.

In this way the interaction time decreases symmetrically when either level decays more rapidly. The measure of saturation, the dimensionless intensity parameter, is then taken to be

$$(\omega_R T)^2 = \frac{\mu^2 E^2}{\hbar^2 \gamma_1 \gamma_2}. \tag{2.8}$$

A more exact definition is given later, but it differs from this only by a numerical constant.

In our considerations of saturation, only the decay of the diagonal elements, the occupation probabilities, has been included. We have seen that the off-diagonal elements, the dipole moments, usually decay faster

$$\gamma_{21} \geq \tfrac{1}{2}(\gamma_1 + \gamma_2) \tag{2.9}$$

because of phase interruption processes. A measure of these is given by the dimensionless coherence parameter

$$\eta = \frac{\gamma_2 + \gamma_1}{2\gamma_{12}} \tag{2.10}$$

which equals one for no phase interruption. Usually it is less than one, but when the Eq. (2.9) holds as an equality, we have $\eta = 1$.

The most rapid time scale in optical resonance problems is the optical frequency† $\Omega/2\pi = 10^{14}$–$10^{15}$ Hz. This is in near resonance with the atomic

†According to the practice in optical spectroscopy we always use *angular frequencies*, but for numerical values we quote the actual frequencies $f$ in hertz.

transition frequency $\omega_{21}$, at which the density matrix element $\rho_{21}$ oscillates in the absence of any couplings. To eliminate this fast component we carry out what is called a *rotating wave approximation*. We are then left only with the detuning $(\omega_{21} - \Omega)$, which usually is found to be of the same order of magnitude as the other rates of our problem: $(\mu E/\hbar)$, $\gamma_1$, $\gamma_2$, $\gamma_{21}$, and so on. This involves an approximation where terms of order $(\mu E/\hbar\Omega)^2$ are neglected. These are the so-called counter-rotating terms. Because for optical frequencies and laser fields usually $\mu E \approx \hbar\gamma$ and the decay times are of the order $\tau = \gamma^{-1} = 1$ ns, we obtain for these corrections an estimate $(\gamma/\Omega)^2 = 10^{-12}$–$10^{-14}$, which can be well neglected. In the radio frequency range, however, $\Omega$ becomes smaller, and fields available in the laboratory have observable effects; see Sec. 2.3. When we investigate sublevels of the ground state, the relaxation times can become very long too.

When each atom sees the same field and the relaxation mechanisms are acting on each separate atom, we call the system *homogeneously broadened*. In a solid the atoms sit well localized, for instance, in the impurity-doped crystals used for solid state lasers. Even then, however, a standing wave pattern makes the atoms unequal; some sit at nodes of the field and saturate hardly at all. The spatial structure of the electromagnetic mode pattern must be included in the analysis.

For a traveling wave, the spatial variation of the field sweeps over the sample and affects all atoms equally. For moving atoms the Doppler shift depends on the rate of change of position, no spatial effects occur. This Doppler shift does, however, change from one atomic individual to the next because of their different velocities. Observations of the spectrum of such an assembly of atoms give a broadened line due to the various resonance positions of the individuals. Such a response is called *inhomogeneously broadened*, because the resonance position is inhomogeneously distributed over the sample. A similar effect can be achieved by a spatially varying crystal field in a solid.

When an atom moves in a standing wave pattern, additional effects emerge. The atom traverses the standing field pattern and tends to average out spatial inhomogeneities. Especially for the fast particles, the averaged description is valid. These are the only particles in resonance with the field when a large detuning requires an appreciable Doppler shift to achieve resonance. The changing field pattern seen by a moving atom induces, however, pulsations in the atomic level populations. These are typical saturation phenomena and require a strong signal theory for their description. Their description within a perturbative expansion is impractical.

There are many types of effects due to strong light intensities. Historically they have acquired different names; terms of order $E^2$ have been designated as *power broadening*, *light shifts* or *Stark-splittings*. All these have very similar origins, and the different names mainly depend on the application where the term appears. Their common origin is a flipping frequency of the type $(\mu E/\hbar)$, which gives rise to a highly nonlinear dependence on $E$. In the following discussion we show many examples of these effects.

In the text it appears as if our solutions would be valid for arbitrary fields $E$ and detunings $\omega_{21} - \Omega$. This is only an illusion, based on our limited model. In real systems there appear restrictions worth remembering. The neglect of the counter-rotating terms will require that $\mu E \ll \hbar\Omega \simeq \hbar\omega_{21}$. When these terms become important in optical spectra, also our restriction to a few levels only breaks down. Transitions to hitherto neglected levels start to take place because their effects will be of order $\mu^2 E^2/\Delta E^2$ approximately equal to $(\mu E/\hbar\Omega)^2$ as many energy level separations $\Delta E$ in an atom are of the same magnitude. Thus for very strong fields, no levels can be neglected, even transitions to the continuum become important.

A similar consideration concerns the detuning. For large enough changes of the light frequency $\Omega$ from its resonance value $\omega_{21}$, it may well approach another resonance with $\omega_{31} = (E_3 - E_1)/\hbar$, and the level $|3\rangle$ can no longer be neglected. The stronger the field is, the sooner the distant levels start to affect the spectroscopy of our truncated system.

In molecules the level structure is even more complicated: The restrictions discussed previously start to be serious even more rapidly, and more care must be exercised when applying simplified theory to real experiments.

**Example.** In a simple two-level system, $|2\rangle \leftrightarrow |1\rangle$ are coupled by the dipole coupling. The state is written

$$|\psi\rangle = \left[c_2 e^{-i\Omega t}|2\rangle + c_1|1\rangle\right] e^{-iE_1 t/\hbar} \tag{2.11}$$

when the field oscillates like $\cos \Omega t$. The Schrödinger equation becomes

$$\begin{aligned} i\hbar\dot{c}_2 &= \hbar(\omega_{21} - \Omega)c_2 - \mu E e^{i\Omega t}\cos\Omega t c_1 \\ i\hbar\dot{c}_1 &= \qquad\qquad\qquad - \mu E e^{-i\Omega t}\cos\Omega t c_2. \end{aligned} \tag{2.12}$$

Using the rotating wave approximation

$$e^{\pm i\Omega t}\cos\Omega t = \tfrac{1}{2} + \tfrac{1}{2}e^{\pm 2i\Omega t} \approx \tfrac{1}{2}, \tag{2.13}$$

we find the equations

$$\begin{aligned} i\dot{c}_2 &= (\omega_{21} - \Omega)c_2 - \frac{\mu E}{2\hbar}c_1 \\ i\dot{c}_1 &= -\frac{\mu E}{2\hbar}c_2 \end{aligned} \tag{2.14}$$

with the two eigenfrequencies

$$\lambda = \frac{\omega_{21} - \Omega}{2} \pm \sqrt{\left(\frac{\omega_{21} - \Omega}{2}\right)^2 + \frac{\mu^2 E^2}{4\hbar^2}}. \tag{2.15}$$

If we use the initial conditions

$$|c_1(0)|^2 = 1, \qquad |c_2(0)|^2 = 0, \tag{2.16}$$

we find for the excited level the solution

$$\begin{aligned} |c_2(t)|^2 &= \frac{\mu^2 E^2}{\hbar^2(\omega_{21} - \Omega)^2 + \mu^2 E^2} \sin^2\left\{\frac{1}{2}\left[(\omega_{21} - \Omega)^2 + \frac{\mu^2 E^2}{\hbar^2}\right]^{1/2} t\right\} \\ &= \frac{\mu^2 E^2}{2\left[\hbar^2(\omega_{21} - \Omega)^2 + \mu^2 E^2\right]} \left\{1 - \cos\left[\left((\omega_{21} - \Omega)^2 + \frac{\mu^2 E^2}{\hbar^2}\right)^{1/2} t\right]\right\}. \end{aligned} \tag{2.17}$$

At exact resonance $\omega_{21} = \Omega$, this displays just the Rabi flipping at the frequency

$$\omega_R = \frac{\mu E}{\hbar} \tag{2.18}$$

corresponding to the field amplitude $\frac{1}{2}E$ [cf. Eq. (2.6)]. This is because one component of the function $\cos \Omega t$ can drive the transition; the other one is counter-rotating, and it has been neglected.

If we have an ensemble of atoms that interact for various times $t$, due to some spontaneous decay process, their interaction times have a probability distribution given by the exponential decay

$$W(t) = \gamma e^{-\gamma t}. \tag{2.19}$$

The averaged population on the upper level is obtained if we weigh the occurrence by $W(t)$ and sum over all values of $t$. We find

$$P_2 = \overline{|c_2|^2} = \int_0^\infty \gamma e^{-\gamma t} |c_2(t)|^2 \, dt. \tag{2.20}$$

The integral can be evaluated from

$$\int_0^\infty e^{-\gamma t}\sin^2\frac{1}{2}\theta t\,dt = \frac{1}{2}\int_0^\infty e^{-\gamma t}\left(1 - \frac{1}{2}e^{i\theta t} - \frac{1}{2}e^{-i\theta t}\right)dt$$

$$= \frac{1}{2\gamma}\frac{\theta^2}{(\theta^2 + \gamma^2)}, \tag{2.21}$$

which from (2.17) and (2.20) gives

$$P_2 = \frac{1}{2}\frac{\mu^2 E^2/\hbar^2}{(\omega_{21} - \Omega)^2 + \gamma^2 + \mu^2 E^2/\hbar^2}. \tag{2.22}$$

This is a Lorentzian of the width

$$\Gamma = \left(\gamma^2 + \frac{\mu^2 E^2}{\hbar^2}\right)^{1/2}, \tag{2.23}$$

which shows that the field causes a *power broadening* of the line shape. This is due to the presence of the Rabi flipping frequency in the sine function.

For a very large field $E \to \infty$ the probability of finding the atom in the upper level becomes

$$P_2 = \tfrac{1}{2}. \tag{2.24}$$

This means that the atom flips so fast that it is found in each level with probability $\frac{1}{2}$. For a strongly saturated system the field tries to make the levels have equal probabilities because then the transitions up and down compensate each other precisely.

## 2.2. STATIONARY TWO-LEVEL ATOMS IN A STANDING WAVE

We consider a two-level system acted upon by the strong standing wave field

$$E(z, t) = E\cos\Omega t\sin kz. \tag{2.25}$$

The atoms are assumed stationary so that each one sees an oscillatory field with the amplitude uniquely determined by its position $z$.

The equation of motion for the density matrix becomes

$$\frac{d}{dt}\rho_{22} = \lambda_2 - \gamma_2\rho_{22} - \frac{i\mu E}{\hbar}\sin kz\cos\Omega t(\rho_{21} - \rho_{12}) \tag{2.26a}$$

$$\frac{d}{dt}\rho_{11} = \lambda_1 - \gamma_1\rho_{11} + \frac{i\mu E}{\hbar}\sin kz\cos\Omega t(\rho_{21} - \rho_{12}) \tag{2.26b}$$

$$\frac{d}{dt}\rho_{21} = -(\gamma_{21} + i\omega)\rho_{21} - \frac{i\mu E}{\hbar}\sin kz\cos\Omega t(\rho_{22} - \rho_{11}), \tag{2.26c}$$

where the energy separation is written $E_2 - E_1 = \hbar\omega$ and incoherent pumping is assumed to occur to both levels. Spontaneous decay between the levels is neglected, and $\gamma_{21} \geq \frac{1}{2}(\gamma_1 + \gamma_2)$ because of phase perturbations.

For the free atom without field, the off-diagonal matrix element $\rho_{21}$ oscillates at a fast optical rate approximately equal to $\omega$. To eliminate this we introduce the new variable

$$\tilde{\rho}_{21} = e^{i\Omega t}\rho_{21}. \tag{2.27}$$

When this ansatz is introduced into the equation of motion we find that we can use the approximation (2.13) because the term oscillating at twice the optical frequency $\Omega$ averages out to zero over a few periods, its effect is small and can be neglected.† Equation (2.26c) then becomes

$$\frac{d}{dt}\tilde{\rho}_{21} = -(\gamma_{21} + i\Delta)\tilde{\rho}_{21} - \frac{i}{2\hbar}\mu E\sin kz(\rho_{22} - \rho_{11}), \tag{2.28}$$

where the detuning parameter

$$\Delta = \omega - \Omega \tag{2.29}$$

has been introduced. If the off-diagonal decay rate $\gamma_{21}$ is large enough, we can assume that $\tilde{\rho}_{21}$ relaxes rapidly to its steady state. It is then possible to use the adiabatic elimination argument introduced in Sec. 1.8 to set $\dot{\tilde{\rho}}_{21}$ equal to zero and solve for

$$\tilde{\rho}_{21}(t) = -\frac{\mu E\sin kz}{2\hbar}\frac{1}{\Delta - i\gamma_{21}}(\rho_{22} - \rho_{11}). \tag{2.30}$$

†This transformation is usually called the rotating wave approximation (RWA). The reason is that in nuclear magnetic resonance it can be shown to correspond to the case of a coordinate system following the rotating motion of the magnetic field.

It must be noticed that the fastest rate in the problem is $\Omega \simeq \omega$, but the substitution (2.27) and the subsequent rotating wave approximation has eliminated it and left only the much smaller rate $\Delta$. With (2.30) we write

$$\begin{aligned}\cos \Omega t(\rho_{21} - \rho_{12}) &= \frac{1}{2}(e^{i\Omega t} + e^{-i\Omega t})(e^{-i\Omega t}\tilde{\rho}_{21} - e^{i\Omega t}\tilde{\rho}_{12}) \\ &= \frac{1}{2}(\tilde{\rho}_{21} - \tilde{\rho}_{12}) = -\frac{i\mu E}{2\hbar\gamma_{21}}L(\Delta)(\rho_{22} - \rho_{11}),\end{aligned} \tag{2.31}$$

where the (dimensionless) Lorentzian is defined by

$$L(\Delta) = \frac{\gamma_{21}^2}{\Delta^2 + \gamma_{21}^2}. \tag{2.32}$$

When (2.31) is inserted into the equations for the occupation probabilities [(2.26a) and (2.26b)] we find the typical rate equations

$$\frac{d}{dt}\rho_{22} = \lambda_2 - \gamma_2\rho_{22} - I\gamma_{21}\zeta \sin^2 kzL(\Delta)(\rho_{22} - \rho_{11}) \tag{2.33a}$$

$$\frac{d}{dt}\rho_{11} = \lambda_1 - \gamma_1\rho_{11} + I\gamma_{21}\zeta \sin^2 kzL(\Delta)(\rho_{22} - \rho_{11}) \tag{2.33b}$$

where the parameters are introduced by

$$\zeta = \frac{\gamma_1\gamma_2}{\gamma_{12}^2} \tag{2.34}$$

$$I = \frac{\mu^2E^2}{2\hbar^2\gamma_1\gamma_2}. \tag{2.35}$$

The last parameter is a dimensionless measure of the field intensity as introduced in (2.8). The rate at which a pure two-level system oscillates between the upper and lower levels is $(\mu E/\hbar)$, the Rabi flipping rate, and when this becomes equal to the average decay rate $\sqrt{\gamma_1\gamma_2}$ of the levels, the system feels the field-induced transitions as often as spontaneous decay, and then saturation sets in. The rate parameter is small in one of the three limits:

1. The field intensity is very low, $I \ll 1$.
2. The system is strongly detuned; $|\Delta| \gg \gamma_{12}$, which gives $L(\Delta) \ll 1$.
3. The phase incoherence is much stronger than spontaneous decay of the populations, $\gamma_{21} \gg \sqrt{\gamma_2 \gamma_1}$, which gives $\zeta \ll 1$.

In all these cases a perturbation solution of Eqs. (2.33) is possible.

When we set the time derivatives in (2.33) equal to zero, that is, we look for steady state solutions, no adiabatic assumption is needed. Then we introduce the quantities

$$\overline{N} = \frac{\lambda_2}{\gamma_2} - \frac{\lambda_1}{\gamma_1} \tag{2.36}$$

and $\eta$ from (2.10). The first quantity is the difference in occupation probability between the levels when no fields act, and the second one is the measure of coherence; when phase perturbations become more important $\eta$ decreases. For no phase perturbations $\eta$ equals one. With these definitions we find

$$\begin{aligned} \rho_{22} - \rho_{11} &= \frac{\overline{N}}{1 + 2I\eta \sin^2 kz L(\Delta)} \\ &= \overline{N}\left[1 - \frac{2I\eta \sin^2 kz \gamma_{12}^2}{\Delta^2 + \gamma_{12}^2(1 + 2I\eta \sin^2 kz)}\right]. \end{aligned} \tag{2.37}$$

The two populations are found to be

$$\rho_{22} = \frac{\lambda_2}{\gamma_2} - \overline{N}\frac{I\gamma_1\gamma_{21}\sin^2 kz}{\Delta^2 + \gamma_{12}^2(1 + 2I\eta \sin^2 kz)} \tag{2.38a}$$

$$\rho_{11} = \frac{\lambda_1}{\gamma_1} + \overline{N}\frac{I\gamma_2\gamma_{21}\sin^2 kz}{\Delta^2 + \gamma_{12}^2(1 + 2I\eta \sin^2 kz)}. \tag{2.38b}$$

The second term in (2.37) denotes the deviation from the unperturbed occupation probability difference $\overline{N}$. This function is found to oscillate between zero and a maximum for $\sin^2 kz = 1$. Thus the atoms are saturated differently depending on their position, see Fig. 2.1. As a function of detuning the deviation from $\overline{N}$ is a Lorentzian function of width

$$\Gamma_p = \gamma_{12}[1 + 2I\eta \sin^2 kz]^{1/2} \tag{2.39}$$

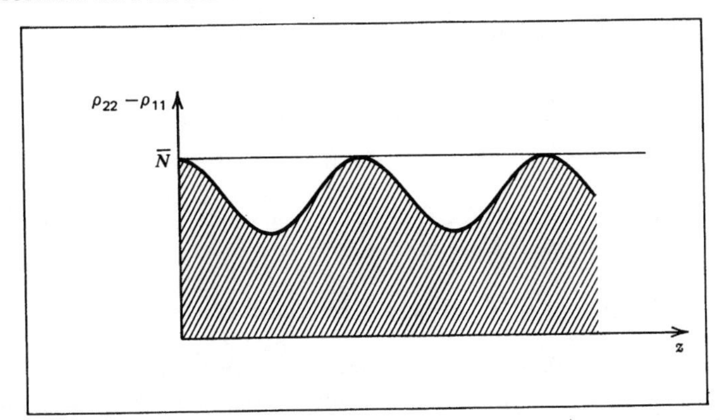


**Fig. 2.1** As a function of position, the population difference mirrors the periodic variation of the saturating field.

see Fig. 2.2. This is found to depend directly on the saturation parameter. For stronger saturation the Lorentzian becomes broader. This effect is called power broadening.

We introduce (2.37) back into (2.30) and find

$$\tilde{\rho}_{21}(t) = -\bar{N}\frac{\mu E \sin kz(\Delta + i\gamma_{21})}{2\hbar\left[\Delta^2 + \gamma_{21}^2(1 + 2I\eta\sin^2 kz)\right]}. \tag{2.40}$$

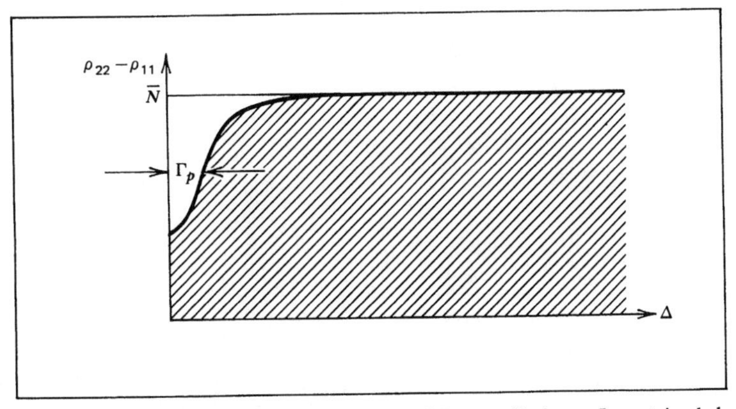


**Fig. 2.2** As a function of detuning, the population difference displays a Lorentzian behavior with the power broadened width $\Gamma_p$. This varies from point to point for stationary atoms.

The total polarization of the sample is, from (1.105),

$$\begin{aligned} P(z,t) &= N_0\langle\mu\rangle = N_0\mu(\rho_{21} + \rho_{12}) \\ &= N_0\mu\left[(\tilde{\rho}_{21} + \tilde{\rho}_{12})\cos\Omega t + i(\tilde{\rho}_{12} - \tilde{\rho}_{21})\sin\Omega t\right]. \end{aligned} \tag{2.41}$$

Because the basic spatial dependence of the field mode is (2.25), we separate the term having the same mode structure by writing

$$P(z,t) = [C\cos\Omega t + S\sin\Omega t]\sin kz + \cdots \tag{2.42}$$

where the terms excluded have the spatial dependence $\cos kz$, $\sin 2kz$, $\cos 2kz$, and so on. We calculate the factors $C$ and $S$ by the projection on the Fourier component $\sin kz$ according to

$$C\cos\Omega t + S\sin\Omega t = \frac{\int_0^L dz\sin kz P(z,t)}{\int_0^L dz\sin^2 kz} \tag{2.43}$$

which gives

$$\begin{aligned} C &= N_0\mu\frac{2}{L}\int_0^L(\tilde{\rho}_{12} + \tilde{\rho}_{21})\sin kz\,dz \\ &= -N_0\frac{\mu^2 E\overline{N}}{\hbar\gamma_{21}^2}\Delta L(\Delta)F(\Delta) \end{aligned} \tag{2.44a}$$

$$\begin{aligned} S &= +iN_0\mu\frac{2}{L}\int_0^L(\tilde{\rho}_{12} - \tilde{\rho}_{21})\sin kz\,dz \\ &= -N_0\frac{\mu^2 E\overline{N}}{\hbar\gamma_{21}}L(\Delta)F(\Delta) \end{aligned} \tag{2.44b}$$

where we have

$$\begin{aligned} F(\Delta) &= \frac{2}{L}\int_0^L\frac{\sin^2 kz}{1 + 2I\eta L(\Delta)\sin^2 kz}dz = \frac{1}{I\eta L(\Delta)}\left[1 - \frac{1}{\sqrt{1 + 2I\eta L(\Delta)}}\right] \\ &= 1 - \tfrac{3}{2}I\eta L(\Delta) + \cdots \end{aligned} \tag{2.45}$$

The last expansion is valid when $I\eta L$ is much less than one.

The factor $S$ is the component of the field that is out of phase with the driving field (2.25). It thus contributes the gain or losses that are needed for absorbtion spectroscopy or laser operation. Its line shape is a symmetric Lorentzian. The component $C$ is in phase with the field and cannot affect the energy. It provides the *dispersion* in the medium, and its shape is the antisymmetric form of the derivative of a Lorentzian.

If we need the energy dissipation in the medium we must, according to Eq. (2.5), calculate the time and space average†

$$\begin{aligned} K &= -\overline{E\frac{dP}{dt}} \\ &= N_0\mu\Omega\frac{2}{L}\overline{\int_0^L E(z,t)\left[(\tilde{\rho}_{21}+\tilde{\rho}_{12})\sin\Omega t - i(\tilde{\rho}_{12}-\tilde{\rho}_{21})\cos\Omega t\right]dz} \\ &= -\frac{1}{2}N_0\mu E\Omega\frac{2}{L}\int_0^L i\sin kz(\tilde{\rho}_{12}-\tilde{\rho}_{21})\,dz \\ &= -\frac{1}{2}E\Omega S \\ &= 2\hbar\Omega\left(\frac{\mu E}{2\hbar}\right)^2\frac{\gamma_{12}}{\Delta^2+\gamma_{12}^2}N_0\overline{N}F(\Delta). \end{aligned} \tag{2.46}$$

If we neglect the intensity dependence in $F(\Delta)$ and introduce the rate

$$W = 2\pi\left(\frac{\mu E}{2\hbar}\right)^2\frac{\gamma_{12}/\pi}{\Delta^2+\gamma_{12}^2}, \tag{2.47}$$

we can write the energy exchange (2.46)

$$K = \hbar\Omega W N_0\overline{N}F(\Delta). \tag{2.48}$$

This equation has a simple interpretation. The energy transfer per process is $\hbar\Omega$, the rate of processes is $W$, and the atomic population difference is $N_0\overline{N}$; $F(\Delta)$ denotes its saturation. Note that (2.47) goes to the "golden rule" of perturbation theory when $\gamma_{12}$ goes to zero. The simple description of energy exchange between the radiation field and matter can be extended into a general theory, which we show briefly in the following section.

† In the time average we set $\overline{\cos^2\Omega t} = \frac{1}{2}$, $\overline{\sin\Omega t\cos\Omega t} = 0$.

**Energy Changes of the Atomic System Due to the Field**

The total energy in the atomic system can be written as a sum of the occupation probability of each level $\rho_{mm}$ times its energy $E_m$

$$E_{\mathrm{at}} = \sum_m E_m \rho_{mm} = \operatorname{Tr} H_{\mathrm{at}} \rho. \tag{2.49}$$

If we calculate the time derivative of this quantity we obtain the energy emission

$$\begin{aligned} K &= -\frac{d}{dt} E_{\mathrm{at}} = -\sum_m E_m \frac{d}{dt} \rho_{mm} \\ &= +i \sum_m \frac{E_m}{\hbar} \langle m | [H, \rho] | m \rangle \\ &= +i \sum_m \frac{E_m}{\hbar} \langle m | [H_{\mathrm{int}}, \rho] | m \rangle \end{aligned} \tag{2.50}$$

because

$$\langle m | [H_{\mathrm{at}}, \rho] | m \rangle = (E_m \rho_{mm} - \rho_{mm} E_m) = 0. \tag{2.51}$$

With the interaction term

$$H_{\mathrm{int}} = -e\mathbf{r} \cdot \mathbf{E}(z, t) \tag{2.52}$$

we find

$$\begin{aligned} K &= -i \sum_{mn} \frac{E_m}{\hbar} E(z, t) \mu (\rho_{nm} - \rho_{mn}) \\ &= -i \left[ \sum_{\substack{mn \\ E_m > E_n}} \frac{E_m}{\hbar} E(z, t) \mu (\rho_{nm} - \rho_{mn}) \right. \\ &\qquad \left. + \sum_{\substack{nm \\ E_n > E_m}} \frac{E_m}{\hbar} E(z, t) \mu (\rho_{nm} - \rho_{mn}) \right] \\ &= -i \left[ \sum_{\substack{nm \\ E_m > E_n}} \frac{E_m - E_n}{\hbar} E(z, t) \mu (\rho_{nm} - \rho_{mn}) \right], \end{aligned} \tag{2.53}$$

where we have interchanged the summation indices $n$ and $m$ in the second sum. The time dependence of the density matrix due to the atomic Hamiltonian gives

$$\rho_{nm}(t) = e^{i(E_m - E_n)t/\hbar}\rho_{nm}(0), \tag{2.54}$$

and consequently we find

$$\begin{aligned} K &= -E(z,t) \sum_{\substack{n,m \\ E_m > E_n}} \mu \frac{d}{dt}(\rho_{nm} + \rho_{mn}) \\ &= -E(z,t)\frac{d}{dt}P(z,t), \end{aligned} \tag{2.55}$$

as is used in the main part of our text. The ordering introduced, $E_m > E_n$, is necessary to avoid counting each transition twice in the sum. For instance, the transition $2 \to 1$ comes in both for ($n = 1$, $m = 2$) and ($n = 2$, $m = 1$). If $E_2 > E_1$, one comes in only for absorption and the other for emission, but our ordering includes each case in a simple notation.

When the electromagnetic field becomes strong it is more difficult to prove equivalence between (2.50) and (2.55) rigorously. A physically convincing argument can be obtained by regarding the atom imbedded in the strong field as one system (the *dressed atom*) with renormalized energy levels $E_m$ and the corresponding eigenstates $|m\rangle$. Then the relation (2.49) holds again and so does (2.54). They result in (2.55), and because this is representation independent we can ignore the fact that the calculation has been carried out in a representation that we do not known how to obtain. In the following sections we assume (2.55) to hold also for strong fields.

## 2.3. STRONG FIELD EFFECTS IN RADIO-FREQUENCY SPECTROSCOPY†

The frequencies of the transitions between the fine structure components of atomic spectra usually lie in the radio-frequency range 1–1000 MHz. The Zeeman splittings in atoms, especially, provide a spectrum with a finite number of components equal to $2J + 1$ where $J$ is the angular momentum quantum number of the level. The energy values are given by

$$E_J = \hbar\omega_m = -g\mu_B B m_J, \tag{2.56}$$

†The rest of this book is independent of the material in this section, which hence can be omitted by the reader.

where $\mu_B$ is the Bohr magneton, $g$ is the $g$-factor, and $B$ is the static magnetic field. The quantum number $m_J$ ranges from $-J$ to $+J$. The levels are coupled by a radio-frequency magnetic field $B_1$ orthogonal to the static field, and the coupling Hamiltonian is

$$H_I = -\mathbf{M}\cdot\mathbf{B}_1\cos\Omega t \tag{2.57}$$

where $M$ is the magnetic dipole operator. The wavelength of the radiation is

$$\lambda = \frac{2\pi}{k} = \frac{2\pi c}{\omega} \simeq 1\ \mathrm{m} \tag{2.58}$$

for $\omega/2\pi = 300$ MHz. A laboratory sample is consequently much smaller than the wavelength, and a factor like $\sin kz$ in Eq. (2.25) can be omitted.

For magnetic systems it is easy to find a pure two-level system because for an atom with zero orbital angular momentum $L = 0$ and one valence electron only, $J = S = \frac{1}{2}$ and only two levels occur. They both decay with the same rate which we denote by

$$\gamma_1 = \gamma_2 = \frac{1}{T_1}, \tag{2.59}$$

in accordance with the terminology of nuclear magnetic resonance theory. The off-diagonal elements decay with the rate

$$\gamma_{12} = \frac{1}{T_2}, \tag{2.60}$$

and from our general considerations it follows that $T_2 \leq T_1$. The time constant $T_1$ is called the *longitudinal decay time*, the constant $T_2$ the *transverse one*. The equations in (2.26) can be used with the replacement

$$\frac{\mu E}{\hbar} \Rightarrow \frac{\langle 1|\mathbf{M}\cdot\mathbf{B}_1|2\rangle}{\hbar} \equiv \omega_1. \tag{2.61}$$

We write

$$\frac{d}{dt}\rho_{22} = \lambda_2 - \frac{1}{T_1}\rho_{22} - i\omega_1\cos\Omega t(\rho_{21} - \rho_{12}) \tag{2.62a}$$

$$\frac{d}{dt}\rho_{11} = \lambda_1 - \frac{1}{T_1}\rho_{11} + i\omega_1\cos\Omega t(\rho_{21} - \rho_{12}) \tag{2.62b}$$

$$\frac{d}{dt}\rho_{21} = -\left(\frac{1}{T_2} + i\omega\right)\rho_{21} - i\omega_1\cos\Omega t(\rho_{22} - \rho_{11}). \tag{2.62c}$$

From these equations we can define new variables in the form

$$R_1 = \rho_{12} + \rho_{21} \tag{2.63a}$$

$$R_2 = -i(\rho_{12} - \rho_{21}) \tag{2.63b}$$

$$R_3 = \rho_{22} - \rho_{11} \tag{2.63c}$$

with interpretations in terms of dipole moments and population differences respectively. We obtain the equations of motion

$$\frac{d}{dt}R_3 = \lambda_2 - \lambda_1 - \frac{R_3}{T_1} - 2\omega_1 \cos \Omega t R_2 \tag{2.64a}$$

$$\frac{d}{dt}R_1 = -\frac{R_1}{T_2} - \omega R_2 \tag{2.64b}$$

$$\frac{d}{dt}R_2 = -\frac{R_2}{T_2} + \omega R_1 + 2\omega_1 \cos \Omega t R_3. \tag{2.64c}$$

If we define the orthogonal unit vectors $\hat{\mathbf{e}}_1$, $\hat{\mathbf{e}}_2$, $\hat{\mathbf{e}}_3$, and the vector

$$\mathbf{R} = R_1\hat{\mathbf{e}}_1 + R_2\hat{\mathbf{e}}_2 + R_3\hat{\mathbf{e}}_3, \tag{2.65}$$

we can write its equation of motion as

$$\frac{d}{dt}\mathbf{R} = -\frac{R_1\hat{\mathbf{e}}_1 + R_2\hat{\mathbf{e}}_2}{T_2} + \frac{(R_3^0 - R_3)\hat{\mathbf{e}}_3}{T_1} + \mathbf{\Omega} \times \mathbf{R}, \tag{2.66}$$

where the external field vector is

$$\mathbf{\Omega} = -2\omega_1 \cos \Omega t \hat{\mathbf{e}}_1 + \omega \hat{\mathbf{e}}_3. \tag{2.67}$$

For no external field the steady state solution is

$$R_1 = R_2 = 0 \tag{2.68a}$$

$$R_3 = R_3^0 = (\lambda_2 - \lambda_1)T_1. \tag{2.68b}$$

Equation (2.66) is the Bloch equation first introduced in nuclear magnetic resonance spectroscopy and widely used to describe magnetic resonance effects. Here the vector **R** is proportional to the magnetic moment of the nucleus for any number of levels, but we can see that Eq. (2.66) also describe the behavior of a general two-level system in optical spectroscopy. Here the vector **R** exists in a ficticious space sometimes called "energy space." Especially for time-dependent phenomena, the *optical Bloch equations* are widely used.

**Example 1.** When $\gamma_2 \neq \gamma_1$ it is not possible to describe the behavior with the vector **R** only because the normalization function also contributes to the time evolution. Show that its equation of motion is

$$\frac{d}{dt}(\rho_{22} + \rho_{11}) = \lambda_1 + \lambda_2 - \tfrac{1}{2}(\gamma_1 + \gamma_2)(\rho_{22} + \rho_{11})$$

$$+ \tfrac{1}{2}(\gamma_1 - \gamma_2)R_3. \tag{2.69}$$

Show that in steady state, Eqs. (2.64) may still be used but with the longitudinal decay time determined by

$$T_1 = \frac{\gamma_1 + \gamma_2}{2\gamma_1\gamma_2}; \tag{2.70}$$

also the pumping term is modified. What happens when $\gamma_1 = \gamma_2$?

In radio-frequency spectroscopy too, one often uses a rotating wave approximation. This is carried out for the density matrix exactly as in Eq. (2.12). It is, however, possible to introduce the rotating wave approximation by the transformation

$$R_1 = \tilde{R}_1 \cos \Omega t - \tilde{R}_2 \sin \Omega t \tag{2.71a}$$

$$R_2 = \tilde{R}_2 \cos \Omega t + \tilde{R}_1 \sin \Omega t, \tag{2.71b}$$

which shows directly that $\tilde{R}_1$ and $\tilde{R}_2$ are the transverse vector components in a coordinate system rotating with the angular frequency $\Omega$ of the field.

**Example 2.** Show that with the approximate replacements

$$\cos^2 \Omega t = \sin^2 \Omega t = \tfrac{1}{2}$$

$$\sin \Omega t \cos \Omega t = 0, \tag{2.72}$$

the vector $\tilde{\mathbf{R}} = \tilde{R}_1 \hat{\mathbf{e}}_1 + \tilde{R}_2 \hat{\mathbf{e}}_2 + \tilde{R}_3 \hat{\mathbf{e}}_3$ satisfies the Bloch Eq. (2.66) with the external field $\mathbf{\Omega} = -\omega_1 \hat{\mathbf{e}}_1 + (\omega - \Omega)\hat{e}_3$.

The effects of the correction terms to the rotating wave approximation are found by making a Fourier expansion of the components of the Bloch vector

$$R_3 = \sum_n d_n e^{in\Omega t} \tag{2.73a}$$

$$R_1 = i\sum_n c_n e^{in\Omega t} \tag{2.73b}$$

$$R_2 = i\sum_n s_n e^{in\Omega t}. \tag{2.73c}$$

Inserting these into (2.66) and equating equal Fourier components, we obtain the steady state equations

$$\left(in\Omega + \frac{1}{T_1}\right)d_n = \frac{R_3^0}{T_1}\delta_{n0} - i\omega_1(s_{n+1} + s_{n-1}) \tag{2.74a}$$

$$\left(in\Omega + \frac{1}{T_2}\right)c_n = -\omega s_n \tag{2.74b}$$

$$\left(in\Omega + \frac{1}{T_2}\right)s_n = \omega c_n - i\omega_1(d_{n+1} + d_{n-1}). \tag{2.74c}$$

We easily find the equations

$$d_n = R_3^0\delta_{n0} + \omega_1 D_1(n)(s_{n+1} + s_{n-1}) \tag{2.75a}$$

$$s_n = \omega_1 D_2(n)(d_{n+1} + d_{n-1}) \tag{2.75b}$$

with

$$D_1(n) = \frac{1}{i\frac{1}{T_1} - n\Omega} \tag{2.76a}$$

$$D_2(n) = \frac{-i}{in\Omega + \frac{1}{T_2} + \frac{\omega^2}{in\Omega + \frac{1}{T_2}}}$$

$$= \frac{1}{2}\left[\frac{1}{i\frac{1}{T_2} - (n\Omega - \omega)} + \frac{1}{i\frac{1}{T_2} - (n\Omega + \omega)}\right]. \tag{2.76b}$$

Equations (2.75) form an infinite set of difference equations with the inhomogeneous term $R_3^0\delta_{n0}$. This makes $d_0$ differ from zero, which in turn acts on $s_1$ and $s_{-1}$, which couple back to $d_0$ and generate $d_{+2}$ and $d_{-2}$. In this manner we can see that $d$ will have even components and $s$ odd ones only. We introduce the variables

$$D(n) = D_1(n); \qquad x_n = \frac{d_n}{R_3^0} \qquad n = \text{even}$$

$$D(n) = D_2(n); \qquad x_n = \frac{s_n}{R_3^0} \qquad n = \text{odd} \tag{2.77}$$

and find the one recurrence relation

$$x_n = \delta_{n0} + \omega_1 D(n)(x_{n+1} + x_{n-1}). \tag{2.78}$$

This equation can be solved in terms of a continued fraction: For $n > 0$ we solve for

$$\frac{x_n}{x_{n-1}} = \frac{\omega_1 D(n)}{1 - \omega_1 D(n)\dfrac{x_{n+1}}{x_n}}. \tag{2.79}$$

We set $n = 1$ and iterate the relationship to obtain

$$\frac{x_1}{x_0} = \frac{s_1}{d_0} = \cfrac{\omega_1 D(1)}{1 - \cfrac{\omega_1^2 D(1) D(2)}{1 - \cfrac{\omega_1^2 D(2) D(3)}{1 - \cdots}}}. \tag{2.80}$$

For $n < 0$ we solve alternatively

$$\frac{x_n}{x_{n+1}} = \frac{\omega_1 D(n)}{1 - \omega_1 D(n)\dfrac{x_{n-1}}{x_n}}, \tag{2.81}$$

which for $n = -1$ gives

$$\frac{x_{-1}}{x_0} = \frac{s_{-1}}{d_0} = \cfrac{\omega_1 D(-1)}{1 - \cfrac{\omega_1^2 D(-1) D(-2)}{1 - \cdots}}. \tag{2.82}$$

Inserting (2.80) and (2.82) into the equation with $n = 0$ we find a solution for $d_0$ in the form

$$d_0 = x_0 R_3^0 = \frac{R_3^0}{1 - \omega_1 D(0)\left[\dfrac{s_1}{d_0} + \dfrac{s_{-1}}{d_0}\right]}$$

$$= R_3^0 \left\{ 1 + i\omega_1^2 T_1 \left[ \frac{D(1)}{1 - \dfrac{\omega_1^2 D(1) D(2)}{1 - \cdots}} + \frac{D(-1)}{1 - \dfrac{\omega_1^2 D(-1) D(-2)}{1 - \cdots}} \right] \right\}^{-1}. \tag{2.83}$$

Because

$$D(-n) = -[D(n)]^* \tag{2.84}$$

for all $n$ we find

$$d_0 = \frac{R_3^0}{1 - 2\omega_1^2 T_1 \mathrm{Im}\left[\dfrac{D(1)}{1 - \dfrac{\omega_1^2 D(1) D(2)}{1 + \cdots}}\right]}. \tag{2.85}$$

In the lowest approximation we replace the continued fraction in the denominator by $D(1)$ and find

$$\mathrm{Im}\, D(1) = -\frac{T_2}{2}\left[\frac{1}{1 + T_2^2(\Omega - \omega)^2} + \frac{1}{1 + T_2^2(\Omega + \omega)^2}\right]. \tag{2.86}$$

Near resonance $\Omega \simeq \omega$ and the second Lorentzian can be neglected. The solution is then

$$d_0 = \frac{R_3^0}{1 + \dfrac{\omega_1^2 T_1 T_2}{1 + T_2^2(\Omega - \omega)^2}}$$

$$= R_3^0\left[1 - \frac{\omega_1^2 T_1}{T_2} \frac{1}{(\Omega - \omega)^2 + 1/T_2^2 + \omega_1^2 T_1/T_2}\right]. \tag{2.87}$$

This is the steady state solution of the rotating wave approximation, showing a Lorentzian deviation from the unperturbed population difference $R_3^0$, but with a power broadened width

$$\Gamma_p = \left[\frac{1}{T_2^2} + \frac{\omega_1^2 T_1}{T_2}\right]^{1/2}. \tag{2.88}$$

In the next approximation we retain the term $D(1)D(2)$ in the continued fraction and find near the resonance $\omega \simeq \Omega \gg 1/T_1$

$$\frac{D(1)}{1 - \omega_1^2 D(1) D(2)} = \frac{T_2}{2}\left[\frac{\dfrac{1}{(\Omega - \omega)T_2 - i}}{1 - \dfrac{T_1 T_2 \omega_1^2/2}{[(\Omega - \omega)T_2 - i](2\Omega T_1 - i)}}\right]$$

$$\approx \frac{T_2}{2}\left[\frac{1}{(\Omega - \omega - \omega_1^2/4\Omega)T_2 - i}\right]. \tag{2.89}$$

When this is inserted instead of the lower-order result (2.86), we find a resonance behavior like that in (2.87) but with the resonance condition given by

$$\Omega = \omega + \frac{\omega_1^2}{4\Omega}. \tag{2.90}$$

The power dependent shift $(\omega_1^2/4\Omega)$ is called the Bloch–Siegert shift and is due to the counter-rotating term of the field.

The counter-rotating term that has been neglected has an amplitude $(\hbar\omega_1/2)$ and is off resonance by the amount $\hbar(\omega + \Omega) \simeq 2\hbar\Omega$. Second-order perturbation theory then gives, see Fig. 2.3,

$$E_2' = E_2 + \frac{(\hbar\omega_1/2)^2}{2\hbar\Omega}$$

$$E_1' = E_1 - \frac{(\hbar\omega_1/2)^2}{2\hbar\Omega}, \tag{2.91}$$

and the resonance transition is shifted to

$$E_2' - E_1' = \hbar\left[\omega + \left(\frac{\omega_1}{2}\right)^2 \frac{1}{\Omega}\right] \tag{2.92}$$

in accordance with (2.90).

From the solution (2.83) we can see that there occurs a resonance behavior for each odd value of $n = 2k + 1$ when

$$(2k + 1)\Omega \simeq \omega \tag{2.93}$$

because the odd functions $D(2k + 1) = D_2(2k + 1)$ behave in a resonant way at such points. These are multiphoton resonances that appear when $(2k + 1)$ photons, each one of energy $\hbar\Omega$, are absorbed to excite the transition of energy difference $\hbar\omega$. As in (2.89), we can calculate a shift for these transitions in the form

$$\begin{aligned}
&\left[1 - \frac{\omega_1^2 D(2k+1) D(2k)}{1 - \omega_1^2 D(2k+1) D(2k+2)}\right]^{-1} \\
&= \frac{[D(2k+1)]^{-1} - \omega_1^2 D(2k+2)}{[D(2k+1)]^{-1} - \omega_1^2 [D(2k) + D(2k+2)]} \\
&= \left[1 + \left(\frac{\omega_1^2}{2k\Omega}\right) \frac{1}{2[(2k+1)\Omega - \omega] - i\dfrac{2}{T_2} - \dfrac{\omega_1^2}{\Omega}\left[\dfrac{1}{2k} + \dfrac{1}{2k+2}\right]}\right] \\
&= \left[1 + \frac{\omega_1^2}{4k\Omega} \frac{1}{\left[(2k+1)\Omega - \omega - \dfrac{\omega_1^2 (2k+1)}{4k(k+1)\Omega} - i\dfrac{1}{T_2}\right]}\right],
\end{aligned} \tag{2.94}$$

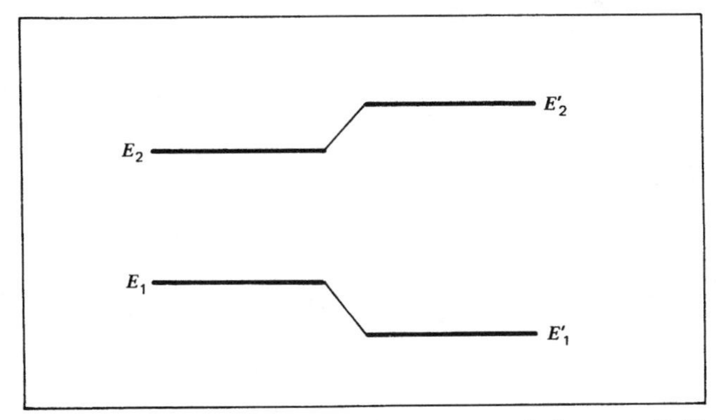


**Fig. 2.3** In second-order perturbation theory the levels $|1\rangle$ and $|2\rangle$ are repelled by the field coupling them.

where we again have assumed $\Omega \gg T_1^{-1}$. We find here also that this resonance has been shifted from (2.93) to

$$(2k+1)\Omega = \omega + \frac{\omega_1^2(2k+1)}{4k(k+1)\Omega}. \tag{2.95}$$

This is a generalized Bloch–Siegert shift of the multiphoton resonances.

In this section we have seen how the effects of counter-rotating terms can be taken into account in an exact way. The result is given in terms of a continued fraction, which shows Bloch–Siegert shifts and multiphoton resonances. Hence the latter appear in our treatment without any need for quantization of the field. It is enough to require the $(2k+1)$ component of the transverse dipole moment components $R_1$ and $R_2$ to resonate with the transition frequency $\omega$ of the two-level system. The only place where the photon concept is needed is in the interpretation of the resonance condition (2.93) as an energy conservation requirement.

## 2.4. MOVING ATOMS IN A TRAVELING WAVE

In this section we turn to consider the case of a moving two-level atom situated in the traveling wave

$$E(z,t) = E\cos(kz - \Omega t). \tag{2.96}$$

The set of equations to be solved is, in analogy with (2.26),

$$\left(\frac{\partial}{\partial t} + v\frac{\partial}{\partial z}\right)\rho_{22} + \gamma_2\rho_{22} = \lambda_2 + i\frac{\mu E}{\hbar}\cos(kz - \Omega t)(\rho_{12} - \rho_{21}) \tag{2.97a}$$

$$\left(\frac{\partial}{\partial t} + v\frac{\partial}{\partial z}\right)\rho_{11} + \gamma_1\rho_{11} = \lambda_1 - i\frac{\mu E}{\hbar}\cos(kz - \Omega t)(\rho_{12} - \rho_{21}) \tag{2.97b}$$

$$\left(\frac{\partial}{\partial t} + v\frac{\partial}{\partial z} + i\omega\right)\rho_{21} + \gamma_{21}\rho_{21} = -i\frac{\mu E}{\hbar}\cos(kz - \Omega t)(\rho_{22} - \rho_{11}). \tag{2.97c}$$

Here again we have a rapid time variation at the frequency $\omega \sim \Omega$, and this can be eliminated by the ansatz

$$\rho_{21} = \tilde{\rho}_{21} e^{i(kz - \Omega t)}. \tag{2.98}$$

and a rotating wave approximation like that in (2.13)

$$e^{\pm i(kz - \Omega t)} \cos(kz - \Omega t) \simeq \tfrac{1}{2}. \tag{2.99}$$

The derivative terms bring down a factor because

$$\left( \frac{\partial}{\partial t} + v \frac{\partial}{\partial z} \right) \rho_{21} \simeq -i(\Omega - kv) \rho_{21}. \tag{2.100}$$

In this section we look for steady state solutions only and neglect all time variation after the rapid one has been eliminated by the explicit exponential (2.98). Then we find

$$\tilde{\rho}_{21} = -\frac{i\mu E}{2\hbar} \frac{1}{i(\Delta + kv) + \gamma_{12}} (\rho_{22} - \rho_{11}) \tag{2.101}$$

and the result

$$\rho_{22} - \rho_{11} = \frac{\overline{N}}{1 + 2I\eta L(\Delta + kv)}, \tag{2.102}$$

where we have used the notation introduced in the preceding section. The quantities $\overline{N}$, $I$, and $\eta$ have exactly the same physical significance, the only change is that the Lorentzian has acquired an additional velocity dependence. The resonance now occurs at the position $\Delta = -kv$ leading to

$$\Omega = \omega + kv, \tag{2.103}$$

which shows the effect of the Doppler shift; see (1.182). For an atom moving in the same direction as the traveling wave, the transition requires a higher frequency to satisfy the resonance condition. We assume that

$$\overline{N}(v) = \Lambda_0 W(v), \tag{2.104}$$

and the field can equalize the atomic populations only for those velocities that satisfy (2.103). The width of the resonance region thus appearing can be seen if we write

$$(\rho_{22} - \rho_{11}) = \overline{N}(v) \left[ 1 - \frac{2I\eta\gamma_{12}^2}{(\Delta + kv)^2 + \gamma_{12}^2(1 + 2I\eta)} \right]. \tag{2.105}$$

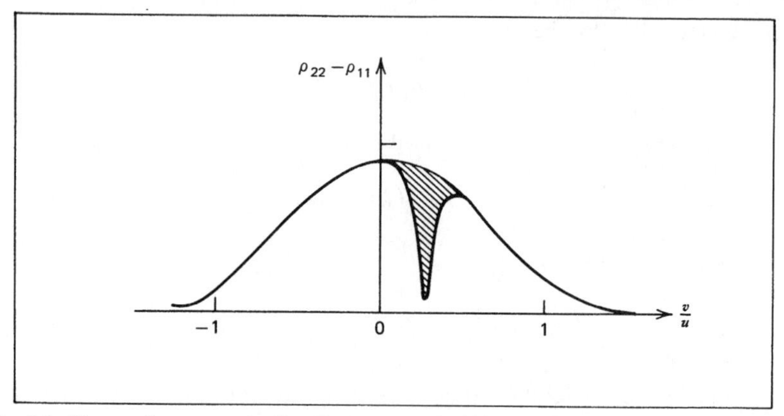


**Fig. 2.4** For moving atoms, the Doppler contour of width $u$ in velocity cannot resonate with the field if $ku \gg \gamma$. Only atoms with a velocity that compensates the detuning can saturate over a velocity range of the order $\gamma/k$.

We can see that the deviation from the unperturbed value $\overline{N}(v)$ is a Lorentzian with the power broadened width

$$\Gamma_p = \gamma_{12}(1 + 2I\eta)^{1/2}. \tag{2.106}$$

A hole appears in the population difference which has been *burned by the interaction* with the field, see Fig. 2.4. This is called the *Bennett hole*. The hole appears because the field transfers population from one level to the other over a limited range

$$\rho_{22}(v) = \frac{\lambda_2(v)}{\gamma_2} - \overline{N}(v)\frac{I\gamma_1\gamma_{21}}{(\Delta + kv)^2 + \gamma_{12}^2(1 + 2I\eta)} \tag{2.107a}$$

$$\rho_{11}(v) = \frac{\lambda_1(v)}{\gamma_1} + \overline{N}(v)\frac{I\gamma_2\gamma_{21}}{(\Delta + kv)^2 + \gamma_{12}^2(1 + 2I\eta)}. \tag{2.107b}$$

These equations are illustrated in Fig. 2.5.

The imaginary part of $\tilde{\rho}_{21}$ becomes

$$\operatorname{Im}\tilde{\rho}_{21} = \frac{1}{2i}(\tilde{\rho}_{21} - \tilde{\rho}_{12}) = -\frac{\mu E}{2\hbar}\frac{\overline{N}\gamma_{12}}{(\Delta + kv)^2 + \gamma_{12}^2(1 + 2I\eta)}. \tag{2.108}$$

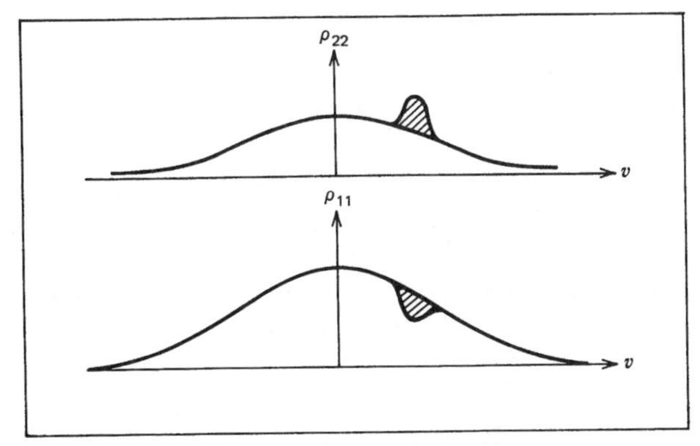


**Fig. 2.5** When atoms of the proper velocity to achieve resonance are removed from the lower level, they appear with the same velocity on the upper level.

This shows a power broadened Lorentzian response to the external driving field. The result comes from all the atoms with a given velocity $v$. To obtain the response of the whole medium we must sum over atoms with all possible velocities.

For a traveling wave we calculate the polarization summed over all velocities in the form

$$P(z,t) = N_0\mu \int \left[ \tilde{\rho}_{21} e^{i(kz-\Omega t)} + \tilde{\rho}_{12} e^{-i(kz-\Omega t)} \right] dv$$

$$= N_0\mu \int \left[ (\tilde{\rho}_{21} + \tilde{\rho}_{12})\cos(kz - \Omega t) + i(\tilde{\rho}_{21} - \tilde{\rho}_{12})\sin(kz - \Omega t) \right] dv. \tag{2.109}$$

We can use these equations to introduce $C$ and $S$ coefficients as in Eq. (2.42)

$$C = N_0\mu \int (\tilde{\rho}_{21} + \tilde{\rho}_{12})\, dv \tag{2.110a}$$

$$S = N_0\mu i \int (\tilde{\rho}_{21} - \tilde{\rho}_{12})\, dv \tag{2.110b}$$

or we can define a susceptibility $\chi$ by setting

$$P(z,t) = \tfrac{1}{2}\left(E\chi e^{i(kz-\Omega t)} + \text{c.c}\right). \tag{2.111}$$

From these definitions we find the relationship

$$\chi = \frac{C - iS}{E} = 2N_0\mu\frac{1}{E}\int\tilde{\rho}_{21}(v)\,dv$$

$$= -i\frac{N_0\mu^2}{\hbar}\int\frac{[\gamma_{21} - i(\Delta + kv)]}{(\Delta + kv)^2 + \gamma_{21}^2(1 + 2I\eta)}\bar{N}(v)\,dv. \qquad (2.112)$$

The susceptibility is complex, and, if we write $\chi = \chi' - i\chi''$, the real part $\chi'$ is related to the dispersion of the medium and the imaginary part $\chi''$ to the losses or gain of the medium. We find also

$$\chi' = \frac{C}{E}, \qquad \chi'' = \frac{S}{E} \qquad (2.113)$$

which connects $\chi$ with the in-phase and out-of-phase components of the polarization.

The concept of a susceptibility can easily be generalized. If the electric field consists of a series of modes

$$E(\mathbf{r}, t) = \sum_n E_n U_n(\mathbf{r})e^{i\Omega_n t}, \qquad (2.114)$$

we can define the susceptibility by expanding the polarization

$$P(\mathbf{r}, t) = \sum_n \chi_n E_n U_n(\mathbf{r})e^{i\Omega_n t}. \qquad (2.115)$$

In general the susceptibility will be a function of all the field amplitudes $\{E_n\}$. If we expand it we obtain a series of coefficients

$$\chi_n(E_1, E_2, \ldots) = \chi_n^{(1)} + \sum_m \chi_{nm}^{(2)}E_m$$

$$+ \sum_{m'} \chi_{nml}^{(3)}E_m E_l + \cdots. \qquad (2.116)$$

In *nonlinear optics* the coefficients $\chi_n^{(1)}$, $\chi_{nm}^{(2)}$, $\chi_{nml}^{(3)}$ and so on are called *nonlinear susceptibilities*.

If we introduce the Maxwellian distribution (1.183) into (2.109) we obtain the convolution between a Gaussian and a Lorentzian line shape. This is called a plasma dispersion function and is tabulated in the literature.†

†The plasma dispersion function is well treated in Abramowitz and Stegun (1970), Section 7, which also contains tables of the function for complex arguments. For numerical work and some asymptotic results the series expansion, Eq. 7.1.8, and the continued fraction representation, Eq. 7.1.15, are useful.

In the limit when the inhomogeneous Doppler broadened width $ku$ much exceeds the homogeneous width $\gamma_{12}$ (with power broadening) the integrals have simple asymptotic expressions.

The imaginary part contains a Lorentzian that samples the Gaussian, and Eq. (2.112) gives

$$\chi'' = \frac{N_0 \Lambda_0 \mu^2}{\hbar} \int \frac{\gamma_{12}}{(\Delta + kv)^2 + \gamma_{21}^2(1 + 2I\eta)} \frac{e^{-v^2/u^2}}{\sqrt{\pi}\, u} dv$$

$$= \frac{N_0 \Lambda_0 \mu^2 e^{-\Delta^2/k^2u^2}}{\sqrt{\pi}\, \hbar k u \sqrt{1 + 2I\eta}} \int \frac{\Gamma_p}{x^2 + \Gamma_p^2} dx$$

$$= \sqrt{\pi} \frac{N_0 \Lambda_0 \mu^2 e^{-\Delta^2/k^2u^2}}{\hbar k u \sqrt{1 + 2I\eta}}. \tag{2.117}$$

The dispersive part of the susceptibility cannot be calculated as easily because the function $x/(x^2 + \gamma^2)$ does not have a convergent integral without the multiplicative Gaussian function. We can, however, obtain an asymptotic estimate for this function, too, in the limit

$$\gamma_{12}(1 + 2I\eta)^{1/2} \ll ku. \tag{2.118}$$

From (2.112) we obtain

$$\chi' = -\frac{N_0 \Lambda_0 \mu^2}{\sqrt{\pi}\, \hbar u} \int \frac{(\Delta + kv) e^{-v^2/u^2}}{(\Delta + kv)^2 + \Gamma_p^2} dv$$

$$= -\frac{N_0 \Lambda_0 \mu^2}{\sqrt{\pi}\, \hbar k u} \int e^{-(x-\Delta)^2/k^2u^2} \frac{x\, dx}{x^2 + \Gamma_p^2}$$

$$= -\frac{N_0 \Lambda_0 \mu^2}{\sqrt{\pi}\, \hbar k u} e^{-\Delta^2/k^2u^2} \int \frac{e^{2\Delta x/k^2u^2} x e^{-x^2/k^2u^2}}{x^2 + \Gamma_p^2} dx$$

$$= -\frac{2\,\Delta N_0 \Lambda_0 \mu^2}{\sqrt{\pi}\, \hbar k^3 u^3} e^{-\Delta^2/k^2u^2} \int \frac{x^2}{x^2 + \Gamma_p^2} e^{-x^2/k^2u^2}\, dx$$

$$= -\frac{2\,\Delta N_0 \Lambda_0 \mu^2}{\hbar (ku)^2} e^{-\Delta^2/k^2u^2}. \tag{2.119}$$

The limit (2.118) is called *the Doppler limit*; its value for qualitative considerations is sometimes great even when its quantitative accuracy is poor.

If we look at the susceptibility $\chi = \chi' - i\chi''$ of the atom in a traveling wave we find the following properties:

1. For low intensities $I \to 0$, both components go to a constant. This means that in this limit we have a linear relationship $P = \chi(0)E$, as expected.

2. For large intensities $I \gg 1$, the imaginary part $\chi'' \propto I^{-1/2}$ that is as $E^{-1}$, whereas the real part $\chi'$ remains unsaturated. The latter fact derives from the dispersive behavior of the integral in (2.119); all atoms contribute, even in the low intensity limit, and power broadening can affect only the imaginary part in (2.117), which essentially is determined only by the resonantly interacting atoms. This saturation behavior differs in an essential way from that of a homogeneously broadened medium, Eqs. (2.44), where the response goes to zero as

$$\chi = \frac{C - iS}{E} \propto \frac{1}{I} \tag{2.120}$$

for asymptotically large values of $I$. Only when the Doppler limit approximation (2.118) breaks down, does the inhomogeneously broadened atomic response start to approach the homogeneously broadened $\chi \propto I^{-1}$, as can be seen directly from (2.112).

3. If we look at the constant value

$$\chi''(I = 0, \Delta = 0) = \sqrt{\pi}\,\frac{N_0 \Lambda_0 \mu^2}{\hbar k u} \tag{2.121}$$

obtained from (2.117) and compare this with the value for the homogeneously broadened case (2.44b)

$$\chi''(I = 0, \Delta = 0) = \frac{S(I = 0, \Delta = 0)}{E} = -\frac{N_0 \bar{N} \mu^2}{\hbar \gamma_{21}} \tag{2.122}$$

we find that the former follows from the latter if we set

$$\bar{N} \sim \sqrt{\pi}\,\frac{\gamma_{21}}{ku} \Lambda_0. \tag{2.123}$$

Because $\gamma_{21} \ll ku$, this result can be understood if we remember that only atoms situated in the Bennett holes of width $\gamma_{12}$ can interact with the radiation. The number of atoms actively participating in the interaction process must hence be decreased by a factor $(\gamma_{12}/ku)$ as the total number of moving atoms is distributed over the Doppler width proportional to $ku$.

4. For asymptotically large detunings, $|\Delta| \gg \gamma_{13}$, the two cases behave differently. Both sample the wings of the interaction resonance function, but whereas the homogeneously broadened case (2.44) leads to a Lorentzian $L(\Delta)$, the inhomogeneous distribution (2.112) follows the Gaussian $\exp(-\Delta^2/k^2u^2)$ for a while. Here we must, however, issue a warning. When $|\Delta|$ becomes *really large*, $|\Delta| \gg ku$, the integral (2.112) also has the asymptotic behavior $\Delta^{-2}$ because the tail of the distant Lorentzians can always find an overlap with the Gaussian function and hence have a contribution from the region $kv \simeq 0$, which will have exactly the weight $(\gamma_{12}/\Delta)^2$. The expression (2.117) can, however, be used in the range $|\Delta| \precsim ku$, which suffices in most practical situations. When $|\Delta|$ becomes larger, either the response goes down too much or hitherto neglected transitions start to approach resonance.

## 2.5. MOVING ATOMS IN A STANDING WAVE

In a laser system the electromagnetic field is situated between two mirrors, and the radiation is reflected back and forth many times through the medium. In this situation the field inside the cavity is well described by a standing wave field of the type (2.25), namely,

$$E(z,t) = E\cos\Omega t \sin kz. \tag{2.124}$$

The equations of motion are identical with those in Eqs. (2.26). The argument used in Sec. 2.2 is again valid for the rapidly varying off-diagonal density matrix elements. We make the ansatz

$$\rho_{21} = e^{-i\Omega t}\tilde{\rho}_{21} \tag{2.125}$$

and apply the approximation discussed after (2.27), the rotating wave approximation. The resulting equations are then for moving atoms

$$\left(\frac{\partial}{\partial t} + v\frac{\partial}{\partial z}\right)\tilde{\rho}_{21} = -(\gamma_{21} + i\Delta)\tilde{\rho}_{21} - \frac{i}{2\hbar}\mu E \sin kz(\rho_{22} - \rho_{11}) \tag{2.126a}$$

$$\left(\frac{\partial}{\partial t} + v\frac{\partial}{\partial z}\right)\rho_{22} = \lambda_2 - \gamma_2\rho_{22} - i\frac{\mu E}{2\hbar}\sin kz(\tilde{\rho}_{21} - \tilde{\rho}_{12}) \tag{2.126b}$$

$$\left(\frac{\partial}{\partial t} + v\frac{\partial}{\partial z}\right)\rho_{11} = \lambda_1 - \gamma_1\rho_{11} + i\frac{\mu E}{2\hbar}\sin kz(\tilde{\rho}_{21} - \tilde{\rho}_{12}). \tag{2.126c}$$

For stationary atoms in Sec. 2.2 we could treat $z$ as a parameter and solve directly for steady state. Because of the atomic motion we cannot do this for

(2.126) because the atomic position depends on time through

$$z(t) = z_0 + v(t - t_0); \tag{2.127}$$

see the representation (1.190). This implicit time dependence gives rise to the drift term $v\partial/\partial z$ on the left-hand side. In the case of a traveling wave, Sec. 2.4., we could eliminate the $z$-dependence together with the time dependence. For a standing wave this is not possible. We can, however, write the standing wave as two traveling waves propagating in opposite directions; from (2.124) we find

$$E(z,t) = \tfrac{1}{2}E[\sin(kz + \Omega t) + \sin(kz - \Omega t)]. \tag{2.128}$$

From physical considerations we can assume the lowest-order effect of the two traveling waves to be a superposition of the effects caused by each wave separately. The induced dipole moment is given by the off-diagonal element of the density matrix. For one traveling wave the off-diagonal element is of the form (2.98). For two waves we can try the ansatz

$$\begin{aligned}\rho_{21} &= \rho_+ e^{i(kz - \Omega t)} + \rho_- e^{-i(kz+\Omega t)} \\ &= e^{-i\Omega t}\left[\rho_+ e^{ikz} + \rho_- e^{-ikz}\right].\end{aligned} \tag{2.129}$$

If we obtain solutions for $\rho_+$ and $\rho_-$ we expect these two to give an acceptable approximation to the exact results. Corrections arise from cross-terms between the two traveling wave components; the accuracy of the approximation is discussed later.

We assume that $\rho_{22}$ and $\rho_{11}$ have no spatial dependence, which means that we replace the level populations with their volume averages. We use an approximation in the spirit of the rotating wave approximation by setting

$$e^{ikz}\sin kz = \frac{i}{2} + \cdots . \tag{2.130}$$

The equations to be solved are then found to be

$$\frac{d}{dt}\rho_+ = -[\gamma_{12} + i(\Delta + kv)]\rho_+ - \frac{\mu E}{4\hbar}(\rho_{22} - \rho_{11}) \tag{2.131a}$$

$$\frac{d}{dt}\rho_- = -[\gamma_{12} + i(\Delta - kv)]\rho_- + \frac{\mu E}{4\hbar}(\rho_{22} - \rho_{11}) \tag{2.131b}$$

$$\frac{d}{dt}\rho_{22} = \lambda_2 - \gamma_2\rho_{22} + \frac{\mu E}{4\hbar}(\rho_+^* + \rho_+ - \rho_- - \rho_-^*) \tag{2.131c}$$

$$\frac{d}{dt}\rho_{11} = \lambda_1 - \gamma_1\rho_{11} - \frac{\mu E}{4\hbar}(\rho_+^* + \rho_+ - \rho_- - \rho_-^*). \tag{2.131d}$$

We now assume that the off-diagonal elements are affected by phase perturbations causing rapid relaxation, and we can assume that $\rho_+$ and $\rho_-$ reach their equilibrium values much faster than the occupation probabilities $\rho_{22}$ and $\rho_{11}$. It is then possible to obtain rate equations along the lines given in Sec. 2.2. The rate is the sum of two contributions† like those in (2.33) but with the Lorentzian at resonance for $\Delta = \pm kv$. In this section we look only for the stationary solution, all time dependence is assumed absent, and then no adiabatic assumption is needed.

In Eqs. (2.131) we set the time derivatives equal to zero and find

$$\rho_{\pm} = \mp \frac{\mu E}{4\hbar} \frac{1}{\gamma_{12} + i(\Delta \pm kv)} (\rho_{22} - \rho_{11}), \tag{2.132}$$

$$\begin{aligned} \rho_+^* + \rho_+ - \rho_- - \rho_-^* &= 2\,\mathrm{Re}(\rho_+ - \rho_-) \\ &= -\frac{\mu E}{2\hbar\gamma_{12}} [L(\Delta + kv) + L(\Delta - kv)](\rho_{22} - \rho_{11}), \end{aligned} \tag{2.133}$$

and

$$\rho_{22} - \rho_{11} = \frac{\overline{N}}{1 + \frac{1}{2} I\eta [L(\Delta + kv) + L(\Delta - kv)]}, \tag{2.134}$$

where we have again used the definitions (2.10), (2.32), and (2.35).

If we compare the result (2.134) with the result (2.105) for one single traveling wave, we find that the population difference will have two holes at velocities $v = \pm\Delta/k$, see Fig. 2.6. Each one will have a magnitude determined by

$$\frac{1}{2} I = \frac{\mu^2 E^2}{4\hbar^2 \gamma_1 \gamma_2}. \tag{2.135}$$

This corresponds to the fact that for each wave of (2.128) the amplitude is $\frac{1}{2}E$, and hence the parameter of (2.105) is given by

$$2\left(\frac{\mu^2}{2\hbar^2\gamma_1\gamma_2}\right)\left(\frac{E}{2}\right)^2 = \frac{1}{2}\left(\frac{\mu^2 E^2}{2\hbar^2\gamma_1\gamma_2}\right), \tag{2.136}$$

as is indeed the case in (2.135).

† Because the result obtained (2.134) corresponds to a rate equation result, the approximation used in this section, namely, the ansatz (2.129), is often called the rate equation approximation (REA) in the literature. For steady state the result is equal to that of a rate equation, but for the time-dependent calculations the validity of our REA is assured if $\gamma_{12}$ is the fastest rate of change in the problem after the RWA has been carried out.

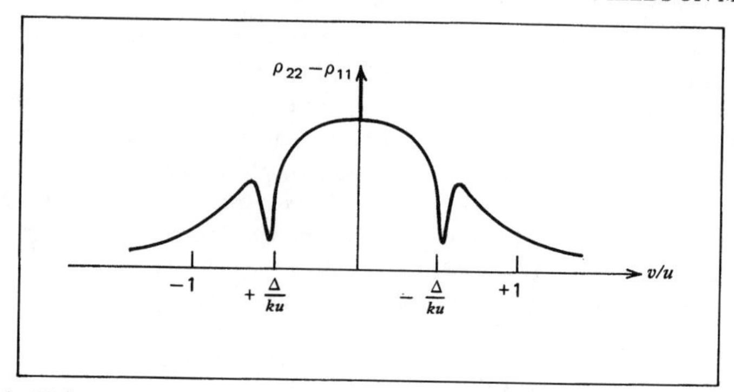


**Fig. 2.6** With a standing wave the simplest physical picture consists of two independent population holes, one burned by each traveling wave component of the standing wave.

The polarization in (2.132) is given by

$$\rho_{\pm} = \mp \frac{\mu E \bar{N}}{4\hbar} \frac{[\gamma_{12} - i(\Delta \pm kv)]}{[\gamma_{12}^2 + (\Delta \pm kv)^2]\{1 + \frac{1}{2}I\eta[L(\Delta - kv) + L(\Delta + kv)]\}}. \tag{2.137}$$

When $|\Delta| \simeq kv$ one of the Lorentzians becomes nearly equal to unity and the other one is given approximately by

$$L(2\Delta) \simeq \left(\frac{\gamma_{12}}{2\Delta}\right)^2. \tag{2.138}$$

If the detuning is dominantly large $|\Delta| \gg \gamma_{12}$, the quantity (2.138) becomes negligible, and we have for that component of $\rho_{21}$ that is nearly at resonance

$$\rho_{\pm} \simeq \mp \frac{\mu E}{4\hbar} \frac{[\gamma_{12} - i(\Delta \pm kv)]}{(\Delta \pm kv)^2 + \gamma_{12}^2(1 + \frac{1}{2}I\eta)} \bar{N}. \tag{2.139}$$

Hence the polarization for large enough detuning $|\Delta|$ becomes Lorentzian with a power broadened line width

$$\Gamma_p = \gamma_{12}(1 + \tfrac{1}{2}I\eta)^{1/2}. \tag{2.140}$$

This corresponds to the Lorentzian of one traveling wave according to (2.108) but with the power broadening determined by the saturation parameter (2.135). For large detunings the two traveling waves constituting the standing wave act independently, and each one digs its own hole in the velocity distribution; compare Fig. 2.6. When

$$|\Delta| \gg \gamma_{12}\left(1 + \tfrac{1}{2}I\eta\right)^{1/2} \tag{2.141}$$

the two holes do not overlap, and no interference takes place between them. When $|\Delta| \lesssim \gamma_{12}$ the holes meet, both traveling waves interact with the same atoms, and our simple approximation of noninteracting waves does not hold any more. We show later that it still constitutes a reasonable approximation in many applications. It gives a qualitatively correct description of the physical processes occurring.

We again define a susceptibility by setting

$$P(z,t) = -\frac{E}{4}\left[i\chi_+ e^{i(kz-\Omega t)} - i\chi_- e^{-i(kz+\Omega t)} + \text{c.c.}\right]; \tag{2.142}$$

see the form (2.111). From (2.129) we find that

$$\begin{aligned}\chi_\pm &= \pm\frac{4i}{E}N_0\mu\int\rho_\pm(v)\,dv\\ &= -\frac{\mu^2 N_0\Lambda_0}{\hbar\sqrt{\pi}\,u}\\ &\quad\times\int\frac{\left[i\gamma_{12} + (\Delta \pm kv)\right]e^{-v^2/u^2}\,dv}{\left[\gamma_{12}^2 + (\Delta \pm kv)\right]^2\left\{1 + \frac{1}{2}I\eta\left[L(\Delta + kv) + L(\Delta - kv)\right]\right\}}.\end{aligned} \tag{2.143}$$

In the Doppler limit $ku \gg \gamma_{12}$, we can approximate this integral by its asymptotic expressions as in Sec. 2.4. Here we consider two limiting cases.

In the limit (2.141) we can neglect the one Lorentzian in the denominator and obtain

$$\chi_\pm = -\frac{\mu^2 N_0\Lambda_0}{\sqrt{\pi}\,\hbar u k}\int\frac{\left[i\gamma_{12} + (\Delta \pm x)\right]e^{-x^2/k^2u^2}dx}{\left[(\Delta \pm x)^2 + \gamma_{12}^2\left(1 + \frac{1}{2}I\eta\right)\right]}. \tag{2.144}$$

As in Eqs. (2.117–2.119) we find

$$\chi''_{\pm} = \frac{\mu^2 N_0 \Lambda_0}{\sqrt{\pi}\,\hbar k u} \int \frac{\gamma_{12} e^{-x^2/k^2u^2}}{(\Delta \pm x)^2 + \gamma_{12}^2\left(1 + \frac{1}{2} I\eta\right)}\, dx$$

$$= \sqrt{\pi}\, \frac{N_0 \Lambda_0 \mu^2 e^{-\Delta^2/k^2u^2}}{\hbar k u \sqrt{(1 + \eta I/2)}} \tag{2.145}$$

and

$$\chi'_{\pm} = -\frac{\mu^2 N_0 \Lambda_0}{\sqrt{\pi}\,\hbar k u} \int \frac{(\Delta \pm x) e^{-x^2/k^2u^2}}{\left[(\Delta \pm x)^2 + \gamma_{12}^2\left(1 + \frac{1}{2} I\eta\right)\right]}\, dx$$

$$= \frac{2\Delta N_0 \Lambda_0 \mu^2}{\hbar (ku)^2} e^{-\Delta^2/k^2u^2}. \tag{2.146}$$

From Eqs. (2.145–2.146) we can see that in the limit of a large detuning the two traveling waves act completely independently. They saturate the medium separately like the one traveling wave treated in Sec. 2.4; only the saturation parameter must be redefined.

When the two resonantly interacting groups of atoms meet, $\Delta \to 0$, the approximation used is no longer valid. It is, however, possible to let the detuning $\Delta$ go to zero in (2.143) and calculate the result, which gives a qualitatively correct picture of the physical properties of the resonant interaction between the radiation and the Doppler broadened atomic system. When $|\Delta| \ll \gamma_{12}$, the two Lorentzians coincide, and we find from (2.143)

$$\chi_{\pm} = -\frac{\mu^2 N_0 \Lambda_0}{\sqrt{\pi}\,\hbar k u} \int \frac{[i\gamma_{12} \pm x] e^{-x^2/k^2u^2}}{\left[x^2 + \gamma_{12}^2(1 + I\eta)\right]}\, dx. \tag{2.147}$$

For the imaginary part of the susceptibility we obtain

$$\chi''_{\pm} = \sqrt{\pi}\, \frac{N_0 \Lambda_0 \mu^2}{\hbar k u \sqrt{(1 + I\eta)}}, \tag{2.148}$$

whereas the real part vanishes with the detuning

$$\chi'_{\pm} = 0 \tag{2.149}$$

even when no assumption about the Doppler limit is made.

## 2.6. CORRECTION TERMS FOR MOVING ATOMS

The approximation presented in the preceding section was based on the assumption that the occupation probabilities $\rho_{22}$ and $\rho_{11}$ have no pulsations in space or time, they equal their average values $\overline{\rho_{22}}$, $\overline{\rho_{11}}$. For optical frequencies there are no corrections from the time variation, but we must include spatial oscillations in the higher-order approximations. Caused by the atomic motion, these lead to pulsations in the populations on the atomic levels; we treat these in the lowest approximation in this section. For a two-level atom with its lower state the ground state, the corrections are easily seen to have a simple physical interpretation. In this case the only decay from the upper level is assumed to be back to the ground state, as no metastable levels are assumed to enter the considerations. As an exercise we treat this case in the present section.

The equations of motion are obtained from Sec. 1.6 in the form

$$\left(\frac{\partial}{\partial t}+v\frac{\partial}{\partial z}\right)\rho_{22}=-\Gamma\rho_{22}-i\frac{\mu E}{2\hbar}\sin kz(\tilde{\rho}_{21}-\tilde{\rho}_{12}) \tag{2.150a}$$

$$\left(\frac{\partial}{\partial t}+v\frac{\partial}{\partial z}\right)\rho_{11}=\Gamma\rho_{22}+i\frac{\mu E}{2\hbar}\sin kz(\tilde{\rho}_{21}-\tilde{\rho}_{12}) \tag{2.150b}$$

$$\left(\frac{\partial}{\partial t}+v\frac{\partial}{\partial z}\right)\tilde{\rho}_{21}=-\left(\frac{1}{2}\Gamma+i\Delta\right)\tilde{\rho}_{21}-\frac{i\mu E}{2\hbar}\sin kz(\rho_{22}-\rho_{11}). \tag{2.150c}$$

In this case the total occupation probability is conserved and can be normalized to the population at that velocity

$$\rho_{11}+\rho_{22}=W(v). \tag{2.151}$$

No decay out of the active levels occurs, and hence no pumping can be allowed into the level pair because no accumulation of probability can be allowed to take place. We can, however, put Eqs. (2.150) in a form familiar from the last section by the use of (2.151) to eliminate $\rho_{22}$ from Eq. (2.150b):

$$\left(\frac{\partial}{\partial t}+v\frac{\partial}{\partial z}\right)\rho_{11}=\Gamma W(v)-\Gamma\rho_{11}+i\frac{\mu E}{2\hbar}\sin kz(\tilde{\rho}_{21}-\tilde{\rho}_{12}). \tag{2.152}$$

Looking back at Eqs. (2.126), we find that we obtained identical equations if we set

$$\lambda_2=0,\qquad \lambda_1=\Gamma W(v),\qquad \gamma_2=\gamma_1=\Gamma. \tag{2.153}$$

In this system an atom stays in interaction with the field for an infinitely long time; no decay out of the interacting levels takes place. In reality the interaction is, of course, terminated when the atom leaves the laser beam in the transverse direction, but we neglect this possibility here. Such a model is sometimes useful, and it can be treated exactly in the same way as our earlier systems, when the replacements (2.153) are used.

In this section we want to include population pulsations in lowest order. It is easily seen that they will pulsate at a spatial frequency twice that of the dipole moment (2.129). Hence we write

$$\rho_{jj}(z) = \overline{\rho_{jj}} + \left[\rho_j e^{2ikz} + \rho_j^* e^{-2ikz}\right] \tag{2.154}$$

where $j = 1$ or 2.

We insert (2.154) together with the ansatz (2.129) into (2.150) and obtain

$$\rho_2 = -\frac{1}{\Gamma + 2ikv}\frac{\mu E}{4\hbar}\left(\rho_+ - \rho_-^*\right)$$

$$\rho_1 = \frac{1}{\Gamma + 2ikv}\frac{\mu E}{4\hbar}\left(\rho_+ - \rho_-^*\right). \tag{2.155}$$

When the relations (2.154) are inserted into (2.150c) we obtain instead of Eqs. (2.131a, 2.131b) the relations

$$\left[\frac{1}{2}\Gamma + i(\Delta + kv)\right]\rho_+ = -\frac{\mu E}{4\hbar}\left(\overline{\rho_{22}} - \overline{\rho_{11}} - \rho_2 + \rho_1\right) \tag{2.156a}$$

$$\left[\frac{1}{2}\Gamma + i(\Delta - kv)\right]\rho_- = \frac{\mu E}{4\hbar}\left(\overline{\rho_{22}} - \overline{\rho_{11}} - \rho_2^* + \rho_1^*\right). \tag{2.156b}$$

From these and Eqs. (2.150a, 2.150b) we find

$$\begin{aligned}
\left(\rho_+ - \rho_-^*\right) &+ \frac{\mu E}{4\hbar}\left[\frac{1}{\frac{1}{2}\Gamma + i(\Delta + kv)} + \frac{1}{\frac{1}{2}\Gamma - i(\Delta - kv)}\right]\left(\overline{\rho_{22}} - \overline{\rho_{11}}\right) \\
&= \frac{\mu E}{4\hbar}\left[\frac{1}{\frac{1}{2}\Gamma + i(\Delta + kv)} + \frac{1}{\frac{1}{2}\Gamma - i(\Delta - kv)}\right]\left(\rho_2 - \rho_1\right) \\
&= -\left(\frac{\mu E}{4\hbar}\right)^2\left[\frac{1}{\frac{1}{2}\Gamma + i(\Delta + kv)} + \frac{1}{\frac{1}{2}\Gamma - i(\Delta - kv)}\right] \\
&\quad \times \frac{2}{\Gamma + 2ikv}\left(\rho_+ - \rho_-^*\right).
\end{aligned} \tag{2.157}$$

For the averaged occupation probabilities $\overline{\rho_{jj}}$ we find equations like (2.131c, 2.131d). Their solution is

$$\overline{\rho_{22}}-\overline{\rho_{11}}=\bar{N}+\frac{\mu E}{2\hbar\Gamma}\left[\left(\rho_{+}-\rho_{-}^{*}\right)+\left(\rho_{+}^{*}-\rho_{-}\right)\right]$$

$$=-W(v)-\left(\frac{\mu E}{2\hbar}\right)^{2}\frac{1}{\Gamma}\operatorname{Re}R\left(\overline{\rho_{22}}-\overline{\rho_{11}}\right) \qquad (2.158)$$

where

$$R=\frac{(\Gamma+2ikv)}{\left[\frac{1}{2}\Gamma+i(\Delta+kv)\right]\left[\frac{1}{2}\Gamma-i(\Delta-kv)\right]+\left(\frac{\mu E}{2\hbar}\right)^{2}\left(\frac{1}{2}\Gamma+ikv\right)\frac{1}{\Gamma+2ikv}}. \qquad (2.159)$$

If we write the denominator in the form

$$\left[\frac{1}{2}\Gamma+i(\Delta+kv)\right]\left[\frac{1}{2}\Gamma-i(\Delta-kv)\right]+\frac{1}{2}\left(\frac{\mu E}{2\hbar}\right)^{2}$$

$$=\left(\frac{1}{2}\Gamma+ikv\right)^{2}+\Delta^{2}+\frac{1}{2}\left(\frac{\mu E}{2\hbar}\right)^{2}\equiv\left[\frac{1}{2}\Gamma+i(kv-\Omega)\right]\left[\frac{1}{2}\Gamma+i(kv+\Omega)\right], \qquad (2.160)$$

we obtain from (2.159)

$$R=\left[\frac{1}{\frac{1}{2}\Gamma+i(kv-\Omega)}+\frac{1}{\frac{1}{2}\Gamma+i(kv+\Omega)}\right], \qquad (2.161)$$

where

$$\Omega=\left[\Delta^{2}+\frac{1}{2}\left(\frac{\mu E}{2\hbar}\right)^{2}\right]^{1/2}. \qquad (2.162)$$

The average population difference becomes from (2.158)

$$\overline{\rho_{11}}-\overline{\rho_{22}}=\frac{W(v)}{1+2(\mu E/2\hbar\Gamma)^{2}[L(kv+\Omega)+L(kv-\Omega)]}, \qquad (2.163)$$

where

$$L(x)=\frac{(\Gamma/2)^{2}}{(\Gamma/2)^{2}+x^{2}}. \qquad (2.164)$$

If we compare this result with that in (2.134) we find that the saturation parameter is

$$\frac{1}{2}I\eta = \frac{1}{2}\left(\frac{\mu^2E^2}{2\hbar^2\gamma_1\gamma_2}\right)\left(\frac{\gamma_1+\gamma_2}{2\gamma_{12}}\right) = \frac{\mu^2E^2}{2\hbar^2\Gamma^2}, \tag{2.165}$$

when (2.153) has been introduced. The Lorentzians are centered at $kv = \pm\Omega$, which agrees with the result $kv = \pm\Delta$ to lowest order in the intensity.

The correction terms we have obtained show that the hole in the population distribution is centered, not at $\pm\Delta$ as in Fig. 2.6 but at

$$kv = \pm\Omega = \pm\left[\Delta + \frac{\mu^2E^2}{16\hbar^2\Delta} + \cdots\right], \tag{2.166}$$

when an expansion in $E$ is used. This means that the distance between the two Bennett holes is larger than in the lowest-order approximation. It is as if the holes tried to push each other apart; there is an apparent repulsion between them. This result can be generalized also to the situation when several modes are present, and each burns its own hole in the population difference.

If we look at the result (2.166) we find that the resonance has been shifted to

$$\omega' = \omega + \frac{\mu^2E^2}{16\hbar^2\Delta}. \tag{2.167}$$

This can be interpreted as a *light shift* of the atomic energy level difference $\hbar\omega$ due to the presence of the two traveling waves; each one sees a level shift due to the other one. The result is correct in second order of perturbation theory if we use the matrix element $(\mu E/4)$ coming from (2.124) and an energy difference† $2\hbar\Delta$. The levels are separated more by the interaction, and we obtain

$$\begin{aligned}\Delta E &= \left[E_2 + \frac{(\mu E/4)^2}{2\hbar\Delta}\right] - \left[E_1 - \frac{(\mu E/4)^2}{2\hbar\Delta}\right] \\ &= \hbar\omega + \frac{(\mu E/4)^2}{\hbar\Delta}\end{aligned} \tag{2.168}$$

in agreement with (2.166).

Another consequence of the result (2.163) is that the two Lorentzians do not coincide for exact resonance but the population difference still contains two holes separated by

$$k\delta v = 2\Omega = \sqrt{2}\,\frac{\mu E}{2\hbar}, \tag{2.169}$$

†Compare the treatment of the Bloch–Siegert shift given in Sec. 2.3.

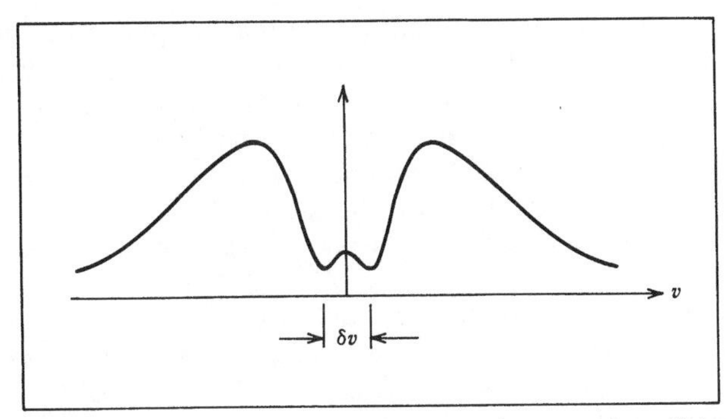


**Fig. 2.7** Higher-order field effects cause the resonant atomic response to split into two holes separated by $\delta v = \mu E/\sqrt{2}\,k\hbar$.

which shows that there is a Stark shift of the resonance just like that in the radio frequency case of Sec. 2.3. The line shape is shown in Fig. 2.7. It is as if the two holes were trying to avoid each other due to their apparent repulsion. They can also saturate atoms over a broader velocity range of the width given by (2.169) and hence polarize more atoms in the sample than they could without the splitting.

The considerations for exact resonance $\Delta \simeq 0$, cannot be regarded as exact. Only for $\Delta \gg \Gamma$ can the result (2.163) be valid. In the next section we show, how one can handle the general case of a two-level atom and generalize the results obtained in this section.

## 2.7. STRONG SIGNAL THEORY FOR MOVING ATOMS

For optical frequencies the rotating wave approximation is usually very good. For a two-level system situated in a standing wave we, however, must include the atomic motion, for instance, in the form given in Eq. (1.199). The field amplitude for a moving atom then takes the form

$$\sin kz = \sin k[z - v(t - t')], \tag{2.170}$$

and thanks to its travel across the stationary pattern of the standing wave the atom sees a field modulation at the frequency $kv$. This is just the nonrelativistic Doppler shift discussed earlier. However, the form of the equations becomes very similar to that of an atom in a strong radio-frequency field without the rotating wave approximation. Instead of the field variation $\cos \Omega t$ in Eq. (2.62) we have the variation (2.170), but the physical situation is very similar. It is possible to achieve a mathematical description identical with the one in Sec. 2.3 by expanding the density matrix elements in a

Fourier series in $e^{ikz}$, the first terms of which are displayed in Eqs. (2.129) and (2.154). These two lowest-order corrections correspond to the case where each traveling wave component affects the polarization in (2.129) independently; their interference effects are neglected. The inclusion of population pulsations in (2.154) does not include real correlations but only a modification of each resonance because of the presence of the other traveling wave. In the case treated in Sec. 2.6 we found a light shift due to the other field.

In this section we want to solve the case of a strong standing wave exactly, but in such a way that the emergence of the correlations between the two traveling wave components is as transparent as possible. To this end we choose to write the field in terms of two independent modes by generalizing (2.128)

$$E(z,t) = \left[E_{-}\sin(kz + \Omega t) + E_{+}\sin(kz - \Omega t)\right]. \qquad (2.171)$$

This field can be considered as a standing wave pattern of the amplitude $E_{-}$ and an additional traveling wave of the amplitude $(E_{+} - E_{-})$. Another possible choice is a standing wave of the amplitude $E_{+}$, but then the traveling wave left over progresses in the opposite direction; other alternatives are easily found. Therefore a field like (2.171) is sometimes called a *quasi-standing wave*.

After introducing the rotating wave approximation as in Sec. 2.5, we obtain from (2.126) the equations

$$\left(\frac{\partial}{\partial t} + v\frac{\partial}{\partial z}\right)\rho_{22} + \gamma_2\rho_{22} = \lambda_2 + \frac{\mu}{2\hbar}\left[\left(E_{+}e^{ikz} - E_{-}e^{-ikz}\right)\tilde{\rho}_{12} - \left(E_{-}e^{ikz} - E_{+}e^{-ikz}\right)\tilde{\rho}_{21}\right] \qquad (2.172a)$$

$$\left(\frac{\partial}{\partial t} + v\frac{\partial}{\partial z}\right)\rho_{11} + \gamma_1\rho_{11} = \lambda_1 - \frac{\mu}{2\hbar}\left[\left(E_{+}e^{ikz} - E_{-}e^{-ikz}\right)\tilde{\rho}_{12} - \left(E_{-}e^{ikz} - E_{+}e^{-ikz}\right)\tilde{\rho}_{21}\right] \qquad (2.172b)$$

$$\left(\frac{\partial}{\partial t} + v\frac{\partial}{\partial z}\right)\tilde{\rho}_{21} + (\gamma_{21} + i\Delta)\tilde{\rho}_{21} = -\frac{\mu}{2\hbar}\left(E_{+}e^{ikz} - E_{-}e^{-ikz}\right)(\rho_{22} - \rho_{11}). \qquad (2.172c)$$

We consider steady state $\partial/\partial t = 0$ and use Fourier series expansions in the form

$$\rho_{ii} = \sum_{\nu}\rho_i(\nu)e^{i\nu kz} \qquad (2.173a)$$

$$\rho_{22} - \rho_{11} = \sum_{\nu}d(\nu)e^{i\nu kz} \equiv \sum_{\nu}\left[\rho_2(\nu) - \rho_1(\nu)\right]e^{i\nu kz} \qquad (2.173b)$$

and

$$\tilde{\rho}_{21} = \sum_{\nu} r(\nu) e^{i\nu kz}, \quad \tilde{\rho}_{12} = \sum_{\nu} r^*(-\nu) e^{i\nu kz} = \tilde{\rho}_{21}^*, \tag{2.174}$$

where in the last equation we have changed the sign of the dummy variable $\nu$. As we have seen in Secs. 2.3 and 2.4, the terms in (2.173) have only *even terms* $\nu = 2k$, and the series in (2.174) has only *odd terms* $\nu = 2k + 1$. Introducing (2.173)–(2.174) into Eqs. (2.172) and assuming all $z$-dependence to occur in the exponents, we find

$$d(\nu) = \bar{N}\delta_{\nu 0} + \frac{\mu}{\hbar} D_1(\nu)\{E_+[r^*(-\nu+1) + r(\nu+1)] - E_-[r^*(-\nu-1) + r(\nu-1)]\} \tag{2.175}$$

$$r(\nu) = -\frac{\mu}{2\hbar} D_2(\nu)[E_+ d(\nu-1) - E_- d(\nu+1)], \tag{2.176}$$

where we have defined the complex Lorentzian functions

$$D_1(\nu) = \frac{1}{2}\left[\frac{1}{\gamma_2 + i\nu kv} + \frac{1}{\gamma_1 + i\nu kv}\right] \tag{2.177a}$$

$$D_2(\nu) = \frac{1}{\gamma_{12} + i(\Delta + \nu kv)}. \tag{2.177b}$$

We combine the equations (2.175)–(2.176) into one equation for the components $d(\nu)$ of the population difference. Owing to the reality condition on the diagonal elements of the density matrix we have

$$d(-\nu) = (d(\nu))^*. \tag{2.178}$$

The resulting equation is

$$\Big\{1 + \frac{1}{2}\frac{\mu^2}{\hbar^2} D_1(\nu)\big[E_+^2\big((D_2(\nu+1) + D_2^*(-\nu+1)\big) + E_-^2\big((D_2(\nu-1) + D_2^*(-\nu-1)\big)\big]\Big\} d(\nu)$$

$$= \bar{N}\delta_{\nu 0} + \frac{1}{2}\frac{\mu^2}{\hbar^2} E_+ E_- \{D_1(\nu)[D_2(\nu+1) + D_2^*(-\nu-1)] d(\nu+2) + D_1(\nu)[D_2(\nu-1) + D_2^*(-\nu+1)] d(\nu-2)\}. \tag{2.179}$$

As we may have expected, only the even components of $d(\nu)$ are coupled by this recurrence equation.

To see the implications of (2.179) we write out the equation with $\nu = 0$, which is the only one with an inhomogeneous term. We find by the use of the definitions of the functions $D_1(\nu)$ and $D_2(\nu)$ that

$$\left\{1 + \left(\frac{\gamma_1 + \gamma_2}{2\gamma_{12}}\right)\frac{\mu^2}{\hbar^2\gamma_1\gamma_2}\left[E_+^2 L(\Delta + kv) + E_-^2 L(\Delta - kv)\right]\right\} d(0)$$
$$= \overline{N} + \left(\frac{\mu^2 E_+ E_-}{\hbar^2}\right)\left(\frac{\gamma_1 + \gamma_2}{2\gamma_1\gamma_2}\right)$$
$$\times\left[\frac{\gamma_{12} + ikv}{(\gamma_{12} + ikv)^2 + \Delta^2} d(2) + \frac{\gamma_{12} - ikv}{(\gamma_{12} - ikv)^2 + \Delta^2} d(-2)\right]. \tag{2.180}$$

From this we can observe several properties of the average occupation probability difference $d(0)$:

1. When one field is exactly equal zero, $E_- = 0$ say, we find that no coupling to other components $d(\nu)$ occurs. The solution obtained is given by

$$d(0) = \frac{1}{1 + \dfrac{\mu^2 E_+^2}{\hbar^2\gamma_1\gamma_2}\left(\dfrac{\gamma_1 + \gamma_2}{2\gamma_{12}}\right) L(\Delta + kv)} \tag{2.181}$$

which agrees exactly with the result (2.102) derived for one traveling wave. Setting $E_+ = 0$ we obtain the result for a wave traveling in the opposite direction with a resonance at $kv = +\Delta$.

2. When both amplitudes are nonvanishing we find that each one has a saturating effect proportional to the intensity $(E_\pm)^2$ with the same Lorentzian as if the waves were not interacting $L(\Delta \pm kv)$. The interference between the waves comes from a cross-term of the form $E_+ E_-$ with the corresponding coupling to higher-order Fourier terms. If this coupling is neglected, we must set $d(\pm 2)$ equal to zero, and obtain the lowest-order approximation to the quasi-standing wave case directly

$$d(0) = \frac{\overline{N}}{1 + \dfrac{\mu^2}{\hbar^2\gamma_1\gamma_2}\left(\dfrac{\gamma_1 + \gamma_2}{2\gamma_{12}}\right)\left[E_+^2 L(\Delta + kv) + E_-^2 L(\Delta - kv)\right]}. \tag{2.182}$$

This corresponds to the case where each traveling wave saturates the atoms independently of the other one. With a proper standing wave $E_+ = E_- = \frac{1}{2}E$ we obtain the result (2.134) previously derived in Sec. 2.5.

It is possible to obtain the exact solution to the set of coupled difference equations (2.179) in terms of a continued fraction. The equation can be written in the form

$$A(n)x_n = \delta_{n0} + B_+(n)x_{n+1} + B_-(n)x_{n-1} \tag{2.183}$$

where

$$x_n = \frac{d(2n)}{\overline{N}}, \tag{2.184}$$

and the functions $A(n)$, $B_{\pm}(n)$ are defined in an obvious way. Using a generalization of the method presented in Sec. 2.3, it is possible to obtain the solution

$$x_0 = \frac{d(0)}{\overline{N}}$$

$$= \left[ A(0) - B_+(0)B_-(0) \left( \cfrac{A(0)}{A^2(0) + \cfrac{B_+(0)B_-(0)A^2(1)}{A^2(1) + \cfrac{B_+(1)B_-(1)A^2(2)}{A^2(2) + \cdots}}} + \cfrac{A(0)}{A^2(0) + \cfrac{B_+(0)B_-(0)A^2(-1)}{A^2(-1) + \cfrac{B_+(-1)B_-(-1)A^2(-2)}{A^2(-2) + \cdots}}} \right) \right]^{-1}. \tag{2.185}$$

The difference between this result and the approximation (2.182) is that the exact function contains oscillating structures due to the interference between the two traveling waves. To lowest order the effect is a light shift for large detunings $\Delta \gg \gamma_{12}$ and a Stark splitting when $\Delta \simeq 0$.

The lowest-order approximation is compared with the exact results in Fig. 2.8 for large and small detunings. The lowest order effects considered in Sec. 2.6 are clearly seen and the additional oscillating interference structure appears most clearly when the detuning $\Delta$ disappears and both traveling waves act on the same atoms. For the detuned case $\Delta \neq 0$, there appears an

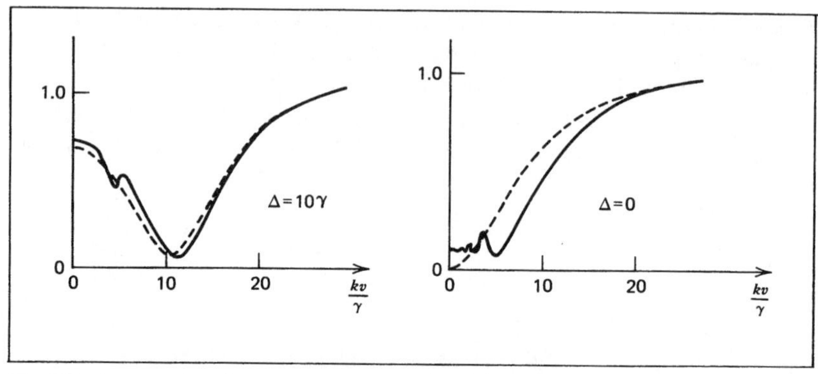


**Fig. 2.8** A comparison between the exact continued fraction treatment (solid line) and the lowest-order approximation (dashed line). To the left a detuned case; to the right the resonant case. The resonance at $\Delta/3$ is clearly seen to the left, and at resonance the split line of Fig. 2.7 breaks up into oscillations. The intensity parameter is given by $(\mu^2 E^2/2\hbar^2\gamma^2) = 30$.

additional resonance at the velocity given by

$$kv = \frac{\Delta}{3} \tag{2.186}$$

due to the resonant behavior of the functions $D_2(\pm 3)$ in the continued fraction. These are three-photon resonances constituting the generalization of the ordinary three-photon resonances (2.95) to the case of a Doppler broadened atomic system; they have been called *multi-Doppleron* resonances.

*Alternative Approach.* For a standing wave $E_+ = E_- = \frac{1}{2}E$ derive the continued fraction result

$$d(0) = \bar{N}\left\{1 + \frac{\mu^2 E^2}{8\hbar^2}\left(\frac{\gamma_1 + \gamma_2}{\gamma_1\gamma_2}\right)\right.$$

$$\left.\times \operatorname{Re}\left[\frac{D_2(1) + D_2^*(-1)}{1 + \dfrac{\dfrac{\mu^2 E^2}{8\hbar^2} D_1(2)\big(D_2(1) + D_2^*(-1)\big)}{1 + \cdots}}\right]\right\}^{-1}$$

*Hint.* Use the equations of motion in the form (2.126). Then insert the Fourier series (2.173)–(2.174) and proceed as in Sec. 2.3 to obtain a relationship between $d(\nu)$ and

$$s(\nu) = r(\nu) - r^*(-\nu).$$

If we want to express the polarization of the medium in terms of the Fourier series (2.173)–(2.174), we have

$$\begin{aligned} P &= N_0 \mathrm{Tr}\, \rho\mu = N_0\mu \int (\rho_{12} + \rho_{21})\, dv \\ &= N_0\mu \int [(\tilde{\rho}_{12} + \tilde{\rho}_{21})\cos \Omega t + i(\tilde{\rho}_{12} - \tilde{\rho}_{21})\sin \Omega t]\, dv \\ &= N_0\mu \Big\{ \cos \Omega t \sum_\nu \int [r(\nu) + r^*(-\nu)]\, dv\, e^{i\nu kz} \\ &\qquad - i \sin \Omega t \sum_\nu \int [r(\nu) - r^*(-\nu)]\, dv\, e^{i\nu kz} \Big\}. \end{aligned} \tag{2.187}$$

If we introduce the coefficients $C$ and $S$ as in (2.42), we obtain

$$\begin{aligned} C &= N_0\mu \frac{\int\int \sin kz \sum_\nu (r(\nu) + r^*(-\nu))\, dv\, e^{i\nu kz}\, dz}{\int \sin^2 kz\, dz} \\ &= \frac{N_0\mu}{i} \int [r(-1) + r^*(1) - r(1) - r^*(-1)]\, dv \\ &= 2N_0\mu \int \mathrm{Im}[r(-1) - r(1)]\, dv \end{aligned} \tag{2.188}$$

where $r(\pm 1)$ is expressed in terms of $d(\nu)$ by the use of (2.176). For the other component we find

$$\begin{aligned} S &= N_0\mu \frac{\int\int \sin kz \sum_\nu i(r(\nu) - r^*(-\nu))\, dv\, e^{i\nu kz}\, dz}{\int \sin^2 kz\, dz} \\ &= N_0\mu \int [r(-1) - r^*(1) - r(1) + r^*(-1)]\, dv \\ &= 2N_0\mu \int \mathrm{Re}(r(-1) - r(1))\, dv. \end{aligned} \tag{2.189}$$

From these expressions we can form an expression for the susceptibility by setting

$$\chi = \frac{C - iS}{E} = \frac{2N_0\mu}{E} \int \frac{1}{i} [r(-1) - r(1)] \, dv. \tag{2.190}$$

## 2.8. COMMENTS AND REFERENCES

For the basis of the interaction between strong light fields and matter, look into Sargent et al. (1974); many interesting aspects are also discussed in Letokhov and Chebotaev (1977). The Lamb dip was first derived by Lamb and reported at the summer school "Enrico Fermi"" (Lamb 1963). It was experimentally verified by McFarlane et al. (1963) and Szöke and Javan (1963). The Bennett hole had been introduced as a concept earlier (Bennett 1962). The strong saturation properties were first treated by continued-fraction techniques by Stenholm and Lamb (1969) and independently by Feldman and Feld (1970). A similar approach was at the same time developed by Holt (1970). The coherence parameter $\eta$ was introduced by Baklanov and Chebotaev (1971).

A continued fraction treatment of RF resonances was early carried out by Autler and Townes (1955). The powerful Floquet theory was applied to the problem of RF resonances by Shirley (1965), and a systematic quantum mechanical approach was presented by Cohen-Tannoudji (1968). The field has later developed considerably both theoretically and experimentally, see the review by Stenholm (1978b), and it now provides one of the best established examples of very strong saturation effects of electromagnetic radiation.

The mathematics of continued fractions has recently been reviewed in a very useful way by Jones and Thron (1980).

The alternative approach to highly nonlinear systems based on higher-ordered susceptibilities has been pioneered by Blombergen (1965). This is called *nonlinear optics* and has been widely applied to a variety of experimental situations. Especially in condensed matter where symmetry properties are of importance, we can extract much valuable information from the susceptibilty coefficients. They are calculable in perturbation theory, and the intermediate states consist of the excitations of the system. For recent reviews of applications see Hellwarth (1977) and Druet and Taran (1981). The monograph by Levenson (1982) also contains a useful review of the applications of nonlinear optics to spectroscopy.

The higher-order resonances at odd fractions of the detuning were first pointed out by Haroche and Hartmann (1972) and discussed as multi-

Doppleron phenomena by Kyrölä and Stenholm (1977 and 1979), and Corbalan et al. (1981). Higher-order resonances in Doppler broadened systems have been observed by Freund et al. (1975) and Read and Oka (1977).

The mechanical manifestations of the photon momentum can be seen in Doppler broadened systems. The theory for nonlinear spectroscopy was introduced by Kol'chenko et al. (1969) and investigated in detail by Stenholm (1974), Aminoff and Stenholm (1976), and Bordé (1976). A new representation of moving atoms was introduced by Shirley (Shirley and Stenholm 1977) and applied to the analysis of recoil effects in saturation spectroscopy by Shirley (1980).

Experimentally, the only system where the recoil effect has been observed is the methane $CH_4$ absorber in a 3.39 $\mu$m He–Ne laser cavity. The resolution of the 2-kHz recoil splitting gives an impressive demonstration of the potentialities of laser spectroscopy. This has been achieved as part of the work on frequency standards in Boulder, Colorado (Hall et al. 1976) and Novosibirsk, USSR (Bagaev et al. 1977); see also the article by Hall and Magyar in Shimoda (1976) and Sec. 10.2.2 of Letokhov and Chebotaev (1977).

A fully quantum mechanical approach to laser spectroscopy of Doppler broadened systems is presented in Stenholm (1978a), where many of the questions discussed here are treated from a slightly different point of view. Many features of the quantum theory of laser light interacting with atomic systems can be found in Haken (1970b).

CHAPTER 3

# Foundation of Laser Theory

## 3.1. GENERAL CONDITIONS FOR LASER OPERATION

The laser is a device that sustains steady state oscillations with a well-defined frequency in the optical range. In electrical engineering such devices are well known, and their operation depends on the properties of amplifiers with frequency-dependent feedback, see Fig. 3.1. The gain must be present to overcome the unavoidable losses in any physical system, and the frequency selectivity is used to fix the oscillational frequency within the frequency range where the amplifier works, the gain bandwidth.

If we cut the feedback loop in Fig. 3.1, we assume the output $Y$ to depend on the input $X$ through

$$Y = A(X) \tag{3.1}$$

and the feedback signal is taken to be

$$X_{\text{fb}} = FY, \tag{3.2}$$

which usually is linear but depends on the frequency of the signal. With the feedback connected, the system satisfies the equation

$$Y = A(X + FY). \tag{3.3}$$

For some parameter ranges it is possible that this equation has an output $Y \neq 0$ for a vanishing input $X = 0$. Here it is essential that the function $A(X)$ is nonlinear. In this condition the system is a self-sustained oscillator, which determines its own amplitude self-consistently. Because this occurs first at some frequency, the oscillational frequency is also fixed by the components, mainly by some resonant element in the feedback circuit. Any textbook in electronics gives numerous examples of such feedback oscillators based on amplifying devices.

In the atomic microwave oscillators, MASERS, the resonant element is a metallic microwave cavity. Only those waves that survive long enough in the

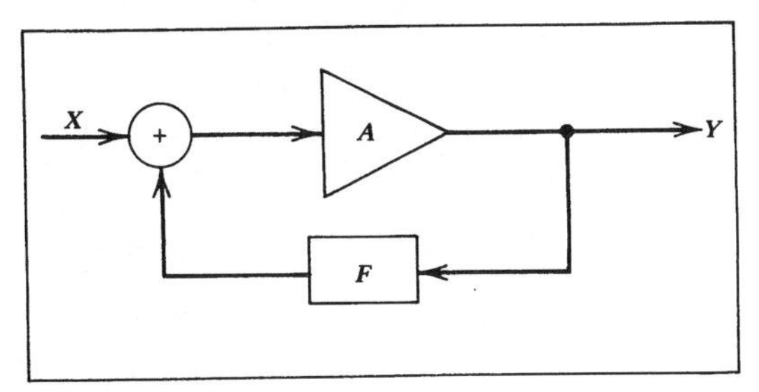


**Fig. 3.1** An amplifier $A$ with an input $X$ and an output $Y$, which is taken back through the feedback loop $F$ to add to the input.

cavity have time to re-interact with the atoms and get amplified. For a laser the situation is similar, only the resonant cavity is an optical resonator, similar to a Fabry–Perot interferometer. The atomic medium in the laser amplifies over a range of frequencies most of which are lost through the diffraction losses and leakage through the mirrors. Only those that fit the resonance conditions, see Sec. 1.2, keep their energy inside the cavity long enough to act back on the medium and get reinforced. Only a few selected frequencies are amplified. The maser and the laser are hence self-sustained oscillators based on the quantum mechanical processes in atomic systems. The name *quantum electronics* seems appropriately chosen for such investigations, even when one is mainly interested in investigating the atomic properties, namely in laser spectroscopy, where we are interested in the nonlinear response of the medium in addition to the linear absorption.

In this book we do not discuss the theory of optical resonators in detail; the main ideas were introduced in Sec. 1.2. For the computations specific to an optical system with considerable diffraction effects we refer to the literature. Here we give only a heuristic discussion of the basic mechanism for losses in the lasers.

In a high-quality cavity the light energy traverses back and forth many times between the end mirrors. If the transmission of the mirror is $\delta$ (assumed $\ll 1$), we have a reduction of the intensity after $N$ passes by a factor $(1-\delta)^N$. For an input $I_0$ we have

$$I_N = (1-\delta)^N I_0. \tag{3.4}$$

If we define a characteristic number of traverses $N_e$ by setting

$$\frac{I_{N_e}}{I_0} = e^{-1}, \tag{3.5}$$

this number corresponds to the lifetime $\tau$ in an exponential decay $e^{-t/\tau}$. From (3.4) and (3.5) we find for a small $\delta$ that

$$-1 = \log e^{-1} = N_e \log(1 - \delta) \approx -N_e\delta, \tag{3.6}$$

and $N_e \approx \delta^{-1}$. The travel time through a cavity of length $L$ is $(2L/c)$, and we define the time constant

$$\tau = N_e \frac{2L}{c} \tag{3.7}$$

and the running time $t$ after $N$ traverses as

$$t = N\frac{2L}{c}. \tag{3.8}$$

Then the basic Eq. (3.4) takes the form

$$\begin{aligned} I_N &= \left(1 - \frac{1}{N_e}\right)^N I_0 \\ &= \left(1 - \frac{t}{N\tau}\right)^N I_0 \to e^{-t/\tau} I_0 \end{aligned} \tag{3.9}$$

for large values of $N$. Thus we can see how a good cavity with many traverses provides a system with nearly an exponential decay; compare the discussion in Sec. 1.2.

For a laser with length $L = 1$ m and 1% losses ($\delta = 0.01$) we find $N_e = 100$ and

$$\tau = 0.7\ \mu s. \tag{3.10}$$

This holds for a radiation frequency satisfying the resonance condition of the cavity which in its crude form, Eq. (1.42), says that twice the resonator length is an integer number of wavelengths, namely,

$$2L = n\lambda \tag{3.11}$$

for some $n$. When $\lambda$ is of the order of 1 $\mu$m we find that $n$ is large, usually $n \sim 10^6$. The nonresonant frequencies cannot remain inside the cavity but are usually lost within a few traverses of its length; their lifetimes can be estimated to be at most $2L/c \sim 10ns$.

Because the mode number $n$ is large, usually many modes fall within the gain band of the laser amplifier, $\Delta\omega$. Thus a laser often oscillates on many modes and special devices are required to achieve single-mode operation with a large amplitude.

For a homogeneously broadened system the frequency range of the response is determined by the line width $\Gamma$. This is usually for atomic systems such that the level lifetime is

$$\tau_{\text{atom}} = \Gamma^{-1} \simeq 10^{-8}\ \text{s},$$

which is much shorter than the characteristic time (3.10) of the cavity. For an inhomogeneously broadened system, the gain band is given by the Doppler width $\Delta\omega \simeq ku$, and is larger than the homogeneous width in the same system.

In the laser the situation is, consequently, the following. The field acts on the atomic system, which can adjust itself in the time $\tau_{\text{atom}}$ to the instantaneous value of the field. The atom acquires a polarization which depends on the field in a nonlinear way. The field, in turn, is driven by the polarization through the Maxwell equations; those components of the field that resonate with the cavity survive for a long enough time to re-interact with the atoms and get amplified. This is the feedback mechanism selecting the frequencies. But the resonant modes can change over a time of length $\tau$ only [see (3.10)], and hence they are slow compared with the atomic processes. The mode changes over the time $\tau$ and during this period samples many dipole moments; each one survives only over $\tau_{\text{atom}} \ll \tau$. It is thus possible to calculate an ensemble averaged dipole moment for a typical atom and assume that the instantaneous field intensity is determined by a sufficient number of atoms that their combined effect approximates the ensemble average.

The situation is suitable for an adiabatic elimination procedure according to the scheme of Sec. 1.8. The atoms are fast enough to reach their equilibrium values in the field intensity $I(t)$ and for times long compared with $\tau_{\text{atom}}$, an equation for only the intensity emerges. If we define a nonlinear gain function $G(I)$, we find the equation to be of the form

$$\frac{dI}{dt} = \left[G(I) - \frac{1}{\tau}\right]I. \tag{3.12}$$

In the empty cavity, $G = 0$, and the correct exponential decay (3.9) of $I(t)$ results. For small intensities we use a linear approximation

$$\frac{dI}{dt} = \left[G(0) - \frac{1}{\tau}\right]I. \tag{3.13}$$

To have a nonzero solution we must have

$$G(0) > \frac{1}{\tau}, \tag{3.14}$$

which gives an exponentially growing intensity. Thus our low intensity approximation will break down, and a steady state ensues when

$$\tau G(I) = 1. \tag{3.15}$$

Because the gain $G(I)$ saturates, its values usually decrease for increasing intensity. The condition (3.14) usually implies that (3.15) is satisfied for some $I = I_0$. In some cases several solutions may exist, but there will always be at least one if (3.14) holds, because $\lim_{I\to\infty} G(I) = 0$ for physical reasons. These considerations have a simple graphic interpretation shown in Fig. 3.2.

If several solutions exist, we must investigate their stability. For this reason we write

$$I = I_0 + i \tag{3.16}$$

and linearize (3.12) by expanding $G(I)$

$$\begin{aligned}\frac{di}{dt} &= \left[G(I_0) + \left(\frac{\partial G}{\partial I}\right)_{I_0} i - \frac{1}{\tau}\right](I_0 + i)\\ &= \left(\frac{\partial G}{\partial I}\right)_{I_0} I_0 i + O(i^2).\end{aligned} \tag{3.17}$$

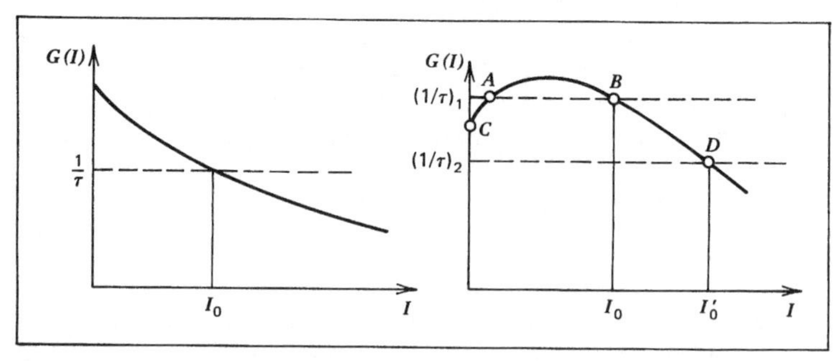


**Fig. 3.2** Two cases of the steady state solution for an oscillator. If the decay time $\tau$ is long enough, there is always only one solution $I_0$, on the left. On the right the gain function is nonmonotonous, and for the losses $(1/\tau)_1$ there are two stable operating points $C$ or $B$. The point $A$ is unstable. For the larger loss $(1/\tau)_2$ there is only the operating point $D$.

Because $I_0 > 0$ we find that the solution is stable if

$$\left(\frac{\partial G}{\partial I}\right)_{I=I_0} < 0. \tag{3.18}$$

Then a spontaneous deviation $i$ will decay exponentially. Thus the crossing $A$ in Fig. 3.2 is an unstable operating point, the point $B$ a stable one. Note also that the point $I = 0$ is a stable nonoscillating operating point for the losses $(1/\tau)_1$.

A simple condition for the threshold in a laser can be obtained in the following way: We consider a cavity of length $L$ and a volume $V$ containing $N_0$ excited atoms. The spontaneous decay rate is $\Gamma$ given in Eq. (1.113). The rate of spontaneous emission by all atoms is $N_0\Gamma$, but only some of this emitted radiation goes into the modes that can be amplified by the laser system. An estimate of this fraction can be obtained from the considerations of radiation cavities.

The cavity can sustain oscillations inside a finite solid angle $\Delta O$ only, and the atomic system has a gain bandwidth $\Delta\omega$. With the two polarizations of the light taken into account, we find the volume element in wave vector space to be $2Vd^3k/(2\pi)^3$. If we define the fraction $\sigma$ of solid angle into which laser radiation must go, we write

$$\Delta O = \sigma 4\pi, \tag{3.19}$$

and requiring this to emit a fraction of the energy roughly of order one into the laser gain bandwidth we set

$$\frac{2Vd^3k}{(2\pi)^3} = V\frac{k^2}{4\pi^3 c}d\omega\, dO \simeq V\frac{k^2}{\pi^2 c}\Delta\omega\sigma \simeq 1, \tag{3.20}$$

when $d\omega$ and $dO$ are replaced by $\Delta\omega$ and $\Delta O$ respectively. From (3.20) we obtain the fraction of useful emission processes $\sigma$, and their total rate of occurrence will be given by $\sigma N_0 \Gamma$. As the radiation energy survives only for a time $\tau$ in the cavity, the useful emission processes must generate energy at a rate equaling its decay rate $\tau^{-1}$, which gives the steady state condition

$$\sigma N_0 \Gamma \tau = 1. \tag{3.21}$$

If $\sigma N_0 \Gamma$ is larger than $\tau^{-1}$, saturation decreases the number of amplifying atoms $N_0$ until (3.21) holds and steady state occurs. The condition (3.21) is a simple example of (3.15); we will meet many similar conditions later.

Inserting (3.20) and $\Gamma = \tau_{\text{atom}}^{-1}$ into (3.21), we obtain the critical threshold density for laser operation

$$\frac{N_0}{V} = \left(\frac{4\Delta\omega}{c\lambda^2}\right)\left(\frac{\tau_{\text{atom}}}{\tau}\right), \tag{3.22}$$

which is the condition originally derived by Schawlow and Townes (1958).

**Example.** In a He–Ne laser $\lambda = 0.633\ \mu\text{m}$, $(\Delta\omega/2\pi) \simeq (ku/2\pi) = 10^9$ Hz, and with (3.22) and (3.10) we obtain for the threshold

$$\frac{N_0}{V} = 5 \times 10^{13}\ \text{m}^{-3}.$$

For future reference it is useful to introduce the expression for $\Gamma$ from (1.113) into the laser threshold condition (3.22) and write it in the form

$$\frac{N_0}{V} = \left(\frac{3}{\pi}\right)\left(\frac{\hbar\varepsilon_0 u}{c\mu^2\tau}\right). \tag{3.23}$$

If we define a quality factor $Q$ for the laser cavity by the ratio of the decay time to the period, we find

$$Q = \frac{\tau\Omega}{2\pi} = \frac{\tau c k}{2\pi}, \tag{3.24}$$

which gives (3.23) the form

$$\frac{N_0}{V} = \frac{3}{2\pi^2}\,\frac{\hbar\varepsilon_0 k u}{\mu^2 Q}. \tag{3.25}$$

At threshold the laser power emitted is given by

$$P_{\text{out}} = \delta P_{\text{in}} = \frac{2L}{\tau c} P_{\text{in}}, \tag{3.26}$$

and, because each emission process involves the energy $\hbar\omega$ and each one has a duration $\tau_{\text{atom}} = \Gamma^{-1}$, we have the relation

$$P_{\text{in}} = \frac{N_0\hbar\omega}{\tau_{\text{atom}}} = \frac{N_0 2\pi\hbar c}{\lambda\tau_{\text{atom}}} = \frac{8\pi\hbar\Delta\omega V}{\lambda^3\tau}, \tag{3.27}$$

where (3.22) has been used. The relations (3.26) and (3.27) can be used to estimate the power obtainable from a laser.

The considerations in this section have been based on simple qualitative arguments; in the following sections we make these quantitative for some cases. We do, however, start with the theory for the laser amplifiers.

## 3.2. THE TRAVELING WAVE AMPLIFIER

In this section we discuss the fate of a traveling wave of the form

$$E(z,t) = E\cos(kz - \Omega t + \varphi) \tag{3.28}$$

traversing a medium of optically active two-level systems. If they are initially in their lower state, we have an absorber, and the nonlinear properties for large fields signify saturation of the absorption. For an inverted medium we have a light amplifier and the inevitable gain saturation. The mathematical treatment of the medium is carried out in Sec. 2.4. There the polarization corresponding to the field (3.28) is written in the form

$$\begin{aligned} P(z,t) &= C\cos(kz - \Omega t + \varphi) + S\sin(kz - \Omega t + \varphi) \\ &= \tfrac{1}{2}E\chi e^{i(kz-\Omega t+\varphi)} + \text{c.c.}\ . \end{aligned} \tag{3.29}$$

In this section we want to derive a relation between the field amplitude $E$ and the components of the polarization $C$ and $S$. We start from the Maxwell's equations of Sec. 1.2. and write them

$$\frac{\partial}{\partial z}E(z,t) = -\mu_0\frac{\partial}{\partial t}H(z,t) \tag{3.30}$$

$$-\frac{\partial}{\partial z}H(z,t) = \frac{\partial}{\partial t}\left[\varepsilon_0 E(z,t) + P(z,t)\right]. \tag{3.31}$$

Eliminating the magnetic field $H$ we find

$$\begin{aligned} c^2\frac{\partial^2}{\partial z^2}E(z,t) &= \frac{\partial^2}{\partial t^2}E(z,t) + \frac{\Omega}{Q}\frac{\partial}{\partial t}E(z,t) \\ &\quad + \frac{1}{\varepsilon_0}\frac{\partial^2}{\partial t^2}P(z,t), \end{aligned} \tag{3.32}$$

where we have added the loss term $(\Omega/Q)(\partial E/\partial t) = (\partial E/\partial t)/\tau$ as in Eq. (1.30).

In the following discussion we assume that $E$, $\varphi$, $C$, and $S$ are slowly varying functions of position $z$ and time $t$. For the polarization we directly neglect the time variation of $C$ and $S$ and write

$$\frac{\partial^2}{\partial t^2}P(z,t) = -\Omega^2 P(z,t). \tag{3.33}$$

For the derivatives of $E(z,t)$ we write

$$\frac{\partial}{\partial t}E(z,t) = \dot{E}\cos(kz - \Omega t + \varphi) - \dot{\varphi}E\sin(kz - \Omega t + \varphi)$$
$$+\Omega E\sin(kz - \Omega t + \varphi) \tag{3.34}$$

$$\frac{\partial^2}{\partial t^2}E(z,t) = 2\Omega\dot{E}\sin(kz - \Omega t + \varphi)$$
$$+2\Omega\dot{\varphi}E\cos(kz - \Omega t + \varphi) - \Omega^2 E\cos(kz - \Omega t + \varphi), \tag{3.35}$$

where we have neglected terms proportional to $\ddot{E}$, $\ddot{\varphi}$ and $\dot{E}\dot{\varphi}$. In a similar way we derive

$$\frac{\partial^2}{\partial z^2}E(z,t) = -2k\frac{\partial E}{\partial z}\sin(kz - \Omega t + \varphi)$$
$$-2k\frac{\partial\varphi}{\partial z}E\cos(kz - \Omega t + \varphi) - k^2 E\cos(kz - \Omega t + \varphi) \tag{3.36}$$

where the second-order terms with respect to spatial derivatives of $E$ and $\varphi$ have been neglected.

The first derivative (3.34) is multiplied by $Q^{-1}$, which is assumed to be a small number. Consequently we neglect also the terms $(1/Q)(\partial/\partial t)$ of $E$ and $\varphi$. Then we note that Eq. (3.32) must hold at all times, and this allows us to identify the coefficients of the cosine and sine terms separately. We obtain in this way the equations

$$\left[\frac{\partial}{\partial t} + c\frac{\partial}{\partial z} + \frac{\Omega}{2Q}\right]E = \frac{\Omega}{2\varepsilon_0}S = \frac{\Omega\chi''}{2\varepsilon_0}E \tag{3.37a}$$

$$2\Omega\dot{\varphi}E + 2kc^2\frac{\partial\varphi}{\partial z}E + (kc + \Omega)(kc - \Omega)E = \frac{\Omega^2}{\varepsilon_0}C. \tag{3.37b}$$

If we remember that for a propagating wave we must have

$$ck \approx \Omega, \tag{3.38}$$

we can divide (3.37b) by $2\Omega$ and obtain

$$\left[\left(\frac{\partial}{\partial t} + c\frac{\partial}{\partial z}\right)\varphi\right]E = \frac{\Omega}{2\varepsilon_0}C = \frac{\Omega}{2\varepsilon_0}\chi' E. \tag{3.39}$$

In Eqs. (3.37) and (3.39) we have also introduced the expressions in terms of the real and imaginary parts of the susceptibility, $\chi'$ and $\chi''$ respectively, from (2.112).

Equation (3.37a) gives the relation between the slowly varying amplitude $E$ and the active medium. We see that the amplitude condition is entirely determined by the out-of-phase component $S \propto \chi''$ of the polarization. Because the intensity of radiation depends on $|E|^2$ this equation determines the propagation of energy through the medium. Looking back at our discussion in Sec. 2.2, Eq. (2.46), we find that $S$ determines the energy exchanged with the medium, which shows why it appears in the propagation of energy in the medium.

In steady state Eqs. (3.37) can describe an amplifying medium if we leave out the time derivatives. To obtain explicit expressions for the amplifier, we introduce the analytic expression (2.117) valid in the Doppler limit $ku \gg \Gamma$; the exact expression can be used for detailed numerical investigations. We find from (3.37a)

$$c\frac{\partial}{\partial z}E = -\frac{\Omega}{2}\left[\frac{1}{Q} - \frac{\sqrt{\pi}\,N_0\Lambda_0\mu^2}{\varepsilon_0\hbar ku\sqrt{1+2I\eta}}e^{-\Delta^2/k^2u^2}\right]E. \tag{3.40}$$

Defining the gain function

$$G_A(I,\Delta) = -\left[\frac{1}{Q} - \sqrt{\pi}\,\frac{\mu^2 N_0\Lambda_0}{\hbar\varepsilon_0 ku\sqrt{1+2I\eta}}e^{-\Delta^2/k^2u^2}\right] \tag{3.41}$$

and noting that the intensity is

$$I = \frac{\mu^2E^2}{2\hbar^2\gamma_1\gamma_2} \tag{3.42}$$

we obtain the equation

$$\frac{\partial I}{\partial z} = kG_A(I,\Delta)I. \tag{3.43}$$

For low intensities the condition for gain at $I = 0$ becomes $G(0, \Delta) > 0$, and this restricts the losses in the medium by

$$\frac{1}{Q} < \sqrt{\pi}\,\frac{\mu^2 N_0 \Lambda_0}{\hbar k u \varepsilon_0} e^{-\Delta^2/k^2u^2}. \tag{3.44}$$

The exponential factor coming from the Doppler broadening makes the loss factor dominate once the field is detuned, so that $|\Delta| \gg ku$. This defines the gain bandwidth of the amplifying medium.

If the laser intensity is small throughout the medium we can use the linear approximation to integrate (3.43) for all values of $z$, namely,

$$I(z) = I(0)e^{G(0,\,\Delta)kz}. \tag{3.45}$$

In particular, this often holds for an absorbing medium $\overline{N} < 0$, where an initially small intensity will decrease further during the propagation. This is an attenuator.

When the intensity is large the gain saturates

$$G(I, \Delta) < G(0, \Delta). \tag{3.46}$$

If the amplification well exceeds the losses we can write (3.40) in the form

$$\frac{\partial I}{\partial z} = \sqrt{\pi}\,\frac{\mu^2 N_0 \Lambda_0}{\hbar \varepsilon_0 u \sqrt{1 + 2I\eta}} e^{-\Delta^2/k^2u^2} I. \tag{3.47}$$

By separating the variables we can integrate this relation analytically, but the ensuing expression is not very instructive. The limiting form for large intensities $\eta I \gg 1$ is easily obtained.

In steady state propagation through a medium we can assume that the phase variable $\varphi$ becomes a linear function of position

$$\varphi = k_{\text{eff}} z - \Omega t. \tag{3.48}$$

The relation (3.39) relates the coefficients by

$$\begin{aligned} k_{\text{eff}} c - \Omega &= \frac{\Omega \chi'}{2\varepsilon_0} \\ &= -\frac{\Omega \Delta N_0 \Lambda_0 \mu^2}{\hbar \varepsilon_0 (ku)^2} e^{-\Delta^2/k^2u^2}, \end{aligned} \tag{3.49}$$

where $\chi'$ is taken from (2.119). The refractive index (dispersion) of the medium is determined by

$$n = \frac{k_{\text{eff}} c}{\Omega} = 1 + \frac{k_{\text{eff}} c - \Omega}{\Omega}$$

$$= 1 - \frac{\Delta N_0 \Lambda_0 \mu^2}{\hbar \varepsilon_0 (ku)^2} e^{-\Delta^2/k^2 u^2}. \tag{3.50}$$

Thus we can see how the real part $\chi'$ of the susceptibility determines the dispersive properties of the medium. In the Doppler limit all atoms of the medium contribute to the dispersion independent of velocity, and no saturation of the dispersion occurs. The gain, Eq. (3.40), depends mainly on the resonant atoms, and saturation acts strongly.

Very often the relation (3.43) is written in terms of an effective cross section for absorption of the incoming radiation. If we omit the loss factor $Q^{-1}$, the absorption or amplification is proportional to the population difference $(n_2 - n_1)$, as was repeatedly found in Chapter 2. We introduce the *optical cross section* $\sigma_{\text{opt}}$ by writing Eq. (3.43) in the form

$$\frac{\partial I}{\partial z} = (n_2 - n_1)\sigma_{\text{opt}} I. \tag{3.51}$$

If the populations $n_i (i = 1, 2)$ are expressed as particle densities, $\sigma_{\text{opt}}$ has the dimension of an area, representing the fraction of the incoming energy flow removed by the atom. Using (3.37a) we can easily relate $\sigma_{\text{opt}}$ to the other loss or gain functions $S$ and $\chi''$, which were calculated in Chapter 2. From Eqs. (2.46) and (2.55) we can easily see that $\chi''$ refers to volume losses of energy, whereas $\sigma_{\text{opt}}$ describes losses of energy flowing across an area. Physically, of course, they carry the same information.

The propagation of strong pulses through an amplifying or absorbing medium is a research field of its own, into which we do not enter here. New phenomena like the self-induced transparency (SIT) and others arise; we refer the readers to the text by Allen and Eberly (1975).

## 3.3. THE TRAVELING WAVE LASER

By the use of mirror arrangements we can make an amplified traveling wave turn back and pass the amplifying medium again. In this way we create a ring cavity which can be made to support laser oscillations because the return of the amplified light provides the feedback mechanism necessary to sustain oscillations; see Fig. 3.3. If the round-trip length of the cavity is $L$,

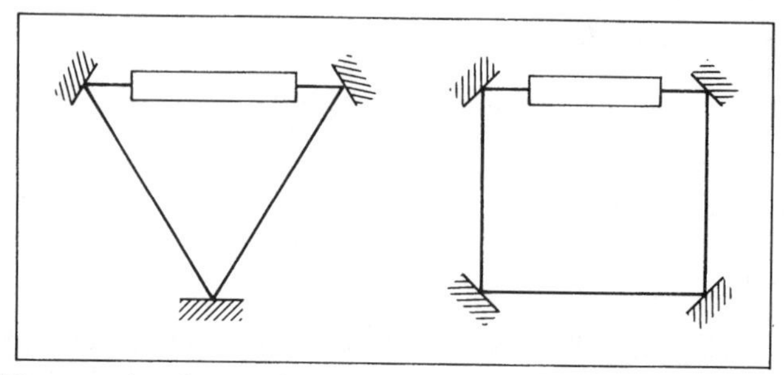

**Fig. 3.3** Ways to make a laser configuration where the beam is turned back onto itself after one round trip. Amplifier tubes can be inserted into each arm of the ring cavity.

only those wavelengths that satisfy [compare Eq. (1.41)]

$$\lambda = \frac{L}{n} \tag{3.52}$$

can return to the amplifying medium in the right phase. Hence the feedback for laser operation is frequency selective.

In general, the ring laser can oscillate on two oppositely traveling waves, as both have the same losses and round-trip path lengths. The degeneracy between the two modes is removed if the laser system is rotating. One traveling wave sees the amplifying medium moving toward it with velocity $v_r = \omega_r R$, where $\omega_r$ is the rotational angular velocity and $R$ is the average radius of the ring cavity. The other wave sees the medium moving with velocity $-v_r$. This leads to a frequency difference $\Delta\Omega \propto \omega_r$ between the two oppositely traveling waves. The frequency difference is in the radio-frequency range and can easily be measured as a beat between the two light frequencies. Thus the laser becomes a sensitive gyroscope for measuring rotations. When the rotation goes to zero, the two frequencies tend toward each other. Any coupling between them leads to a strongly resonant coupling, and the modes lock to the same frequency. Atomic saturation and especially linear backscattering make the laser gyro useless for very small rotation rates. This phenomenon belongs to the area of multimode operation with strong competition, which we do not treat in this book.

It is also possible to achieve single-traveling-mode operation by including an asymmetric optical element into the cavity. This can be based, for example, on polarization properties and the Faraday effect. In this Section we assume that single-mode operation prevails, and proceed with the analysis from there.

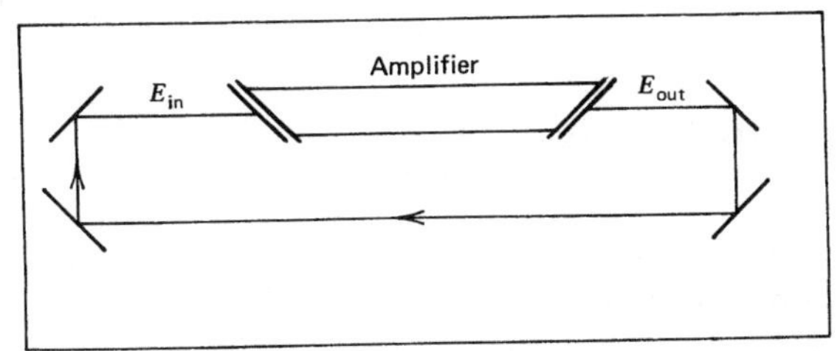


**Fig. 3.4** A linear amplifier is made into a laser oscillator when the outgoing light is used as a feedback signal to sustain the oscillation. To prevent loss-scattering in the windows of the amplifier cell they are oriented at the Brewster angle of the light. This also causes the light to be polarized.

After the amplifying medium the field is given by $E_{out}$, see Fig. 3.4, and if we assume the amplification during one pass of the amplifiers to be small

$$(E_{out} - E_{in})_{\text{one pass}} \ll E_{out}, \tag{3.53}$$

we can require self-consistency by setting $E_{in} = E_{out}$. Then there appears only one laser field amplitude $E$, and no spatial variation along the light pass need be considered.

From Eq. (3.37) we then find

$$\frac{\partial E}{\partial t} = -\frac{\Omega}{2Q}\left[1 - \frac{Q\chi''}{\varepsilon_0}\right]E, \tag{3.54}$$

where the $Q$-factor is presumed to represent all the losses around the light path of the cavity. For low intensities $E^2$, the susceptibility $\chi''$ is evaluated from (3.37) at $I = 0$. If

$$\chi''(0) > \frac{\varepsilon_0}{Q}, \tag{3.55}$$

the laser intensity starts to grow exponentially in time; in the opposite case the losses will dominate, and the intensity decreases to zero. The condition (3.14) is equivalent with (3.55), and in accordance with Eq. (3.13) we introduce

$$\frac{\partial E}{\partial t} = \frac{\Omega}{2}\left[G_L(I) - \frac{1}{Q}\right]E \tag{3.56}$$

where the laser gain function $G_L(I)$ is taken analogously with that of Eq.

(3.47) to be

$$G_L(I) = \sqrt{\pi}\,\frac{\mu^2\Lambda_0 N_0}{\hbar\varepsilon_0 ku\sqrt{1+2I\eta}}\,e^{-\Delta^2/k^2u^2}. \tag{3.57}$$

For low intensities $I = 0$ and central tuning $\Delta = 0$, we obtain the threshold inversion $\Lambda_T$ from the condition (3.55)

$$\sqrt{\pi}\,\frac{\mu^2 N_0 \Lambda_T}{\hbar\varepsilon_0 ku} = \frac{1}{Q} \tag{3.58}$$

giving

$$N_0\Lambda_T = \frac{\hbar\varepsilon_0 ku}{\sqrt{\pi}\,\mu^2 Q}. \tag{3.59}$$

Apart from the numerical factor, this is the result (3.25) of our heuristic derivation in Sec. 3.1.

Using $\Lambda_T$ we can write the gain function (3.57) in the form

$$G_L(I) = \frac{N}{Q}\,\frac{e^{-\Delta^2/k^2u^2}}{\sqrt{1+2\eta I}}, \tag{3.60}$$

where

$$N = \frac{\Lambda_0}{\Lambda_T} \tag{3.61}$$

is a normalized pumping parameter telling how much the excitation exceeds threshold for laser operation; $N = 1$ gives threshold for lasing at $\Delta = 0$. The time-dependent amplitude in the laser is then described by the equation

$$\frac{dE}{dt} = \frac{1}{2\tau}\left[\frac{Ne^{-\Delta^2/k^2u^2}}{\sqrt{1+2\eta I}} - 1\right]E. \tag{3.62}$$

In steady state, the time derivative is zero, and the intensity is given by

$$\eta I = \tfrac{1}{2}\left[N^2 e^{-2\Delta^2/k^2u^2} - 1\right]. \tag{3.63}$$

For those values of $N$ or $\Delta$ where $I$ becomes zero, oscillation ceases. If

$|\Delta| \ll ku$ and $N \simeq 1$, we find the result

$$\eta I = \tfrac{1}{2}(N-1)(N+1) \simeq N-1. \tag{3.64}$$

Near threshold the intensity is then linear in the amount by which the pumping exceeds threshold. For $N > 1$, cavity tuning decreases the intensity until we have

$$e^{\Delta^2/k^2u^2} = N, \tag{3.65}$$

after which no oscillations take place. Thus the ring laser is a self-sustained oscillator that determines its own amplitude.

The laser also determines its own frequency. From the relation (3.39) we find

$$\dot{\varphi} = \frac{\Omega}{2\varepsilon_0}\chi'. \tag{3.66}$$

In the steady state case, we again assume that the phase has a linear part $\varphi(t) = -\Delta\Omega t + \cdots$. This gives [from (3.66) and (2.119)] the result

$$\Delta\Omega = (\Omega_L - \Omega) = \frac{\Omega\Delta N_0 \Lambda_0 \mu^2}{\hbar\varepsilon_0 (ku)^2} e^{-\Delta^2/k^2u^2}. \tag{3.67}$$

Thus for the laser oscillational frequency we have the condition

$$\frac{\Omega_L - \Omega}{\omega - \Omega_L} = \frac{\Omega N_0 \Lambda_0 \mu^2}{\hbar\varepsilon_0 (ku)^2} e^{-\Delta^2/k^2u^2} \tag{3.68}$$

where we have replaced $\Delta$ by $\omega - \Omega_L$, as the field used to calculate the polarization must oscillate at just the frequency generated by the laser, and hence the detuning $\Delta$ must consist of the difference between the actual laser frequency and the atomic transition frequency. For small intensities and zero detuning, we use the definition of the threshold parameter (3.59) and (3.61) to write

$$\frac{\Omega_L - \Omega}{\omega - \Omega_L} = \frac{\Omega N}{\sqrt{\pi}\, Qku} \simeq \frac{1/\tau}{\sqrt{\pi}\, ku} \ll 1. \tag{3.69}$$

Because the cavity line width $1/\tau$ is much smaller than the gain band of the medium $ku$, the laser oscillational frequency is much closer to the cavity frequency $\Omega$ than to the atomic transition frequency $\omega$. Thus there is no need to distinguish between $\Omega$ and $\Omega_L$ on the right-hand side of (3.68). Because this right-hand side is always positive, we find that the sign of $(\Omega_L - \Omega)$ is the same as that of $(\omega - \Omega_L)$, which shows that the laser oscillates somewhere between the cavity eigenfrequency $\Omega$ and the atomic transition frequency $\omega$, as we may expect on physical grounds. The laser, however, tunes mainly as the cavity frequency.

When the intensity is finite, we use the steady state condition (3.63) to write (3.68) in the form

$$\frac{\Omega_L - \Omega}{\omega - \Omega_L} = \frac{\sqrt{1 + 2I\eta}}{\sqrt{\pi}\,\tau k u}. \tag{3.70}$$

For increasing intensity $I$, the lasing frequency $\Omega_L$ is pushed away from the cavity eigenfrequency $\Omega$. This mode pushing makes the laser frequency shift with the intensity of the output.

In this section we have given a simple view of the operation of a laser. In the following sections of this chapter we present a more detailed formulation of laser theory. In Sec. 3.5 we consider laser operation in the common standing wave cavity, and in Sec. 3.6 we discuss a laser with a nonlinear absorber inside the laser cavity. Readers who are satisfied with the understanding of laser operation gained so far can, without loss of continuity, leave the rest of this chapter and proceed directly to Chapter 4. When a more detailed understanding of gas laser operation is needed they can return to this chapter.

## 3.4. MORE GENERAL LASER THEORY

In this section we formulate the basis of a general description of oscillations in a laser cavity. The cavity is assumed to be described by a set of low-loss eigenmodes satisfying the differential equation

$$\nabla^2 \mathbf{U}_n(\mathbf{r}) + \mathbf{k}_n^2 \mathbf{U}_n(\mathbf{r}) = 0 \tag{3.71}$$

and the appropriate boundary conditions. The modes are taken to be transverse

$$\nabla \cdot \mathbf{U}_n = 0,$$

and hence two orthogonal eigenmodes belong to each eigenfrequency $\Omega_n = ck_n$; see the discussion in Sec. 1.2.

The eigenmode problem for optical cavities is treated by diffraction theory and differs slightly from that of metallic, microwave cavities. The modes always display diffraction losses and are hence sometimes called quasi-modes; in the literature they are often referred to as Fox–Li modes because Fox and Li were the first to carry out detailed diffraction calculations for an optical cavity.

Following the treatment in Sec. 1.2 we expand the electric field as

$$\mathbf{E}(\mathbf{r}, t) = \sum_n E_n(t)\mathbf{U}_n(\mathbf{r}) \tag{3.72}$$

with $E_n(t)$ given in (1.24). The polarization is expanded in the same way as

$$\mathbf{P}(\mathbf{r}, t) = \sum_n P_n(t)\mathbf{U}_n(\mathbf{r}) \tag{3.73a}$$

with

$$P_n(t) = \frac{\int d^3\mathbf{r}\,\mathbf{P}(\mathbf{r}, t)\cdot\mathbf{U}_n(\mathbf{r})}{\int d^3\mathbf{r}[\mathbf{U}_n(\mathbf{r})]^2}. \tag{3.73b}$$

Combining Maxwell's equation (1.14) with the loss term from (1.30) we have the result

$$\left(c^2\nabla^2 - \frac{\partial^2}{\partial t^2}\right)\mathbf{E} - \frac{1}{\tau}\frac{\partial}{\partial t}\mathbf{E} = \frac{1}{\varepsilon_0}\frac{\partial^2}{\partial t^2}\mathbf{P}. \tag{3.74}$$

Inserting (3.72) and (3.73), using the equation (3.71) and an approximation like (3.33), we find

$$\frac{d^2}{dt^2}E_n(t) + \frac{1}{\tau_n}\frac{d}{dt}E_n(t) + \Omega_n^2 E_n(t) = \frac{\Omega_n^2}{\varepsilon_0}P_n(t). \tag{3.75}$$

Here the index $n$ is added to the decay time $\tau_n$ to allow for different losses in different modes. The time derivatives in $P_n$ can be replaced by $-\Omega^2$ because we assume that the rest of the time dependence in $P_n$ is much slower than $\Omega$.

For laser operation we extract the time dependence of the cavity eigenmode from $E_n(t)$ by setting

$$E_n(t) = E_n \cos(\Omega_n t + \varphi_n). \tag{3.76}$$

The amplitude $E_n$ and the phase $\varphi_n$ are still assumed to be capable of slow time variation compared with the cavity frequency $\Omega_n$. We calculate

$$\frac{d}{dt}E_n(t) = -(\Omega_n + \dot{\varphi}_n)E_n \sin(\Omega_n t + \varphi_n) + \dot{E}_n \cos(\Omega_n t + \varphi_n)$$

$$\frac{d^2}{dt^2}E_n(t) = -2\Omega_n \dot{E}_n \sin(\Omega_n t + \varphi_n) - (\Omega_n + \dot{\varphi}_n)^2 E_n \cos(\Omega_n t + \varphi_n)$$

$$(3.77)$$

where we have neglected terms proportional to $\ddot{\varphi}$, $\ddot{E}$ and $\dot{\varphi}\dot{E}$ as small.

The polarization is written in the same way as (3.76) but we must assume a phase shift $\theta_n$ between the electric field and the driving polarization

$$\begin{aligned} P_n(t) &= P_n \cos(\Omega_n t + \varphi_n - \theta_n) \\ &= P_n \cos\theta_n \cos(\Omega_n t + \varphi_n) + P_n \sin\theta_n \sin(\Omega_n t + \varphi_n) \\ &\equiv C_n \cos(\Omega_n t + \varphi_n) + S_n \sin(\Omega_n t + \varphi_n). \end{aligned} \tag{3.78}$$

We insert (3.77) and (3.78) into Eq. (3.75) and require that the sine and cosine functions have separately equal amplitudes. This is necessary to satisfy the equation at each instant of time. In addition, we assume that the loss term $(1/\tau)(dE_n/dt)$ is small and the terms $\dot{\varphi}/\tau$ and $\dot{E}/\tau$ are neglected. This is necessary for the cavity to be of the low-loss type needed in lasers. We find the equations

$$\dot{E}_n + \frac{1}{2\tau_n}E_n = -\frac{\Omega_n}{2\varepsilon_0}S_n \tag{3.79}$$

$$\left[\Omega_n^2 - (\Omega_n + \dot{\varphi}_n)^2\right]E_n = \frac{\Omega_n^2}{\varepsilon_0}C_n. \tag{3.80}$$

Writing for the laser oscillational frequency

$$\Omega_n + \dot{\varphi}_n \simeq \Omega_n \tag{3.81}$$

we find

$$\Omega_n^2 - (\Omega_n + \dot{\varphi}_n)^2 \simeq -2\Omega_n\dot{\varphi}_n, \tag{3.82}$$

and Eq. (3.80) becomes

$$\dot{\varphi}_n E_n = -\frac{\Omega_n}{2\varepsilon_0}C_n. \tag{3.83}$$

Equations (3.79) and (3.83) determine the mode amplitude $E_n$ and phase $\varphi_n$ self-consistently. The components of the polarization $S_n$ and $C_n$ will be functions of all amplitudes $E_1, E_2, \ldots$ and the phases $\varphi_1, \varphi_2, \ldots$ giving coupled nonlinear equations. For laser operation it is often sufficient to solve the equations iteratively, to assume the oscillational frequencies to coincide with the cavity eigenfrequencies $\Omega_n$. Then the amplitudes are solved from the nonlinear Eq. (3.79). The proper oscillational frequencies then follow from (3.83). For multimode operation this procedure often breaks down. The relative phases adjust themselves to each other, and mode-locked operation of the laser ensues. In this book we do not consider the complicated phenomena connected with multimode operation but refer the reader to the references listed at the end of this chapter.

In this section we illustrate the use of the general laser equation for the case of a homogeneously broadened single-mode laser. This case may represent the situation in an ideal solid state laser where each active atom is surrounded by the same environment. For simplicity we drop the mode index $n$ from the equations.

The coefficients $C$ and $S$ are obtained from Eqs. (2.44a) and (2.44b). It is instructive for the reader to compare our treatment in Sec. 2.2 of the ansatz (2.25) and (2.42) with the general approach of this section.

For the amplitude we find the equation

$$\dot{E} + \frac{\Omega}{2Q}E = \frac{\Omega\mu^2 \overline{N} N_0}{2\varepsilon_0 \hbar \gamma_{21}} L(\Delta) F(\Delta) E, \tag{3.84}$$

which is of the type (3.54) encountered in the preceding section. The threshold condition determines the oscillational requirement at resonance $\Delta = 0$,

$$\overline{N} N_0 > \overline{N}_T N_0 = \frac{\varepsilon_0 \hbar \gamma_{21}}{Q\mu^2}. \tag{3.85}$$

If we compare this threshold pumping with that of the inhomogeneously broadened case of the preceding section we find from (3.59)

$$\frac{\overline{N}_T\,(\text{homogeneous})}{\overline{N}_T\,(\text{inhomogeneous})} = \sqrt{\pi}\,\frac{\gamma_{21}}{ku} < 1. \tag{3.86}$$

In the inhomogeneously broadened case only a fraction of the order $\gamma_{21}/ku$ of all atoms can contribute to the gain, and hence the pumping must be that much more efficient to achieve lasing than in the corresponding homogeneously broadened case.

Using the expansion in (2.45) and (3.85) we write for the weakly saturating system the equilibrium condition

$$\begin{aligned} 1 &= NL(\Delta)F(\Delta, I) \\ &\approx NL(\Delta)[1 - \tfrac{3}{2}I\eta L(\Delta) + \cdots], \end{aligned} \tag{3.87}$$

where $N = \overline{N}/\overline{N}_T$. Solving for the intensity we find

$$\begin{aligned} \eta I &= \frac{2}{3}\frac{NL(\Delta) - 1}{NL^2(\Delta)} \\ &\approx \frac{2}{3L^2(\Delta)}[NL(\Delta) - 1]. \end{aligned} \tag{3.88}$$

Near resonance we find $I \propto (N - 1)$ in analogy with the earlier result (3.64). When the intensity increases, $N$ grows well above 1, we must use the exact form of the function $F(\Delta)$ in (2.45). At resonance $\Delta = 0$ we solve for $I$ and find

$$\begin{aligned} \eta I &= (N - \tfrac{1}{4}) - [\tfrac{1}{2}(N + \tfrac{1}{8})]^{1/2} \\ &\approx \tfrac{2}{3}(N - 1) + \tfrac{2}{27}(N - 1)^2 - \tfrac{8}{243}(N - 1)^3 + \cdots \end{aligned} \tag{3.89}$$

The lowest term is in agreement with our lowest-order calculation (3.88). The term $(N - 1)^2$ increases the output, whereas the next term in $(N - 1)^3$ makes the output saturate, that is, the intensity turns downward.

From (3.88) we can see that in an oscillating laser, $N > 1$, the oscillation ceases when the cavity is detuned so that

$$L(\Delta) \approx \frac{1}{N}. \tag{3.90}$$

Because this happens when $|\Delta| \gtrsim \gamma_{21}$ we can see that the oscillational bandwidth is of the order $\gamma_{21}$, as expected from physical considerations.

To determine the oscillational frequency we insert the coefficient $C$ from (2.44a) into (3.83) and set

$$\dot{\varphi} = (\Omega_L - \Omega) \tag{3.91}$$

as in the preceding section. We find

$$(\Omega_L - \Omega) = \frac{\Omega N_0 \mu^2 \overline{N}\Delta}{2\varepsilon_0 \hbar \gamma_{21}^2} L(\Delta) F(\Delta). \tag{3.92}$$

Writing again $\Delta = \omega - \Omega_L$ and using (3.87) we find

$$\frac{\Omega_L - \Omega}{\omega - \Omega_L} = \frac{1/\tau}{2\gamma_{21}}, \tag{3.93}$$

where the cavity bandwidth $1/\tau$ again is much less than the homogeneous linewidth $\gamma_{21}$: Thus we again find that the laser oscillates close to the cavity eigenfrequency $\Omega_L \simeq \Omega$.

Solving for the frequency we find

$$\Omega_L = \frac{2\gamma_{21}\Omega + (1/\tau)\omega}{2\gamma_{21} + 1/\tau}, \tag{3.94}$$

which shows that the laser frequency is the weighted average of the cavity eigenfrequency $\Omega$ and the atomic transition frequency $\omega$.

In contrast with the inhomogeneously broadened case the result (3.93) is exact here. Thus the laser oscillational frequency is independent of the oscillational amplitude, and no mode pushing occurs. The reason is that the same saturation function $F(\Delta)$ occurs both in the amplitude condition (3.87) and the frequency condition (3.92). For the inhomogeneously broadened case the dispersion was unsaturated, hence there will appear a saturation function in the frequency condition when the pumping parameter is eliminated.

## 3.5. THE STANDING WAVE LASER

The most common laser configuration is that of an amplifier cell situated between the nearly totally reflecting mirrors of an optical cavity, Fig. 3.5. In this case the field consists of two oppositely traveling waves of equal

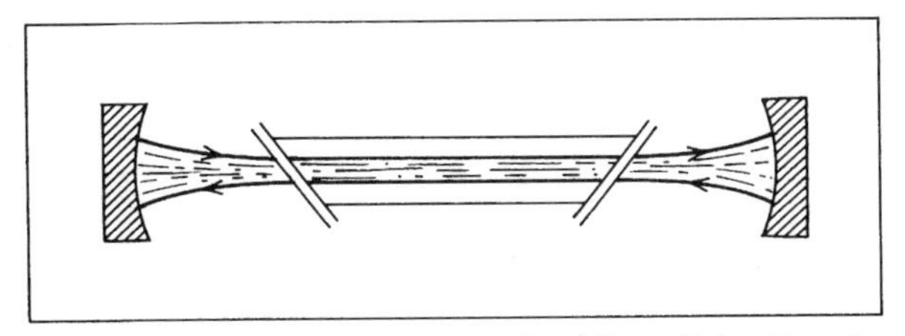

**Fig. 3.5** In a linear laser the amplifier cell is introduced into a Fabry–Perot laser cavity. The laser output is taken through transmission at one of the mirrors.

amplitude, which make up a standing wave. The field can be written as

$$\begin{aligned} E(z,t) &= \tfrac{1}{2}E\left[\sin(kz-\Omega t)+\sin(kz+\Omega t)\right] \\ &= -\frac{E}{4}\left[ie^{i(kz-\Omega t)}-ie^{-i(kz+\Omega t)}+\text{c.c.}\right] \\ &= E\cos\Omega t\sin kz, \end{aligned} \tag{3.95}$$

where the cavity eigenmode is taken to be represented by a single sine function. The field (3.95) is treated in Sec. 2.5 and its polarization is written in the form (2.142), which, according to (3.72-73), must be projected onto the cavity eigenmode

$$P(t)=\frac{\int_0^L P(z,t)\sin kz\,dz}{\int_0^L \sin^2 kz\,dz}=C\cos\Omega t+S\sin\Omega t \tag{3.96}$$

where we have

$$S=\tfrac{1}{2}E\,\mathrm{Im}(\chi_+ + \chi_-)=-\tfrac{1}{2}E(\chi_+'' + \chi_-'') \tag{3.97}$$

$$C=\tfrac{1}{2}E\,\mathrm{Re}(\chi_+ + \chi_-)=\tfrac{1}{2}E(\chi_+' + \chi_-') \tag{3.98}$$

with $\chi_\pm$ given by (2.143). Combining we find

$$\frac{1}{2}(\chi_+ + \chi_-) = -\frac{\mu^2 N_0\Lambda_0}{2\hbar\sqrt{\pi}\,u\gamma_{12}^2}$$

$$\times\int\frac{\left[(i\gamma_{12}+\Delta+kv)L(\Delta+kv)+(i\gamma_{12}+\Delta-kv)L(\Delta-kv)\right]e^{-v^2/u^2}\,dv}{\left[1+\tfrac{1}{2}I\eta(L(\Delta+kv)+L(\Delta-kv))\right]}. \tag{3.99}$$

The laser amplitude is now determined from Eq. (3.79) in the form

$$\dot{E}+\frac{1}{2\tau}E=-\frac{\Omega}{4\varepsilon_0}E\,\mathrm{Im}(\chi_+ + \chi_-)=\frac{\Omega}{4\varepsilon_0}E(\chi_+'' + \chi_-''). \tag{3.100}$$

When we use (3.99) we obtain the steady state operation condition

$$1 = \frac{\Omega\tau\mu^2 N_0 \Lambda_0}{2\varepsilon_0 \hbar\sqrt{\pi}\, ku\gamma_{12}} \int \frac{[L(\Delta + x) + L(\Delta - x)]\, e^{-x^2/k^2u^2}\, dx}{[1 + \frac{1}{2}I\eta(L(\Delta + x) + L(\Delta - x))]}. \tag{3.101}$$

When $\Delta = 0$ and $I = 0$, we obtain the threshold pumping $N_0\Lambda_T$ from (3.101) in the form

$$\begin{aligned} 1 &= \frac{\Omega\tau\mu^2 N_0 \Lambda_T}{\varepsilon_0 \hbar\sqrt{\pi}\, ku\gamma_{12}} \int L(x) e^{-x^2/k^2u^2}\, dx \\ &= \frac{\Omega\tau\mu^2 N_0 \Lambda_T}{\varepsilon_0 \hbar\sqrt{\pi}\, ku\gamma_{12}} \pi\gamma_{12}, \end{aligned} \tag{3.102}$$

where we have assumed $\gamma_{12} \ll ku$. Together with $\Omega\tau = Q$, (3.102) gives

$$N_0\Lambda_T = \frac{\hbar\varepsilon_0 ku}{\sqrt{\pi}\, Q\mu^2} \tag{3.103}$$

which agrees with (3.59). The same physical discussion holds again when we compare this with the threshold for the homogeneously broadened case (3.85). Inserting (3.103) into (3.101) we obtain

$$2\pi\gamma_{12} = N \int \frac{(L(\Delta + x) + L(\Delta - x))e^{-x^2/k^2u^2}}{[1 + \frac{1}{2}I\eta(L(\Delta + x) + L(\Delta - x))]}\, dx. \tag{3.104}$$

From this result we can obtain various limiting cases.

Equation (3.104) is a nonlinear relationship between the laser output intensity $I$, the cavity detuning $\Delta$, and the pumping rate $N_0\Lambda_0$. It can be solved numerically for $I$, or if the fourth-order polynomial in the denominator is expanded in partial fractions, we can express the result in terms of the plasma dispersion function, but the output intensity $I$ can, in any case, only be obtained in an implicit way. The result (3.101) is sometimes called the REA (= rate equation approximation) in the literature. This name derives from the fact that (3.99) results from an approximation where each traveling wave of the standing wave saturates the medium independently. This was discussed in Sec. 2.5. Together with a rate approximation, see Sec. 1.8, this leads to (3.99) for the steady state. The same result is, however, of a more general validity; when the steady state result is taken for the ansatz (2.129) of Sec. 2.5 the result is (3.99). No adiabatic assumptions are needed for the steady state result.

**Example.** If we assume the fast dephasing limit $\gamma_{21} \gg \mu E/\hbar, \Delta$, show that we have the rate equations

$$\frac{d}{dt}\rho_{22} = \lambda_2 - \gamma_2\rho_{22} + W(\rho_{11} - \rho_{22})$$

$$\frac{d}{dt}\rho_{11} = \lambda_1 - \gamma_1\rho_{11} - W(\rho_{11} - \rho_{22}),$$

where the rate $W$ is given by

$$W = \frac{\mu^2 E^2}{8\hbar^2\gamma_{12}}[L(\Delta + kv) + L(\Delta - kv)].$$

In the Doppler limit

$$\gamma_{21} \ll ku \tag{3.105}$$

we can approximate the result for several limiting cases. The corresponding approximations for $\chi_\pm$ were obtained in Sec. 2.5. At exact resonance $\Delta = 0$ the approximation (2.148) can be used in (3.100). The same result follows directly from (3.104) in the form

$$\pi\gamma_{12} = N\int \frac{L(x)}{1 + I\eta L(x)} e^{-x^2/k^2u^2}\,dx$$

$$\simeq \gamma_{12}^2 N \int \frac{1}{x^2 + \gamma_{12}^2(1 + I\eta)}\,dx = \frac{\pi\gamma_{12}N}{\sqrt{1 + I\eta}}, \tag{3.106}$$

which gives the output intensity in the form

$$\eta I = N^2 - 1 = 2(N - 1) + \cdots. \tag{3.107}$$

This is the quadratic dependence $I \sim N^2$, which we already found in the traveling wave case (3.63). It is a consequence of the inhomogeneous broadening in the Doppler limit. For a traveling wave, the result (3.63) is an exact consequence of this approximation, here it only follows from the lowest-order approximation, the REA result.

We have for a traveling wave of amplitude $E_0$ the intensity $I = (\mu^2 E_0^2 / 2\hbar^2 \gamma_1 \gamma_2)$, and the traveling wave amplitude of (3.95) is $E_0 = \frac{1}{2}E$. At resonance two such waves contribute to the intensity, and we can estimate the saturation parameter to be

$$\frac{\mu^2}{2\hbar^2 \gamma_1 \gamma_2}\left[\left(\frac{E}{2}\right)^2 + \left(\frac{E}{2}\right)^2\right] = \frac{1}{2}I. \tag{3.108}$$

When this is inserted into the traveling wave result (3.63) at resonance, we obtain (3.107). From (3.108) we can see that the two waves do add independently to the saturation in the present approximation.

Another useful limit is that of a very small intensity $\eta I \ll 1$. This concerns laser operation near threshold, the pumping parameter $N$ is close to unity or the laser is detuned nearly until operation ceases. We then expand the integrand in (3.104) and obtain

$$\begin{aligned} 2\pi\gamma_{12} &= N\int [L(\Delta + x) + L(\Delta - x)] \\ &\quad \times \{1 - \tfrac{1}{2}I\eta[L(\Delta + x) + L(\Delta - x)]\} e^{-x^2/k^2u^2}\, dx \\ &= Ne^{-\Delta^2/k^2u^2}\int \{2L(\Delta + x) - \eta I[L(\Delta + x)]^2 \\ &\quad - \eta I L(\Delta + x) L(\Delta - x)\}\, dx. \end{aligned} \tag{3.109}$$

Using the relations

$$\int L(y)^2\, dy = \gamma_{12}\int \frac{\gamma_{12}^3}{(y^2 + \gamma_{12}^2)^2}\, dy = \frac{\pi\gamma_{12}}{2} \tag{3.110}$$

$$\begin{aligned} \int L(x - \Delta) L(x + \Delta)\, dx &= \gamma_{12}\int \frac{\gamma_{12}^3}{\left((x - \Delta)^2 + \gamma_{12}^2\right)\left((x + \Delta)^2 + \gamma_{12}^2\right)}\, dx \\ &= \frac{\pi}{2}\gamma_{12} L(\Delta) \end{aligned} \tag{3.111}$$

we obtain from (3.109)

$$1 = Ne^{-\Delta^2/k^2u^2}\left[1 - \frac{\eta I}{4}(1 + L(\Delta))\right] \tag{3.112}$$

which can be solved for the intensity

$$\eta I = 4\frac{1 - N^{-1}e^{\Delta^2/k^2u^2}}{1 + L(\Delta)}$$

or differently

$$\eta I \simeq 4\frac{N - e^{\Delta^2/k^2u^2}}{1 + L(\Delta)}, \tag{3.113}$$

where we have taken into account the fact that near resonance (3.112) can hold only when $N \simeq 1$. This is the result of the third-order perturbation theory derived by Lamb. At resonance the intensity is

$$\eta I \simeq 2(N - 1) \tag{3.114}$$

in accordance with (3.107). When $|\Delta| \simeq \gamma_{12}$ but $|\Delta| \ll ku$ we find

$$\eta I = 4\frac{N - 1}{1 + L(\Delta)}, \tag{3.115}$$

which shows that the intensity grows when $|\Delta|$ increases. This is because when $\Delta = 0$, the two Bennett holes of the atomic population inversion overlap and all utilize the same atoms near zero velocity. When $|\Delta| > \gamma_{21}$ the two traveling waves start to utilize different atomic velocity groups, and the number of atoms effectively involved with amplifying the signal goes up by nearly a factor of two, hence the result (3.115). This is the Lamb dip at the center of the single-mode laser detuning curve. When $|\Delta| \simeq ku$, the number of atoms goes down because the Maxwellian velocity distribution goes to zero. Then the intensity drops to zero as

$$\eta I = 4\left(N - e^{\Delta^2/k^2u^2}\right). \tag{3.116}$$

Thus the full tuning curve for the Doppler limit has the shape shown in Fig. 3.6.

A similar shape follows from the full relationship (3.101) when we evaluate $I$ as a function of $N$ by numerical integration. In reality, both the series expansion result (3.113) and the Doppler limit result (3.107) are of rather limited validity. To illustrate the basic features of an inhomogeneously broadened laser they are very useful.

The result (3.113) is derived by assuming both the Doppler limit and a perturbation expansion in $I$. In the general case there are correction terms of order $I$ to the result (3.109), but their contribution is proportional to $(\gamma_{12}/ku)$ which makes them disappear in the Doppler limit.

To obtain an expression for the laser frequency we must take the expressions of (3.99) and insert them into (3.98). These determine the laser

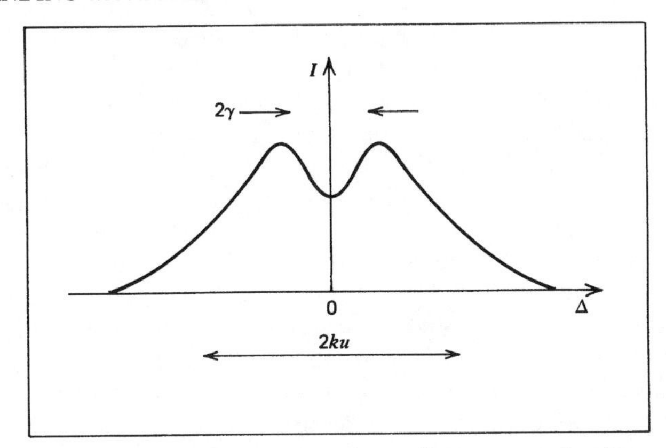


**Fig. 3.6** The single-mode laser output intensity oscillates over a detuning range of the magnitude of the inhomogeneous gain profile $\simeq ku$. At resonance there is Lamb's tuning dip of a width determined by the power broadened homogeneous atomic line width.

frequency from expression (3.83) in the form

$$\Delta\Omega = -\frac{\Omega}{2\varepsilon_0 E}C = -\frac{\Omega}{4\varepsilon_0}(\chi'_+ + \chi'_-). \tag{3.117}$$

We find

$$\begin{aligned}
\Omega_L - \Omega &= \Delta\Omega \\
&= \frac{N}{4\pi\gamma_{12}^2\tau}\int[(\Delta + x)L(\Delta + x) \\
&\quad + (\Delta - x)L(\Delta - x)]e^{-x^2/k^2u^2}\,dx + O(I) \\
&= \frac{N}{4\pi\gamma_{12}^2\tau}\int yL(y)\left[e^{-(y-\Delta)^2/k^2u^2} - e^{-(y+\Delta)^2/k^2u^2}\right]dy \\
&= \frac{Ne^{-\Delta^2/k^2u^2}}{2\pi\gamma_{12}^2\tau}\int yL(y)\sinh\left(\frac{2\Delta y}{k^2u^2}\right)e^{-y^2/k^2u^2}\,dy \\
&= \frac{N\Delta e^{-\Delta^2/k^2u^2}}{\pi\tau(ku)^2}\int\frac{y^2}{y^2 + \gamma_{12}^2}e^{-y^2/k^2u^2}\,dy \\
&= \frac{N\Delta e^{-\Delta^2/k^2u^2}}{\sqrt{\pi}\,\tau ku}.
\end{aligned} \tag{3.118}$$

This is in agreement with the result (2.146) inserted directly into (3.117); it is also in agreement with the result for a traveling wave (3.68). Its main conclusions are like those following (3.69), namely, that the laser frequency $\Omega_L$ nearly coincides with that of the cavity eigenmode $\Omega$.

The results of this section are based on the lowest-order results for a standing wave given in Sec. 2.5. Its correction terms are derived in Sec. 2.6, and their physical significance is discussed. For one single mode, we can solve the strong signal case exactly in terms of continued fractions. This solution is derived in Sec. 2.7. To obtain a strong signal theory for the single-mode laser we must take the coefficients $C$ and $S$ from (2.188) and (2.189) and insert them into Eqs. (3.79) and (3.83). For the ensuing conclusions concerning the laser operation we refer the reader to the sources discussed at the end of this chapter in Sec. 3.7.

## 3.6. A LASER WITH A SATURABLE ABSORBER

So far we have considered the nonlinear response of the atomic medium as an amplifier in an oscillating laser. This requires an inverted population; the upper level is initially more populated than the lower level. Saturation implies that the populations become nearly equal and that there is approximately the same number of transitions up and down. The medium no longer amplifies. If we, on the other hand, start with the population on the lower level, we have an absorber. When the field is strong enough to transfer a considerable fraction of the population to the upper level its absorption decreases. Saturation makes the absorber less efficient; it is bleached by the action of the strong field. This effect gives rise to new phenomena in the operation of a laser.

We can discuss the main features of a saturable absorber by looking at the laser system shown in Fig. 3.7. The laser is kept oscillating by the amplifier cell, which gives enough gain to overcome the total losses of the system. These include the absorption in the absorber cell. If this is strongly saturated, its net effect may, however, be rather small. For other frequencies near the operating one the absorber retains its full absorbing power, and hence it acts to suppress these modes from oscillating. This is the basis for its use to achieve single-mode operation in a laser. In general, the effect of absorption on the mode interaction in a laser is rather multifarious but lies outside the scope of this book. We refer to the references in the final Section. Because of the standing wave structure in the laser cavity the exact positions of the cells are of importance. In gas lasers the density of the active medium is so low that the influence of the gas is mainly determined by the bulk of atoms. The end effects are small and can be ignored in our

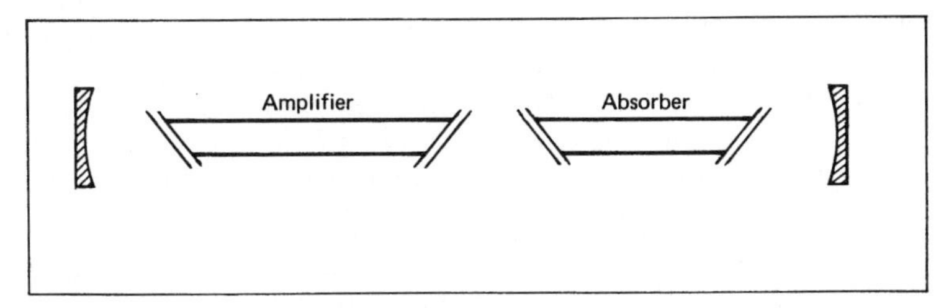


**Fig. 3.7** A laser with an intracavity absorber is kept oscillating by the amplifier tube, and the losses in the absorber are partly saturated (bleached) by the light intensity.

discussion. For solid and liquid laser media this assumption cannot be taken for granted but must be checked.

The Lamb dip gives an indication of the center of the atomic transition frequency, with a width determined by the line width $\gamma$; see Fig. 3.6. This is much less than the gain bandwidth $ku$, but it is not enough for high-precision stabilization of the laser. Such applications occur, for example, when the laser is used to construct a time standard. Then it is useful to be able to make a frequency mark on the laser output that is much sharper than the amplifier line width which is affected both by pressure effects and power broadening. By introducing a saturable absorber into the laser cavity we can achieve a much sharper resolution.

To understand the origin of the effect we write down the low-intensity condition (3.112) for a laser with a saturating absorber included. For simplicity we assume $|\Delta| \ll ku$ and add the effect of the absorber directly to the right-hand side

$$1 = N\left\{1 - \frac{\eta}{4} I\left[1 + L(\Delta)\right]\right\}$$

$$-N_{\text{abs}}\left\{1 - \tfrac{1}{4} I_{\text{abs}}\left[1 + L_{\text{abs}}(\Delta)\right]\right\}. \tag{3.119}$$

Here the absorber pumping constant $N_{\text{abs}}$ is provided with a negative sign to indicate losses and

$$I_{\text{abs}} = \left(\frac{\mu_{\text{abs}}^2 E^2}{2\hbar^2 \gamma_1^{\text{abs}} \gamma_2^{\text{abs}}}\right) \frac{\gamma_2^{\text{abs}} + \gamma_2^{\text{abs}}}{2\gamma_{12}^{\text{abs}}}$$

$$\equiv \xi \eta I, \tag{3.120}$$

where the parameters $\gamma^{\text{abs}}$ refer to the gas in the absorber cell. The field amplitude $E$ is the same in both cells. To make the absorber saturate more strongly than the amplifier we must choose it so that its line widths are less than those of the amplifier, $\gamma^{abs} < \gamma$. Then the parameter $\xi$ defined by (3.120) is larger than one. The absorber Lorentzian $L_{abs}(\Delta)$ also is then narrower than the amplifier Lorentzian $L(\Delta)$. From (3.119–3.120) we solve

$$\eta I = \frac{4}{N} \frac{N - N_{abs} - 1}{1 + L(\Delta) - M\left[1 + L_{abs}(\Delta)\right]} \tag{3.121}$$

where

$$M = \frac{\xi N_{abs}}{N}. \tag{3.122}$$

When we have

$$N > \xi N_{abs} \quad \text{and} \quad N > N_{abs} + 1 \tag{3.123}$$

the intensity in (3.121) is positive; the laser is above threshold. In the denominator we have two Lorentzians: $L(\Delta)$ is of width $\gamma_{12}$, and when $\Delta$ grows it increases the intensity. The other, $L_{abs}(\Delta)$, is narrower and is of width $\gamma_{12}^{abs}$. It signifies the saturation of the absorber when the two oppositely traveling waves in the laser use the same absorber atoms. When $\Delta$ exceeds $\gamma_{12}^{abs}$ two holes appear in the absorber, and the saturation decreases. Then the laser intensity goes down. Thus in the laser output there appears a

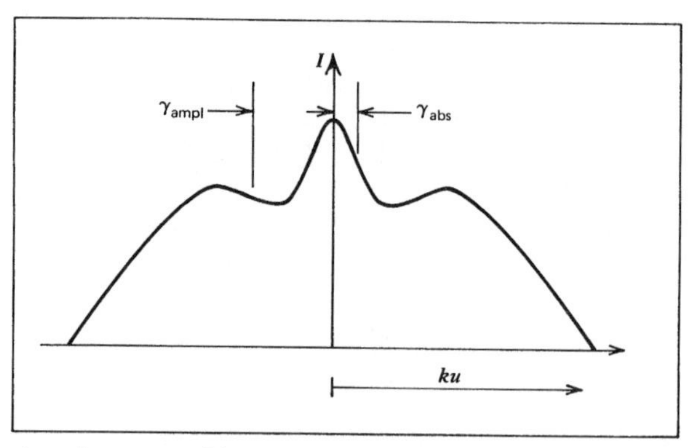


**Fig. 3.8** The intensity output with a saturable absorber shows an inverted Lamb dip in the middle of the ordinary one. This emerges when the absorber saturates strongly inside its homogeneous line width $\gamma_{abs}$.

peak, an inverted Lamb dip at the center of the ordinary Lamb dip. This is narrow, of width $\gamma_{12}^{\text{abs}}$, and can be used to stabilize the laser frequency. The output is of the type indicated in Fig. 3.8.

In general, the amplifier and the absorber have their resonance frequencies centered at different points. Then the two Lorentzians have their peaks at different points, and the curve in Fig. 3.8 becomes asymmetric.

In a He–Ne laser, absorption can be achieved by the insertion of a low-pressure Ne cell. Its line is narrower because it is less pressure broadened than is the amplifier gas. But the latter is also pressure shifted, and the two resonances do not coincide.

For time standards a He–Ne laser oscillating at the infrared wavelength $\lambda = 3.39\ \mu\text{m}$ is equipped with a methane, $CH_4$, absorber. The extremely narrow molecular lines make it possible to achieve a frequency accuracy down to about 1 Hz, which is about one part in $10^{14}$.

When we calculate the effect of the absorber to lowest order in the intensity, we miss some very essential features in the behavior of the laser. These include bistable operation and hysteresis effects. Such effects occur to higher orders in the intensity. They are easy to understand because of the action of the absorber.

When the laser intensity is small, the absorber acts with full power. If the gain is not large enough to overcome the losses of the absorber and the cavity, no oscillation is possible. But when the intensity grows, the absorber saturates, and the gain need only overcome the cavity losses. If the laser is oscillating it can go on doing so in a stable way.

To see how the optically bistable laser operates, let us assume a laser with an intracavity absorber tuned to exact resonance with both amplifier and absorber. No difference between their resonance frequencies is assumed. We also utilize the Doppler limit $\gamma_{12} \ll ku$. Then the result (3.106) becomes

$$1 = \frac{N}{\sqrt{1 + \eta I}} - \frac{N_{\text{abs}}}{\sqrt{1 + \xi \eta I}} \tag{3.124}$$

where we have already introduced (3.120). If we write this equation in terms of a gain function $G(I)$ as in Sec. 3.1, we have the steady state condition

$$G(I) = 1. \tag{3.125}$$

The function $G(I)$ consists of two parts shown separately in Fig. 3.9. Because $\xi > 1$ the negative part decreases faster than the positive one. If $N_{\text{abs}} < N$ this leads to a gain curve with a maximum like that in Fig. 3.2. Equation (3.124) can then be solved graphically by drawing the line at the cavity loss and looking for intersections. There are three possibilities as

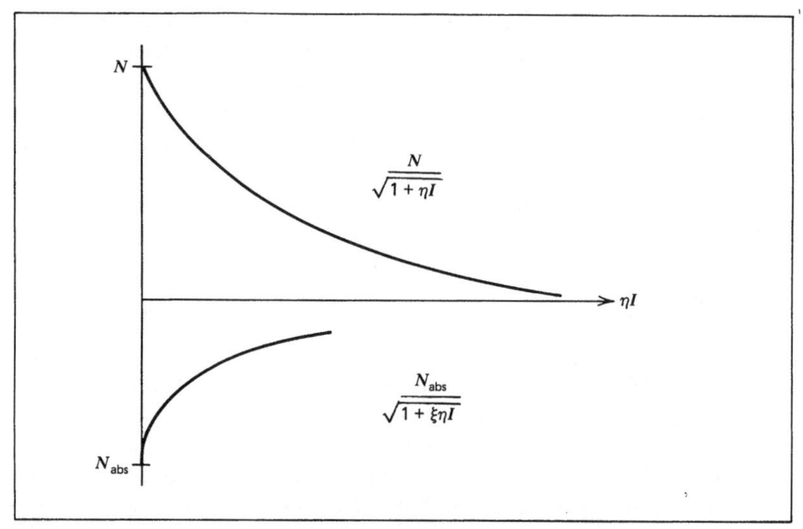


**Fig. 3.9** This figure shows how absorption saturates much faster than amplification because of the factor $\xi(>1)$. The drawing shows the resonant case $\Delta = 0$.

shown in Fig. 3.10. In case 1 there is no stable solution; the laser is below threshold. In case 3 there is only one stable solution; the laser is well above threshold. Case 2 is the interesting one. We have two crossings, but also the point $I = 0$ gives a stable solution. There the losses exceed the gain, and small intensity fluctuations are rapidly damped out.

The crossing indicated by a cross is unstable. A fluctuation down makes the losses dominate, and the intensity rapidly drops to zero. An upward fluctuation in intensity makes the gain dominate and takes the intensity to the operating point indicated by a dot. This is stable, as can easily be seen by a similar argument. Thus for case 2 the operation is bistable. Letting the value $N$ vary, we easily find a hysteresis in the intensity as shown by Fig. 3.11, which can be constructed by inspection from Fig. 3.10.

Following the method in Sec. 3.1 we can write, as in (3.13),

$$\frac{dI}{dt} = \frac{1}{\tau}[G(I) - 1]I \tag{3.126}$$

where $G(I)$ is defined by (3.124). A detailed mathematical analysis can confirm the conclusions we just obtained by simple heuristic reasoning. The stability of the operating point, especially, as given by (3.18), agrees with our conclusions from Fig. 3.10. The laser with an intracavity saturable absorber thus gives an example of the behavior discussed in Sec. 3.1.

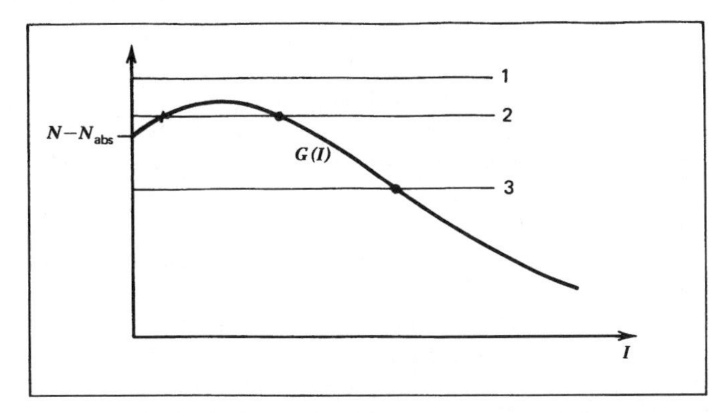


**Fig. 3.10** The net gain at resonance as a function of intensity. The three loss lines, 1, 2, and 3, indicate the cases of no stable operation, a bistable situation, and one single stable operating point, respectively. At zero intensity the gain is $N - N_{abs}$.

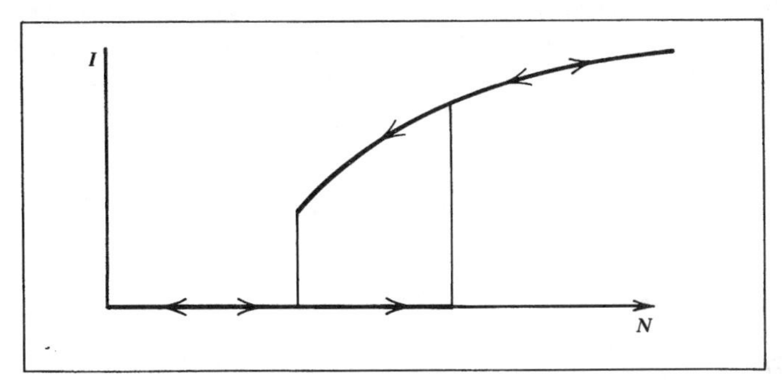


**Fig. 3.11** Laser intensity as a function of amplifier-cell pumping. For a certain range bistable operation is possible, and a hysteresis effect can be seen.

Bistability and hysteresis also appear as a function of detuning near the threshold for oscillations. These can be discussed along the lines indicated here, but the argument becomes more complicated. The qualitative features are easily understood from asymptotic expressions like the result in (2.145), but the discussion can be omitted here.

## 3.7. COMMENTS AND REFERENCES

The basic condition for the oscillation of a laser system, as given in Sec. 3.1, was presented by Schawlow and Townes (1958). A good introduction to laser physics is found in Svelto (1976). The self-consistent semiclassical theory was forged by W. E. Lamb Jr. (1964), and its further developments are presented in Sargent et al. (1974). Simultaneously, laser theory was developed by H. Haken in Stuttgart, West Germany (Haken 1970a, b) and M. Lax at Bell Telephone Laboratories, New Jersey (Lax 1968). For an early version of Lamb's views consult his lectures in De Witt et al. (1964), which is an excellent early source for quantum electronics theories.

The semiclassical traveling wave amplifier was treated by Close (1967) and the traveling wave laser by Smith (1966). For the other aspects of gas laser theory consult Sargent et al. (1974), which treats multimode lasers, and Zeeman and ring lasers semiclassically.

The laser with an intracavity saturable absorber was introduced experimentally by Lee et al. (1968) and Lisitsyn and Chebotaev (1968a, b). The theory was considered by Beterov et al. (1971a, b) and further continued by Salomaa and Stenholm (1973a, b). Many later considerations and suggestions are treated in Letokhov and Chebotaev (1977) and the references therein. This was the first optically bistable nonlinear element to emerge in quantum electronics. Later many others have attracted a great theoretical and experimental interest; see, for example, Bowden et al. (1981).

CHAPTER 4

# Topics in Laser Spectroscopy

## 4.1. INTRODUCTION

We have now shown how the influence of a strong coherent field modifies the properties of an atomic system. In Chapter 3 we demonstrated how these modifications can be utilized to build up a self-sustained oscillator, the laser.

Many spectroscopic applications must, however, use strong light sources to achieve effects not observable in ordinary linear spectroscopy. Features due to saturation effects or multiphoton processes, especially, are essentially nonlinear phenomena.

In this chapter we consider a few applications of the basic concepts to various situations of interest in laser spectroscopy. The various sections are to a large extent independent, and the reader can choose to proceed from this introduction to any one he chooses. Together they do, however, complement each other and give different points of view on the same phenomena.

The Doppler broadened system, especially, hides many of its atomic features behind the large inhomogeneous width. With nonlinear methods these can be brought forward and investigated. In Sec. 4.2 we consider the effect on a weak probe by the saturating power of a strong counterpropagating field. This is a situation similar to that inside a laser cavity, where, however, both fields must have the same amplitude. For spectroscopy in a cell, it is advantageous to keep one wave weak.

It is also possible to investigate a strongly saturated two-level system by coupling one level weakly to a third level. The spectroscopy of such three-level systems has played an important role in the development of laser applications. In Sec. 4.3 we discuss this situation and several special cases of importance in applications.

One old way to observe features inside the Doppler profile has been the level-crossing technique discovered by Hanle in 1923. With laser signals, several types of level crossings are possible. These are discussed and compared in Sec. 4.4.

For recent work on multiphoton excitation of molecular systems, the general $N$-level case has become of interest. In isotope separation work and laser-induced chemistry such considerations are of the utmost importance. The case rapidly becomes very complicated, and in Sec. 4.5 we can only give a few basic ideas and some approximate treatments.

The choice of topics here is highly subjective. There are many important applications missing. One that is rather obvious is based on the saturation of dispersion and not of absorption. From Eq. (3.39) we can see that the phase of light will be modified by the active medium owing to the dispersive part of the susceptibility $\chi'$: This will depend on both velocity and intensity, and hence it is possible to cause different phase shifts for different waves in passing through the atomic medium. With elliptically polarized light, the two components may experience different phase shifts in passing through the same medium, and hence polarization rotation (Faraday effect) can be observed. Owing to cross-saturation, when several waves affect the same atoms, the polarization rotation shows features similar to those of absorption spectroscopy. This has been utilized by Hänsch and Schawlow to investigate spectra by the polarization properties of light; see Sec. 4.2.c.

When the dielectric constant of an optical Fabry–Perot cavity depends on the light intensity, its optical length can be tuned in such a way that it is resonant at a high intensity but lossy for low intensities. We then have a bistable device which can either be filled with radiation at a given intensity or strives to stay free of light energy. Such a dispersive optical bistability has given rise to many theoretical investigations over the last years. Experimentally, the device is of interest because of its potentiality as a component in optical data processing systems. For this we refer the readers to the recent conference, Bowden et al. (1981).

In these sections we regard the light as impinging on the system from an external (laser) source. Then there appears a difference as compared with the laser case described in Chapter 3.

In an active laser, the $k$-vector is determined by the geometry of the cavity, see Section 1.2, and the oscillational frequency is determined self-consistently by the oscillator as explained in Chapter 3.

In an absorber the light comes from the outside, and its frequency is determined by its source. If the phase is modified by the active medium, this appears as a change of wavelength, and hence optical bistability becomes possible. In conditions characteristic of normal laser operation, dispersion does not lead to bistability. The absorptive bistability due to an intracavity saturable absorber as discussed in Sec. 3.6 has its counterpart in a passive cavity too. Then the source of energy is the externally injected laser signal.

In the equation of evolution of the phase (3.39) there is a time variation $\dot{\varphi}$ in the laser (the frequency is shifted) and a change with respect to position

$(c\partial\varphi/\partial z)$ in an external cavity (a change of wavelength and $k$-vector). Both can be modified by the light intensity, but the physical manifestations will be different.

Other applications of lasers involving scattering, parametric effects, laser field-induced collisional phenomena, and laser ionization are omitted. All these fields have recently become very important, but their treatments fall outside the general point of view of this book. So does a treatment of laser action based on free electrons. Proper mathematical treatment would take us too far, and a brief review of the present situation would rapidly become outdated.

In some very recent works it has been found that optical systems under certain conditions display a chaotic behavior. These developments are, at the moment of writing, very new and cannot be properly discussed within the framework of this book.

## 4.2. SATURATION SPECTROSCOPY

### 4.2.a. General Ideas

When a strong traveling wave interacts with a two-level system it modifies the state in a way that depends on the parameters of the light field as well as on those of the atomic system. In linear spectroscopy many atomic features of the spectra are overwhelmed by the Doppler width $ku$ which is larger than the line widths and fine details of the spectrum. Only by using nonlinear methods can we bring forth the hidden information residing in the holes burned in the population differences and their accompanying induced polarizations.

When we direct one weak probe beam against the original strong beam, it ordinarily feels the full absorbing power of the atomic system over the whole Doppler profile. However, when its detuning is such that its resonantly interacting atoms also are acted on by the strong field, their absorbing power is already saturated, and there is only little change due to the probe. The sample appears to become nearly transparent to the probe.

The simplest way to achieve the situation described is to reflect part of the strong beam back onto itself and let it act as the probe. The situation is shown in Fig. 4.1. Both beams then have the same frequency, and they interact with atomic velocity groups situated symmetrically with respect to the line center. The hole due to the strong beam is, however, much larger because of high intensity effects. The situation for the detuned case is shown in Fig. 4.2. When we tune the laser beam to line center, the two holes meet and the probe absorption effectively gives an image of the strong field hole.

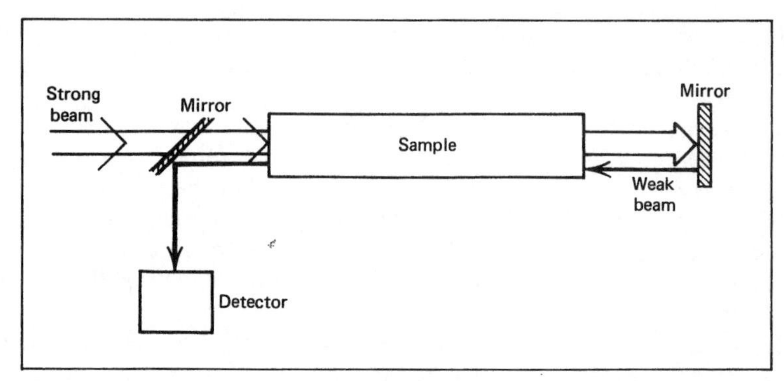


**Fig. 4.1** A typical setup for saturation spectroscopy. A strong beam passes the sample, and a fraction of it is returned to probe the ensuing saturation.

Because this is an absorber, there appears a maximum in the transmission. The line is power broadened, but because the probe is weak it feels the full effect of the strong field saturation. A similar effect was earlier shown in the laser in the form of the Lamb dips (Figs. 3.6 and 3.8). In the operating laser, however, both beams are necessarily strong, and the additional saturation when the holes overlap is smaller than in the present case of a weak probe.

With the advent of tunable dye lasers, the present method has become an efficient way to investigate atomic structure. In the next section we give a more mathematical formulation of the situation described here.

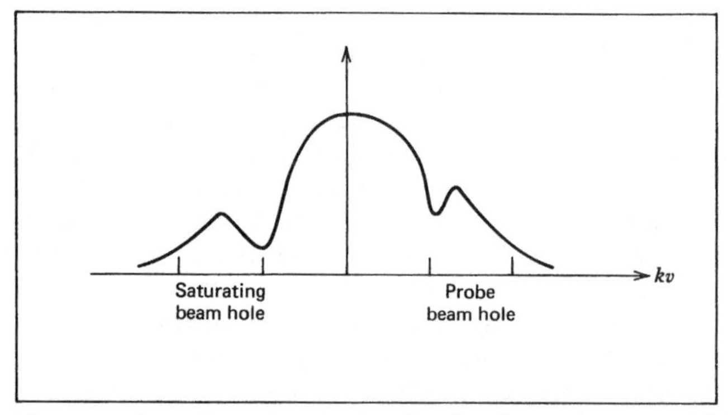


**Fig. 4.2** The strong beam burns a power broadened and strongly saturated hole in the population. This is probed by the counterpropagating weak beam.

### 4.2.b. The Theoretical Description

In this section we consider the general case of two traveling waves interacting with the two-level system in Fig. 4.3. The field is taken to be in the form

$$E(z,t) = E_1 \cos \varphi_1 + E_2 \cos \varphi_2, \tag{4.1}$$

where the phases are

$$\varphi_i = \Omega_i t + k_i z \qquad (i = 1, 2). \tag{4.2}$$

For the simple case of backscattering of the initial laser beam we have $k_1 = -k_2$. Following our treatment in Chapter 2, we introduce a rotating wave approximation by neglecting rapidly varying terms. We also use the notation

$$\alpha_i = \frac{\mu_{12} E_i}{2\hbar}$$

$$\Delta\omega_i = \omega_0 - \frac{d\varphi_i}{dt} = \omega_0 - \Omega_i - k_i v. \tag{4.3}$$

First we treat the strong field $E_1$ only and set $E_2 = 0$. Then with

$$\tilde{\rho}_{21} = e^{i\varphi_1} \rho_{21} \tag{4.4}$$

we obtain directly from Sec. 2.4 the results: The population modification is given by

$$\Delta n = (\rho_{22} - \rho_{11}) = \frac{\overline{N}_{21}}{1 + I \dfrac{\gamma_{12}^2}{\Delta\omega_1^2 + \gamma_{12}^2}}$$

$$= \overline{N}_{21} \left[ 1 - \frac{I \gamma_{12}^2}{\Delta\omega_1^2 + \gamma_{12}^2 (1 + I)} \right], \tag{4.5}$$

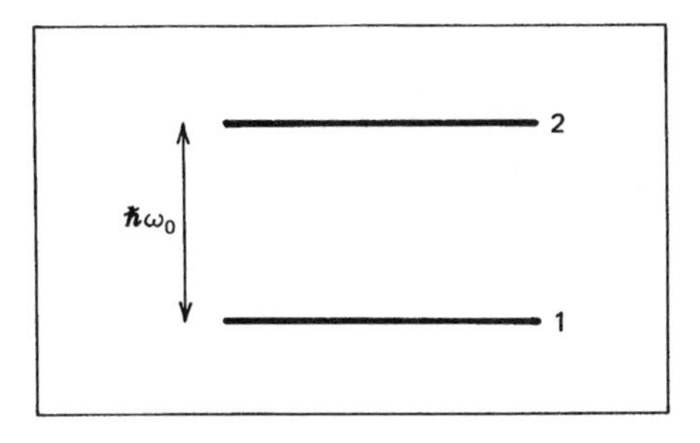


**Fig. 4.3** The two-level system for saturation spectroscopy with a weak probe.

and the induced dipole is determined by

$$\tilde{\rho}_{21} = -i\alpha_1 \frac{\rho_{22} - \rho_{11}}{\gamma_{12} + i\,\Delta\omega_1}, \tag{4.6}$$

where we have

$$I = \frac{\mu_{12}^2 E_1^2}{2\gamma_{12}\hbar^2}\left(\frac{1}{\gamma_1} + \frac{1}{\gamma_2}\right)$$

$$\overline{N}_{21} = \frac{\lambda_2}{\gamma_2} - \frac{\lambda_1}{\gamma_1}. \tag{4.7}$$

For an absorber ordinarily $\lambda_2 = 0$ and $\overline{N}_{21} < 0$.

To perform the perturbation calculation to lowest order in $E_2$ we write the density matrix $\rho = \rho^{(0)} + \rho^{(1)}$ and let $\rho^{(0)}$ be given by (4.5)–(4.6). We then obtain the perturbation equations

$$\dot{\rho}_{22}^{(1)} + \gamma_2\rho_{22}^{(1)} = -i\alpha_1\left[e^{i\varphi_1}\rho_{21}^{(1)} - e^{-i\varphi_1}\rho_{12}^{(1)}\right]$$
$$-i\alpha_2\left(e^{i\varphi_2}\rho_{21}^{(0)} - e^{-i\varphi_2}\rho_{12}^{(0)}\right) = -\dot{\rho}_{11}^{(1)} - \gamma_1\rho_{11}^{(1)} \tag{4.8}$$

$$\dot{\rho}_{21}^{(1)} + (\gamma_{21} + i\omega_0)\rho_{21}^{(1)} = -i\alpha_1\left(\rho_{22}^{(1)} - \rho_{11}^{(1)}\right)e^{-i\varphi_1} - i\alpha_2\left(\rho_{22}^{(0)} - \rho_{11}^{(0)}\right)e^{-i\varphi_2}. \tag{4.9}$$

Terms of order $\alpha_2\rho^{(1)}$ have been neglected. We introduce

$$\Delta = \varphi_2 - \varphi_1 \tag{4.10}$$

and write

$$e^{i\varphi_2}\rho_{21}^{(0)} - e^{-i\varphi_2}\rho_{12}^{(0)} = e^{i\Delta}\tilde{\rho}_{21} - e^{-i\Delta}\tilde{\rho}_{12}$$

$$\rho_{ii}^{(1)} = \rho_{ii}^{+}e^{i\Delta} + \rho_{ii}^{-}e^{-i\Delta} \qquad (i = 1, 2)$$

$$\rho_{21}^{(1)} = \rho_{21}^{I}e^{-i\varphi_2} + \rho_{21}^{II}e^{-i(\varphi_1 - \Delta)} = \left[\rho_{12}^{(1)}\right]^*. \tag{4.11}$$

With

$$\Delta\Omega = \frac{d}{dt}\Delta = \Omega_2 - \Omega_1 + v(k_2 - k_1) \tag{4.12}$$

we have

$$(\gamma_2 + i\Delta\Omega)\rho_{22}^{+} = -i\alpha_1(\rho_{21}^{II} - \rho_{12}^{I}) - i\alpha_2\tilde{\rho}_{21} \tag{4.13a}$$

$$(\gamma_1 + i\Delta\Omega)\rho_{11}^{+} = i\alpha_1(\rho_{21}^{II} - \rho_{12}^{I}) + i\alpha_2\tilde{\rho}_{21} \tag{4.13b}$$

and

$$(\gamma_{21} - i\,\Delta\omega_2)\rho_{12}^{\mathrm{I}} = i\alpha_1(\rho_{22}^{+} - \rho_{11}^{+}) + i\alpha_2\,\Delta n \tag{4.14a}$$

$$[\gamma_{21} + i(\Delta\omega_1 + \Delta\Omega)]\,\rho_{21}^{\mathrm{II}} = -i\alpha_1(\rho_{22}^{+} - \rho_{11}^{+}). \tag{4.14b}$$

If we solve for the induced dipole with the phase $\varphi_2$ we find

$$\rho_{12}^{\mathrm{I}} = \frac{i\alpha_2}{\gamma_{21} - i\,\Delta\omega_2}\Delta n + \frac{i\alpha_1}{\gamma_{21} - i\,\Delta\omega_2}(\rho_{22}^{+} - \rho_{11}^{+}). \tag{4.15}$$

The first term is a pure rate term; the probe absorption is modified by the modification of the population difference $\Delta n$. The second term in (4.15) is a coherence term. The quantity $\rho_{22}^{+} - \rho_{11}^{+}$ is induced by $\alpha_2$ but still contains coherence in $\alpha_1$ which combines with the explicit $\alpha_1$ in (4.15).

If we observe the probe absorption we have, in accordance with the result of Sec. 2.2, the result

$$W_2 \propto \alpha_2\langle \mathrm{Im}\,\rho_{12}^{\mathrm{I}}\rangle, \tag{4.16}$$

where the bracket denotes a velocity average. Introducing the first term of (4.15) we observe

$$\begin{aligned} W_2 &= C\left\langle \frac{\alpha_2^2\gamma_{21}}{\gamma_{21}^2 + \Delta\omega_2^2}\Delta n \right\rangle \\ &= C\gamma_{21}\alpha_2^2\left\langle \left[\frac{1}{\gamma_{21}^2 + \Delta\omega_2^2} - \frac{I\gamma_{21}^2}{(\gamma_{21}^2 + \Delta\omega_2^2)(\Delta\omega_1^2 + \gamma_{12}^2(1 + I))}\right]\right\rangle \overline{N}_{21}. \end{aligned} \tag{4.17}$$

Here $C$ is the scale constant of proportionality for the measuring system.

The first term in (4.17) just maps the Doppler contour with a resolution of $\gamma_{21}$; this is the result of linear spectroscopy. The second term is the nonlinear response. It is clearly a product of two rates with Lorentzian line shapes; only the first rate is power broadened. The product involves two velocity groups defined by $\Delta\omega_i = 0$. When these coincide $\Delta\omega_1 = \Delta\omega_2$ we have a modification in the behavior, and a resonance appears, as is easily proved in the Doppler limit. This occurs at the velocity

$$v_0 = \frac{\Omega_1 - \Omega_2}{k_2 - k_1}. \tag{4.18}$$

When we have counterpropagating beams $|k_2 - k_1| \simeq 2k$, and for identical frequencies, the holes meet in the middle of the Doppler curve at $v = 0$. It is, however, possible to probe other velocity groups by using beams of different frequencies. This can be used to check the line width dependence on velocity because a large $v_0$ ($\gg u$) in the direction of the beam is likely to dominate over the transverse velocities, which are mostly of the order $u$.

To include the coherence effects in our problem we must solve the coupled Eqs. (4.13) and (4.14). We introduce the notation

$$D_i = (\gamma_i + i\,\Delta\Omega)^{-1} \qquad (i = 1,2)$$

$$D_{\mathrm{I}} = (\gamma_{21} - i\,\Delta\omega_2)^{-1}$$

$$D_{\mathrm{II}} = [\gamma_{21} + i(\Delta\omega_1 + \Delta\Omega)]^{-1} \tag{4.19}$$

and find straightforwardly

$$(\rho_{22}^{+} - \rho_{11}^{+}) = [1 + \alpha_1^2(D_{\mathrm{I}} + D_{\mathrm{II}})(D_1 + D_2)]^{-1}$$

$$\times[-i\alpha_2(D_1 + D_2)\tilde{\rho}_{21} - \alpha_1\alpha_2 D_{\mathrm{I}}(D_1 + D_2)\,\Delta n]. \tag{4.20}$$

The original strong field $\alpha_1$ has created both population modifications ($\propto \Delta n$) and an induced dipole ($\propto \tilde{\rho}_{21}$). Both affect the value of $(\rho_{22}^{+} - \rho_{11}^{+})$ according to (4.20). To simplify the expression we introduce $\tilde{\rho}_{21}$ in terms of $\Delta n$ from (4.6) and find

$$(\rho_{22}^{+} - \rho_{11}^{+}) = -\alpha_1\alpha_2(D_1 + D_2)(D_{12} + D_{\mathrm{I}})\,\Delta n\,H^{-1} \tag{4.21}$$

where

$$H = 1 + \alpha_1^2(D_{\mathrm{I}} + D_{\mathrm{II}})(D_1 + D_2) \tag{4.22a}$$

$$D_{12} = (\gamma_{12} + i\Delta\,\omega_1)^{-1}. \tag{4.22b}$$

Introducing this back into the second term of (4.15) we find the observable

$$W_2 = C\alpha_2 \langle \operatorname{Im} \rho_{12}^{\mathrm{I}} \rangle$$
$$= C\alpha_2^2 \gamma_{12} \left\langle \frac{\Delta n}{\gamma_{12}^2 + \Delta\omega_2^2} (1 - I \operatorname{Re} B) \right\rangle. \tag{4.23}$$

The term proportional to unity gives the rate contribution discussed earlier, and $\operatorname{Re} B$ is consequently a coherence contribution. Defining the function $F(v)$ by setting

$$D_1(v) + D_2(v) = \frac{2}{\gamma F(v)}, \tag{4.24}$$

where

$$\gamma^{-1} = \frac{1}{2}\left(\frac{1}{\gamma_1} + \frac{1}{\gamma_2}\right) \tag{4.25}$$

we can write

$$B = \frac{\gamma_{21} + i(\omega_0 - \Omega_1 - k_1 v)\left[\dfrac{1}{\gamma_{21} + i(\omega_0 - \Omega_1 - k_1 v)} + \dfrac{1}{\gamma_{21} - i(\omega_0 - \Omega_2 - k_2 v)}\right]}{2F(v) + I\gamma_{12}\left[\dfrac{1}{\gamma_{21} - i(\omega_0 - \Omega_2 - k_2 v)} + \dfrac{1}{\gamma_{21} + i(\omega_0 - 2\Omega_1 + \Omega_2 - (2k_1 - k_2)v)}\right]} \tag{4.26}$$

The velocity dependence is seen to be extremely complicated and difficult to interpret physically. Some features are, however, noteworthy: The function $F(v)$ experiences resonance behavior at

$$\Delta\Omega = \Omega_2 - \Omega_1 + v(k_2 - k_1) = 0. \tag{4.27}$$

For coinciding frequencies this resonance always occurs at zero velocity independent of all other parameters. The other feature is the resonance in the last term in the denominator. For $\Omega_1 = \Omega_2 = \Omega$ and $k_1 = -k_2 = k$ we have resonance at the velocity group

$$v_1 = \frac{\omega_0 - \Omega}{3k}. \tag{4.28}$$

This is a hole-burning effect at one-third of the velocity characteristic of the main Bennett hole. It is a typical coherence feature; more exact calculations show that there will be higher-order holes at all the velocities

$$v_l = \frac{\omega_0 - \Omega}{(2l + 1)k}. \tag{4.29}$$

Their origin can be understood from the continued fraction treatment in Sec. 2.7 by looking at the resonance denominators lower down in the continued fraction. These features have been called "multi-Doppleron" resonances. See also Sec. 2.8.

### 4.2.c. Comments and References

Lasers rapidly found a wide use in spectroscopy. A general review of many applications can be found in the book by Corney (1977). Once it was realized that the Lamb dip could be used for spectroscopy outside the laser cavity (Bagaev et al. 1968 and Letokhov and Chebotaev 1969), the method rapidly became very useful. Many applications are found in the article by Letokhov in Shimoda (1976) and Letokhov and Chebotaev (1977). A detailed theoretical investigation of the weak probe-beam response was undertaken by Baklanov and Chebotaev (1971 and 1972) and Haroche and Hartmann (1972). When tunable lasers became available, this method could be extended to many atomic and molecular investigations. One very successful early application was the high-precision work by Th. W. Hänsch. He measured the values of the Rydberg constant and the Lamb shift with saturation spectroscopy (Hänsch 1973). This was the first measurement of the quantum electrodynamic shift using optical radiation only. The use of intensity-dependent polarization properties to probe inside the Doppler profile was initiated by Wieman and Hänsch (1976); the later developments of the method can be followed from Dabkiewicz et al. (1981).

Saturation spectroscopy selecting a velocity group different from zero was introduced by Javan (1977) and his collaborators Mattick et al. (1973). They utilized the method to investigate collisional effects.

Later many applications of nonlinear laser spectroscopy have emerged. A recent discussion of some theoretical principles has been given by Sargent (1978). A review of applications is found in Levenson (1982).

## 4.3. THE THREE-LEVEL SYSTEM

### 4.3.a. General Remarks

Our theoretical treatment has been obsessed with two-level systems. In real atomic systems, of course, a multitude of levels and sublevels are encountered. To treat these analytically would involve us in a hopeless maze of algebra. The next simplest case, the three-level system can, however, be discussed while retaining some intuition for the physics of the processes. Because laser beams are extremely monochromatic, they do actually select only a few levels out of the atomic and molecular manifolds, and our intuition can be transferred to real observations with some confidence.

We consider the situation depicted in Fig. 4.4. The lower level pair forms a two-level system acted on by a traveling wave

$$\mathscr{E}_1(z, t) = E_1\cos(\Omega_1 t + k_1 z) \tag{4.30}$$

by the dipole moment $\mu_{12}$. The level 2 is, in its turn, coupled to the uppermost level 3 by the field

$$\mathscr{E}_2(z, t) = E_2\cos(\Omega_2 t + k_2 z) \tag{4.31}$$

and dipole $\mu_{23}$. There is no coupling assumed between levels 3 and 1 ($\mu_{13} = 0$), and each field is taken to affect only one transition not the other.

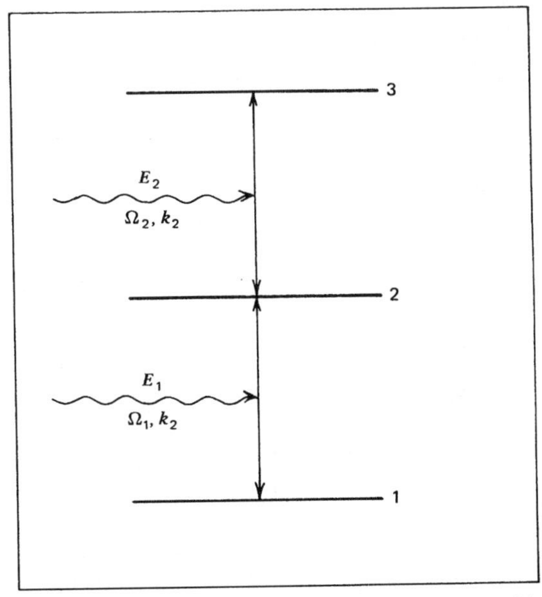


**Fig. 4.4** The three-level systems. The strong field $E_1$ acts on the first transition $|1\rangle \to |2\rangle$, the weak field $E_2$ probes the modifications of the lower two-level system.

The latter situation can be achieved by choosing the polarization of the light suitably or by choosing levels so that $|\omega_{21} - \omega_{32}|$ is large enough. Then only one field can interact resonantly with each transition.

This type of system can serve as a model for many cases of atomic spectroscopy. The experimental investigations on three-level atomic systems form a field of spectroscopy in themselves. Frequency-selective excitation and ionization of atoms and molecules is by now a standard tool to detect, identify, or separate nearly identical species, for example, isotopes or rare earth atoms. For these processes multiphoton ladders give the simplest model; the three-level system is their prototype.

Many Doppler-free laser techniques are based on two-photon processes, and here again the three-level system is our model. Such cases are discussed in Sec. 4.3.c.

We discuss the cascade situation depicted in Fig. 4.4. With very slight modifications our equations can be made to describe the folded situations in Fig. 4.5; they are called the V-configuration (Fig. 4.5a) and the Λ- or inverted V-figuration (Fig. 4.5b). We refer to these later in the discussion.

### 4.3.b. The Theoretical Description

If we introduce the obvious notation

$$\alpha_i = \frac{\mu_i E_i}{2\hbar} \qquad (i = 1, 2) \tag{4.32}$$

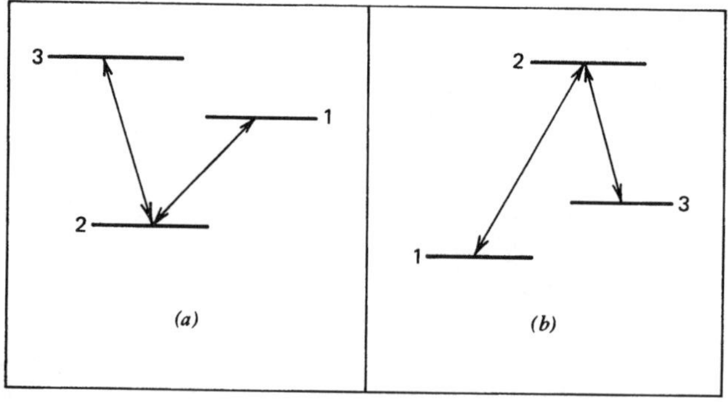


**Fig. 4.5** Other three-level configurations: (*a*) The V-configuration; (*b*) The inverted-V, or Λ-configuration.

(where $\mu_i$ is the appropriate dipole matrix element), we obtain the equations

$$\dot{\rho}_{33} = -\gamma_3\rho_{33} + 2i\alpha_2\cos(\Omega_2 t + k_2 z)(\rho_{32} - \rho_{23})$$

$$\dot{\rho}_{22} = -\gamma_2\rho_{22} - 2i\alpha_2\cos(\Omega_2 t + k_2 z)(\rho_{32} - \rho_{23})$$
$$+2i\alpha_1\cos(\Omega_1 t + k_1 z)(\rho_{21} - \rho_{12})$$

$$\dot{\rho}_{11} = \lambda_1 - \gamma_1\rho_{11} - 2i\alpha_1\cos(\Omega_1 t + k_1 z)(\rho_{21} - \rho_{12}) \tag{4.33}$$

and

$$\dot{\rho}_{32} = -(\gamma_{32} + i\omega_{32})\rho_{32} + 2i\alpha_2\cos(\Omega_2 t + k_2 z)(\rho_{33} - \rho_{22})$$
$$+2i\alpha_1\cos(\Omega_1 t + k_1 z)\rho_{31}$$

$$\dot{\rho}_{31} = -(\gamma_{31} + i\omega_{31})\rho_{31} + 2i\alpha_1\cos(\Omega_1 t + k_1 z)\rho_{32}$$
$$-2i\alpha_2\cos(\Omega_2 t + k_2 z)\rho_{21}$$

$$\dot{\rho}_{21} = -(\gamma_{21} + i\omega_{21})\rho_{21} + 2i\alpha_1\cos(\Omega_1 t + k_1 z)(\rho_{22} - \rho_{11})$$
$$-2i\alpha_2\cos(\Omega_2 t + k_2 z)\rho_{31}. \tag{4.34}$$

We have assumed no pumping except to the lowest level and hence we have

$$\overline{N} = -\frac{\lambda_1}{\gamma_1}. \tag{4.35}$$

If we now again apply a rotating wave approximation we set

$$\rho_{21} = e^{-i(k_1 z + \Omega_1 t)}\tilde{\rho}_{21} \tag{4.36}$$

$$\rho_{32} = e^{-i(k_2 z + \Omega_2 t)}\tilde{\rho}_{32} \tag{4.37}$$

$$\rho_{31} = \exp\{-[i(k_1 + k_2)z + (\Omega_1 + \Omega_2)t]\}\tilde{\rho}_{31}. \tag{4.38}$$

We choose to apply a perturbation approach by setting $\alpha_2 = 0$, and then we obtain the solution $\tilde{\rho}_{21}^{(1)}$, $\rho_{22}^{(1)}$, and $\rho_{11}^{(1)}$ given in Sec. 2.4. We here want to use the index 1 instead of zero to indicate the presence of field $E_1$ only. These solutions are now

$$\rho_{22}^{(1)} = -2\alpha_1^2 \bar{N} \frac{\gamma_{12}}{\gamma_2} \frac{1}{\left[(\Delta_1 - k_1 v)^2 + \Gamma_1^2\right]} \tag{4.39}$$

$$\rho_{11}^{(1)} = \frac{\lambda_1}{\gamma_1} + 2\alpha_1^2 \bar{N} \frac{\gamma_{12}}{\gamma_1} \frac{1}{\left[(\Delta_1 - k_1 v)^2 + \Gamma_1^2\right]} \tag{4.40}$$

$$\tilde{\rho}_{21}^{(1)} = \frac{\bar{N}\alpha_1(\Delta_1 - k_1 v + i\gamma_{12})}{(\Delta_1 - k_1 v)^2 + \Gamma_1^2} \tag{4.41}$$

where

$$\Delta_1 = \omega_{21} - \Omega_1 \tag{4.42}$$

and

$$\Gamma_1^2 = \gamma_{12}^2 + 2\alpha_1^2 \frac{\gamma_{12}(\gamma_1 + \gamma_2)}{\gamma_1 \gamma_2}. \tag{4.43}$$

Introducing these results into the equations for $\rho_{33}$, $\rho_{32}$, and $\rho_{31}$ and retaining terms to lowest order in $E_2$, we find that only two new functions $\tilde{\rho}_{32}$ and $\tilde{\rho}_{21}$ enter this approximation. They are coupled and give the equations

$$(\Delta_2 - k_2 v - i\gamma_{32})\tilde{\rho}_{32}^{(2)} - \alpha_1 \tilde{\rho}_{31}^{(2)} = -\alpha_2 \rho_{22}^{(1)} \tag{4.44a}$$

$$\left[\Delta_1 + \Delta_2 - (k_1 + k_2)v - i\gamma_{31}\right]\tilde{\rho}_{31}^{(2)} - \alpha_1 \tilde{\rho}_{32}^{(2)} = -\alpha_2 \tilde{\rho}_{21}^{(1)}, \tag{4.44b}$$

where we set

$$\Delta_2 = \omega_{32} - \Omega_2. \tag{4.45}$$

From this we can see that both the *population modification* $\rho_{22}^{(1)}$ and the *induced dipole* $\tilde{\rho}_{21}^{(1)}$ act as driving terms for the upper transition.

Solving for $\tilde{\rho}_{32}^{(2)}$ we obtain the dipole moment, the imaginary part of which gives the absorption on the transition $2 \rightarrow 3$, which is the observable of the probe beam. This is, as we know, equivalent with a measure of the population reaching level 3.

If we solve Eqs. (4.44)–(4.45) we find

$$\tilde{\rho}_{32} = -\alpha_2 \frac{[\Delta_1 + \Delta_2 - (k_1 + k_2)v - i\gamma_{31}]\rho_{22}^{(1)} + \alpha_1 \tilde{\rho}_{21}^{(1)}}{(\Delta_2 - k_2 v - i\gamma_{32})[\Delta_1 + \Delta_2 - (k_1 + k_2)v - i\gamma_{31}] - \alpha_1^2}. \tag{4.46}$$

Here we can see two contributions: one proportional to the population modification on level 2, the other to the induced dipole $\tilde{\rho}_{21}^{(1)}$. The roles of these two can most easily be seen if we resort to the perturbation theory limit; $\alpha_1$ becomes small. Then it disappears from the denominator of (4.46), and $\Gamma_1$ becomes equal to $\gamma_{12}$ according to (4.43).

The population modification induced term then becomes

$$\left(\tilde{\rho}_{32}^{(2)}\right)_{\text{rate}} = \left(\frac{\alpha_2}{\Delta_2 - k_2 v - i\gamma_{32}}\right) \frac{2\gamma_{12}\overline{N}}{\gamma_2} \frac{\alpha_1^2}{\left[(\Delta_1 - k_1 v)^2 + \gamma_{12}^2\right]}. \tag{4.47}$$

The observed absorption on the second transition is then

$$W = C\alpha_2 \langle \operatorname{Im} \tilde{\rho}_{32}^{(2)} \rangle \tag{4.48a}$$

where $C$ is the scale factor of proportionality in the measuring system. We find with (4.47)

$$W = \frac{2C\overline{N}}{\gamma_2} \left\langle \left[\frac{\alpha_2^2 \gamma_{32}}{(\Delta_2 - k_2 v)^2 + \gamma_{32}^2}\right] \left[\frac{\alpha_1^2 \gamma_{21}}{(\Delta_1 - k_1 v)^2 + \gamma_{21}^2}\right] \right\rangle. \tag{4.48b}$$

The bracket denotes a velocity average. This result has an interpretation in terms of the product of two rates. This is an *incoherent rate contribution*. It is also called the two-step process because the intermediate level population probability $\rho_{22}$ must become nonzero for the process to be operative. If some strongly incoherent agent acts to destroy the coherence either $\tilde{\rho}_{21}^{(1)}$ or alternatively $\tilde{\rho}_{31}^{(2)}$, this is the only contribution that survives. Such agents may be collisions, see Sec. 1.6, or laser field fluctuations, see Chapter 5.

To lowest order in $\alpha_1^2$ the coherent contribution becomes

$$\left(\tilde{\rho}_{32}^{(2)}\right)_{\text{coh}} = -\frac{\alpha_1^2 \alpha_2 \overline{N}}{(\Delta_2 - k_2 v - i\gamma_{32})} \left[\frac{1}{(\Delta_1 - k_1 v) - i\gamma_{13}}\right] \times \left[\frac{1}{\Delta_1 + \Delta_2 - (k_1 + k_2)v - i\gamma_3}\right]. \tag{4.49}$$

The imaginary part of this shows a complicated dependence on the detuning. When both $\Delta_1$ and $\Delta_2$ approach zero, both transitions are tuned to resonance, also $\Delta_1 + \Delta_2$ goes to zero, and all three denominators behave resonantly at the same point. A complicated behavior ensues. This is a *totally coherent process*; it is induced by $\tilde{\rho}_{21}^{(1)}$ and goes through $\tilde{\rho}_{31}^{(2)}$. It is sometimes called a two-quantum process even if this is slightly misleading; both processes involve two quanta (photons).

It is possible to develop a diagrammatic perturbation theory for the density matrix elements. We start with the lowest level $\rho_{11}$ and draw two lines following the history of each index of $\rho_{ij}$ out to the observable $\rho_{33}$. We indicate the way this is done in Fig. 4.6; the rate process in $(a)$ and the coherent process in $(b)$. The resonance denominators can be read off at the position of the dashed lines. The interaction structure of the three-level system can also be represented differently by the way we pass the density matrix elements to reach $\rho_{33}$ from $\rho_{11}$. This is shown in Fig. 4.7. We see that to lowest order in the $\alpha$'s we must have the factor $\alpha_1^2\alpha_2^2$ just as in Fig. 4.6.

Finally, we want to stress that our present treatment is simplified by the assumption that only the lowest level is pumped. If there occurs a popula-

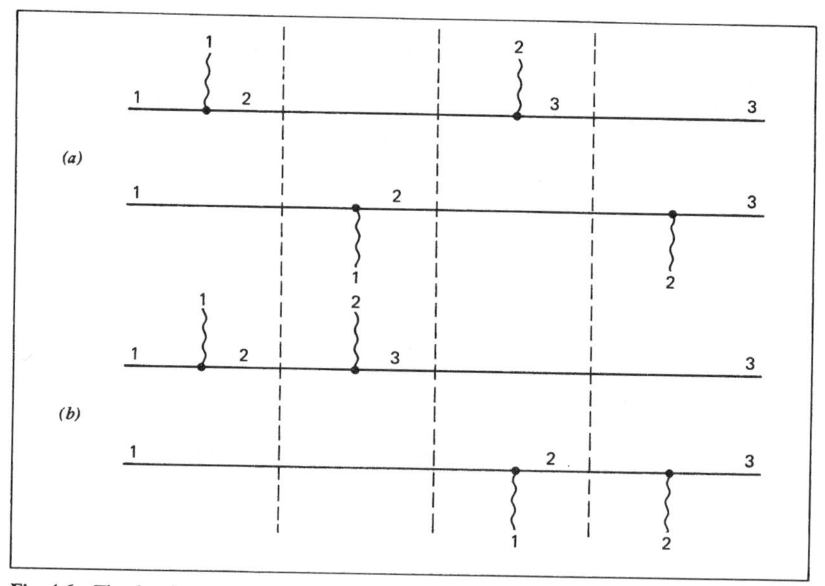


**Fig. 4.6** The density matrix element $\rho_{11}$ is taken to the excited state $\rho_{33}$ to lowest order in the fields (wavy lines). Each field, 1 and 2, is seen to act twice, but the rate process $(a)$ goes through the intermediate state population $\rho_{22}$, whereas the coherent process $(b)$ utilizes the coherence $\rho_{31}$.

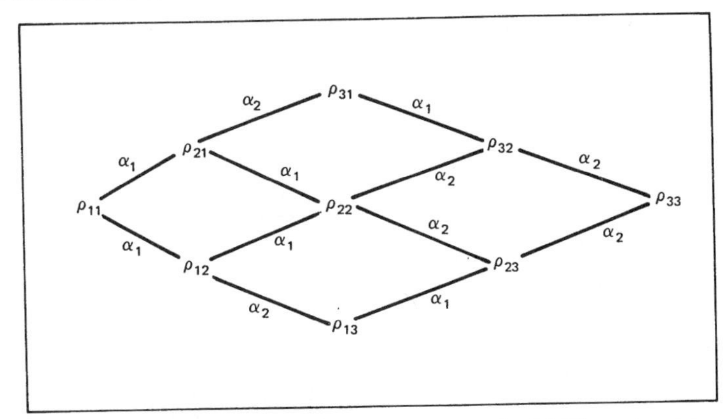


**Fig. 4.7** The same couplings by the fields $\alpha_1$ and $\alpha_2$ as in Fig. 4.6. The various intermediate density matrix elements are clearly seen. To get from $\rho_{11}$ to $\rho_{33}$ we must, to lowest order, get the factor $\alpha_1^2\alpha_2^2$. To the same order there appears a modification of $\rho_{11}$ only through the intermediate elements $\rho_{31}$ and $\rho_{13}$.

tion on the levels 2 and 3, also without any fields, there appear additional terms in our equations, and these can be observed to show resonant behavior with laser tuning.

When we take the expressions $\rho_{22}^{(1)}$ and $\tilde{\rho}_{21}^{(1)}$ for a two-level system in a strong field from Sec. 2.4 we obtain from (4.46)

$$\tilde{\rho}_{32} = \frac{\alpha_1^2\alpha_2\bar{N}\gamma_{21}}{\gamma_2\left[(\Delta_1 - k_1 v)^2 + \Gamma_1^2\right]} \times \left\{ \frac{2[\Delta_1 + \Delta_2 - v(k_1 + k_2) - i\gamma_{31}] - (\gamma_2/\gamma_{21})(\Delta_1 - k_1 v + i\gamma_{21})}{(\Delta_2 - k_2 v - i\gamma_{32})[\Delta_1 + \Delta_2 - (k_1 + k_2)v - i\gamma_{31}] - \alpha_1^2} \right\}. \tag{4.50}$$

This expression contains three resonances: $\Delta_1 = 0$, $\Delta_2 = 0$, and the two-photon resonance $\Delta_1 + \Delta_2 = 0$. Each has its own physical implications. The presence of the strong field $\alpha_1^2$ in the denominator leads to additional physical phenomena which cannot be found in perturbation theory. In the following sections we discuss some of these features in detail.

### 4.3.c. Doppler-Free Two-Photon Spectroscopy

In this case we look at the situation where both intermediate steps are well off-resonance, $\Delta_1 \neq 0$, $\Delta_2 \neq 0$, but the two-photon transition near resonance, $\Delta_1 + \Delta_2 \simeq 0$. In this case we can write (4.50) approximately as

$$\tilde{\rho}_{32} \simeq \frac{\alpha_1^2 \alpha_2 \bar{N}}{\left[(\Delta_1 - k_1 v)^2 + \Gamma_1^2\right](\Delta_2 - k_2 v - i\gamma_{32})} \times \frac{(\Delta_1 - k_1 v + i\gamma_{21})}{\left[\Delta_1 + \Delta_2 - (k_1 + k_2)v - i\gamma_{31} - \dfrac{\alpha_1^2}{\Delta_2 - k_2 v - i\gamma_{32}}\right]}. \tag{4.51}$$

The incoherent two-step process does not give any contribution because the detuning prevents population buildup on the intermediate level.

The atomic velocity is, at most, of the order $u$, and if the detunings are large enough so that

$$ku \ll |\Delta_1| \simeq |\Delta_2|, \tag{4.52}$$

we find that the detunings $\Delta$ dominate all other terms except in the combination $\Delta_1 + \Delta_2$. Hence we obtain from (4.51)

$$\tilde{\rho}_{32} = \frac{\alpha_1^2 \alpha_2 \bar{N}}{\Delta_1 \Delta_2} \left[\frac{1}{\Delta_1 + \Delta_2 - (k_1 + k_2)v - i\gamma_{31} - \alpha_1^2/\Delta_2}\right]. \tag{4.53}$$

In this expression we have a single-resonance denominator with the width $\gamma_{31}$. This is the line width of a dipole forbidden transition, and it is often rather insensitive to perturbations, for example, by atomic collisions. Hence the resonance may be very narrow.

The resonance occurs for the velocity group with

$$v = \frac{\Delta_1 + \Delta_2}{k_1 + k_2}; \tag{4.54}$$

if both fields derive from the same laser but turned back to propagate in the opposite direction, $k_1 = -k_2$, and the resonance becomes velocity independent. All atoms within the Doppler contour experience the same excitation and a fully Doppler-free resonance is seen. Its resonance condition

$$\Delta_1 + \Delta_2 = \omega_{31} - 2\Omega = 0 \tag{4.55}$$

shows the two-photon nature of the transfer $1 \to 3$. In many practical applications our assumption that one field acts selectively on each transition becomes unrealistic. A realistic analysis becomes a little more complicated, but the narrow two-photon resonance remains. In addition, a broad Doppler profile appears as a background.

The term $\alpha_1^2/\Delta_2$ in the denominator gives a shift of the resonance, but in this approximation we have no power broadening. This will come from terms of order $\alpha_1^2\alpha_2^2/|\Delta|^2$.

The Doppler-free two-photon spectroscopy has proven to be an efficient way to obtain high resolution spectra. The experiments must be carried out with as low an intensity as possible to avoid power distortion of the resonances. Then their intensity tends to become small because of the factor $\Delta_1\Delta_2 \simeq |\Delta_1|^2$ in the denominator. Thus there may be an advantage to tune the laser into resonance with the intermediate step. The Doppler shifts become important but so do also the two resonance conditions

$$\begin{aligned} \Delta_1 - k_1 v &= 0 \\ \Delta_2 - k_2 v &= 0. \end{aligned} \tag{4.56}$$

These affect the same atoms when the detunings satisfy

$$\frac{\Delta_1}{k_1} - \frac{\Delta_2}{k_2} = 0. \tag{4.57}$$

If $k_1 \simeq -k_2$, this resonance nearly coincides with the two-photon one. Such resonances have the product of two $\gamma$'s instead of the $\Delta_1\Delta_2$ factor in (4.53). Then, however, the incoherent two-step process must be included too. We return to this question in the next section.

### 4.3.d. The Two-Photon Transition Near Resonance

Returning again to the perturbation limit we find from (4.50)

$$\tilde{\rho}_{32} = -\frac{\alpha_1^2\alpha_2\overline{N}\gamma_{21}}{\gamma_2}\left\{\frac{2}{\left[(\Delta_1 - k_1 v)^2 + \gamma_{12}^2\right](\Delta_2 - k_2 v - i\gamma_{32})}\right.$$
$$\left. - \frac{(\gamma_2/\gamma_{21})}{[(\Delta_1 - k_1 v) - i\gamma_{21}][(\Delta_2 - k_2 v) - i\gamma_{32}][\Delta_1 + \Delta_2 - (k_1 + k_2)v - i\gamma_{31}]}\right\}. \tag{4.58}$$

At exact resonance the absorbed power is given by

$$W \propto \alpha_2 \mathrm{Im}\, \tilde{\rho}_{32} \simeq \frac{2\alpha_1^2\alpha_2^2\overline{N}}{\gamma_2\gamma_{12}\gamma_{32}}\left(1 + \frac{\gamma_2}{2\gamma_{13}}\right). \tag{4.59}$$

The second term in the parentheses is the contribution from the two-quantum coherent process. Even at exact resonance its relative contribution is given by the factor

$$\frac{\gamma_2}{\gamma_{31}} = \frac{\text{life-time of coherence } \rho_{31}}{\text{life-time of population } \rho_{22}}. \tag{4.60}$$

The coherent process goes through $\rho_{31}$, the incoherent process through $\rho_{22}$. Which path dominates is determined simply by their relative durability. This simple consideration holds even when we go beyond perturbation theory.

The same types of arguments can be applied also to the three-level configurations V and inverted V of Fig. 4.5. In Fig. 4.5a where both 1 and 3 are coupled to the lower level 2, we expect $\gamma_{31} \gtrsim \frac{1}{2}(\gamma_{33} + \gamma_{11}) \gg \gamma_{22}$, and then the incoherent rate process is expected to dominate the transitions $1 \to 3$. In Fig. 4.5b the level 2 can decay spontaneously to levels 1 and 3, and hence we expect $\gamma_{22} \gg \gamma_{31}$. In this case the coherent processes dominate, and the rate approximation is expected to be invalid. These conclusions are in agreement with numerical calculations on multilevel systems.

A more detailed calculation of the V and inverted-V must be done from the beginning. The pumping situation, especially, becomes different. It is, however, simple to carry out the modifications necessary to describe these cases.

All perturbations of the phases tend to increase $\gamma_{31}$ faster than $\gamma_{22}$. Hence they will decrease the contribution from the coherent process. Such effects can arise from collisions or laser light fluctuations, see Secs. 1.6 and 5.2. To call on such processes to make the system operate in the rate approximation is, however, a self-destructive approach. The overall rate is proportional to the factor $(\gamma_{12}\gamma_{23})^{-1}$ [see Eq. (4.59)], and this makes the destruction of coherence imply that the total transfer rate becomes very small. This is due to the inevitable intermediate states $\rho_{21}$ and $\rho_{23}$. To obtain the optimal transfer of atoms to the topmost level, we must try to utilize the coherent as well as the rate transitions.

To obtain the observed absorption from Eq. (4.58) we must integrate over the velocity distribution. This introduces a Gaussian weight function, but if we consider detunings such that the resonances occur only close to its center and $|\Delta| \ll ku$, we can replace it by a constant. Then the velocity integrals can be carried out by closing the velocity integrations in the complex $v$-plane. The poles following from (4.58) are

$$v = \frac{\Delta_1}{k_1} \pm \frac{i\gamma_{12}}{k_1}$$

$$v = \frac{\Delta_2}{k_2} - \frac{i\gamma_{32}}{k_2}$$

$$v = \frac{\Delta_1 + \Delta_2}{k_1 + k_2} - i\frac{\gamma_{31}}{k_1 + k_2}. \tag{4.61}$$

If $k_1 + k_2 > 0$ we can close in the upper half-plane and obtain the result

$$\langle \tilde{\rho}_{32} \rangle = \left( \frac{2\pi\alpha_1^2\alpha_2 \bar{N}}{\gamma_2} \right) \left[ \frac{1}{\left( \Delta_2 - \frac{k_2\Delta_1}{k_1} \right) - i\left( \gamma_{32} + \frac{k_2\gamma_{12}}{k_1} \right)} \right]. \tag{4.62}$$

The imaginary part of this gives a simple Lorentzian resonance at the position given by (4.57). In the limit $\gamma \simeq |\Delta| \ll ku$ this result is exact.

When, however,the sign of $k_1 + k_2$ is changed the imaginary part ($\gamma_{31}/k_1 + k_2$) crosses to the upper half-plane, and a coherent contribution from the second term enters too. This sensitivity to the sign of the Doppler shift $(k_1 + k_2)v$ is characteristic of three-level resonances, and it contributes to the complicated line shape encountered in these systems.

Some special features of interest can be seen from the perturbation result (4.58) directly. With copropagating waves $k_1 \| k_2$, all factors containing $v$ have the same sign in the second, coherent contribution. If we integrate over velocity with a broad, nearly constant velocity distribution, the result of the integration gives zero because all the complex $v$-poles are in the same half-plane. In this situation only the incoherent, rate term contributes, and the signal probes the perturbations of the population $\rho_{22}$ and not those of the coherence $\rho_{31}$ at all. Again, these can be investigated if the change of the ground state occupation $\rho_{11}$ appearing to order $\alpha_1^2\alpha_2^2$ can be separated. Using modulation of the laser beams at different frequencies one makes this experimentally possible. From Fig. 4.7 we can observe that the only way to reach back to $\rho_{11}$ after acting twice with $\alpha_1$ and twice with $\alpha_2$ is to go through $\rho_{31}$ or $\rho_{13}$; no pass through $\rho_{22}$ takes us back to $\rho_{11}$ to this order. Thus we can use nonlinear laser spectroscopy experimentally to separate the relaxation processes affecting various density matrix elements.

In an interesting example the lower level is the ground state and the upper level is ionized by a special rate out to a continuum. This situation occurs in the case of selective ionization by two-photon transitions to a state near the continuum of the spectrum followed by ionization due to a static electric field or a third radiation field. If this is characterized by a rate $\gamma_0$ out of level three, we have the equations

$$\dot{\rho}_{33} = -\gamma_0 \rho_{33} + \cdots$$

$$\dot{\rho}_{22} = -\Gamma \rho_{22} + \cdots$$

$$\dot{\rho}_{11} = \lambda_1 + \Gamma \rho_{22} + \cdots$$

$$\dot{\rho}_{21} = -\left(i\omega_{21} + \tfrac{1}{2}\Gamma\right)\rho_{21} + \cdots \tag{4.63}$$

The spontaneous emission terms are taken from Sec. 1.6.

In this situation $\gamma_2/\gamma_{12} = 2$ and the relation (4.50) becomes in the limit $\alpha_1 \to 0$

$$\tilde{\rho}_{32} = \alpha_1^2 \alpha_2 \overline{N} \left( \frac{1}{(\Delta_1 - k_1 v)^2 + \gamma_{12}^2} \right) \times \left( \frac{1}{\Delta_1 + \Delta_2 - (k_1 + k_2) v - i\frac{1}{2}\gamma_0} \right) \tag{4.64}$$

because $\gamma_1 = 0$ and hence $\gamma_{31} + \gamma_{21} = \gamma_{32}$. We see that in this special case the incoherent process disappears completely and only the coherent one survives. Incoherence and higher-order processes tend to restore the resonance at $\Delta_2 \simeq 0$, and hence an exchange of intensity between the two resonances at $\Delta_1 + \Delta_2 \simeq 0$ and $\Delta_2 \simeq 0$ for fixed $\Delta_1$ can take place. This is sometimes called "line-inversion."

### 4.3.e. Saturation Effects on the Probe

The effects of the strong field $\alpha_1$ acting on the first transition $1 \to 2$ are seen in two places of Eq. (4.50). The first is the power broadening of the transition $1 \to 2$ as manifested in $\Gamma_1$. The second one is the $\alpha_1^2$ in the denominator. The latter effect is of more interest. In the Doppler limit, the resonance structure observed is determined by the singularities of $\tilde{\rho}_{32}$ as we saw in the preceding section.

We thus want to consider the roots of the equation

$$(\Delta_2 - k_2 v - i\gamma_{32})[\Delta_1 + \Delta_2 - (k_1 + k_2)v - i\gamma_{31}] = \alpha_1^2. \quad (4.65)$$

To discuss its properties let us put $\gamma_{32} = \gamma_{31} = \gamma$ and neglect the velocity dependence $v = 0$. We obtain the roots

$$\Delta_2 = i\gamma - \frac{1}{2}\Delta_1 \pm \sqrt{\left(\frac{\Delta_1}{2}\right)^2 + \alpha_1^2}\,. \quad (4.66)$$

When the first transition is tuned to resonance $\Delta_1 = 0$, the probe sees its transition split into a doublet at $\pm\alpha_1$, see Fig. 4.8. This is called the *dynamic Stark effect*, the AC *Stark effect* or the *Autler–Townes effect* (see comments in Sec. 4.3.f).

If $\Delta_1$ is nonzero, the line shape becomes asymmetric but the doublet structure remains. We look at the two limits:

1. The amplitude dominates the detuning $\alpha_1 \gg |\Delta_1/2|$. Then we can expand in $\Delta_1$ and find the resonance positions

$$\Delta_2 = \begin{cases} \alpha_1 - \dfrac{1}{2}\Delta_1 + \dfrac{\Delta_1^2}{8\alpha_1} \\ -\alpha_1 - \dfrac{1}{2}\Delta_1 - \dfrac{\Delta_1^2}{8\alpha_1}. \end{cases} \quad (4.67)$$

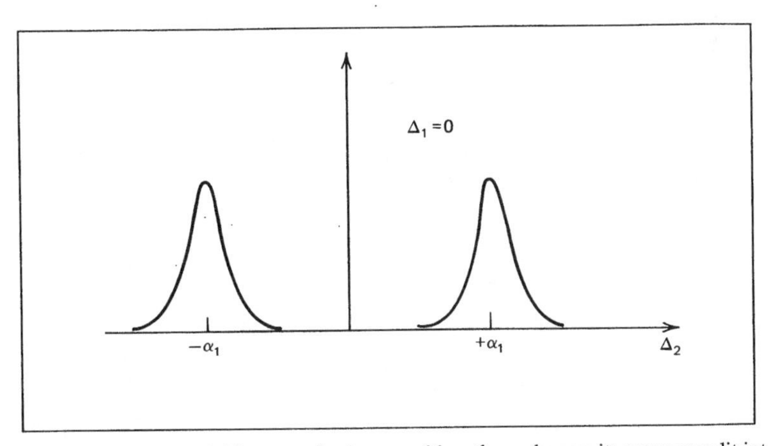


**Fig. 4.8** When a strong field acts on the first transition, the probe sees its response split into a doublet separated by $2\alpha_1$.

We can see the lowest-order Stark shift $\pm\alpha_1$ to be modified by $(\Delta_1^2/8\alpha_1)$. The terms $-\Delta_1/2$ cause an essential asymmetry.

2. The detuning dominates the amplitude, $\alpha_1 \ll |\Delta_1/2|$. We find the asymmetric resonances

$$\Delta_2 = \begin{cases} 0 + \dfrac{\alpha_1^2}{\Delta_1} \\ -\Delta_1 - \dfrac{\alpha_1^2}{\Delta_1}. \end{cases} \tag{4.68}$$

The first is the resonance position for the second transition, the second one is the position of the two-photon resonance $\Delta_1 + \Delta_2 = 0$. Both experience a *light shift* of magnitude $\alpha_1^2/\Delta_1$.

At this point it is interesting to compare our results with those for the Bloch–Siegert shift discussed in Sec. 2.3.

To see how the dynamic Stark effect manifests itself in a Doppler broadened system we look at the simple situation $\gamma_{32} = \gamma_{31}$ and $k_1 = -k_2$. We also tune the first transition to resonance $\Delta_1 = 0$, which is the situation displaying the symmetric Stark splitting of Fig. 4.8. We find from (4.50)

$$\tilde{\rho}_{32} = \frac{\alpha_1^2\alpha_2 \bar{N}\,[2(\Delta_2 - i\gamma) - (kv + i\gamma)]}{\left[(\Delta_2 - kv - i\gamma)(\Delta_2 - i\gamma) - \alpha_1^2\right]\left[(kv)^2 + \Gamma_1^2\right]}. \tag{4.69}$$

As a function of velocity no splitting is seen, but for small field amplitudes $\alpha_1$ there is a light shift of the resonance to

$$kv = \Delta_2 - \frac{\alpha_1^2}{\Delta_2}.$$

If we again assume the Doppler limit, we can perform the Doppler integration in the upper complex half-plane, and the only pole which contributes is the one at $kv = i\Gamma_1$. The result is

$$\langle\tilde{\rho}_{32}\rangle \propto \frac{2\Delta_2 - i(3\gamma + \Gamma_1)}{[\Delta_2 - i(\gamma + \Gamma_1)][\Delta_2 - i\gamma] - \alpha_1^2}. \tag{4.70}$$

The observed line shape when $\Delta_2$ is tuned has resonances at the points

$$\begin{aligned} \Delta_2 &= \frac{i}{2}\left[2\gamma + \Gamma_1 \pm \sqrt{(2\gamma + \Gamma_1)^2 - 4\gamma(\gamma + \Gamma_1) - 4\alpha_1^2}\right] \\ &= \frac{i}{2}\begin{cases} 3\gamma + \Gamma_1 \\ \gamma + \Gamma_1 \end{cases}; \end{aligned} \tag{4.71}$$

here the definition of $\Gamma_1$ in Eq. (4.43) has been used. These appear as resonances in the observed response. Thus there are two Lorentzians centered at $\Delta_2 = 0$, but one of the width approximately $2\gamma$ and the other approximately equal to $\gamma$. Both are power broadened by $\alpha_1$. In this case both fall at top of each other, but they are still present if $\Delta_1 \neq 0$ and then they are no longer at the same position.

The observed quantity becomes

$$W \propto \alpha_2 \text{Im}\langle\tilde{\rho}_{32}\rangle \propto \alpha_1^2 \alpha_2^2 \frac{1}{\Delta_2^2 + \frac{1}{4}(\gamma + \Gamma_1)^2}. \tag{4.72}$$

Surprisingly enough, the assumption $\gamma_{32} = \gamma_{31}$ implies that the weight of the broad resonance vanishes and only the narrow resonance is seen. In the general case both resonances survive and display their different line widths.

In conclusion, we note that the velocity-averaged response (4.72) does not show the AC Stark splitting (4.67) which is seen as a function of $\Delta_2$. When we relax the simplifying assumptions of this section we will find that the resonances become doublets again. These doublets are, however, not straightforward images of the AC Stark splittings seen when no Doppler broadening is present. To see the further complications occurring in three-level spectroscopy the reader is referred to the references in Sec. 4.3.f.

### 4.3.f. Comments and References

The method to probe a two-level system by coupling to a third level has been known for a long time. It was considered for laser spectroscopy in a thorough way by Feld and Javan (1969), Hänsch and Toschek (1970), Popova et al. (1970a, b) and Baklanov and Chebotaev (1971). Strong signal aspects were further discussed by Feldman and Feld (1972). The general situation is outlined by Chebotaev in an article in Shimoda (1976) and discussed in Sargent (1978). General considerations are also found in Schenzle and Brewer (1978). The Doppler-free technique was suggested by Vasilenko et al. (1970) and soon observed experimentally by several groups; see the article by Blombergen and Levenson in Shimoda (1976). The effect of the intermediate resonance was discussed in Salomaa and Stenholm (1975, 1976a) and Salomaa (1977), where approximate expressions are discussed also. The contributions from coherent processes at resonance were discussed in Salomaa and Stenholm (1978). These questions were clarified experimentally by the work of Bjorkholm and Liao (1976). Further interest has been devoted to the saturated state of the two-level system as seen by the probe (Kyrölä and Salomaa 1981). Salomaa and Stenholm (1976b) suggested observing two-photon resonances on an atomic beam, which has been carried out successfully by Poulsen and Winstrup (1981).

Many aspects of three-level spectroscopy and its use in investigating collisional properties of atoms were discussed by Berman at the 1982 Les Houches summer school. The proceedings are due to appear in the near future.

The disappearence of the incoherent contribution from the ground state is explained by strict energy conservation in Berman (1978), Sec. 4.1. As the first step takes the atom from a sharp state by absorbing one quantum of radiation energy, the second can only bridge the difference between this energy and the topmost level. The width of the intermediate state plays no role. The line-inversion phenomenon is discussed and explained in Dixit et al. (1980).

The AC Stark shift was discussed by Autler and Townes (1955), and the light shift was observed first in optical pumping, see Cohen-Tannoudji (1968).

## 4.4. COHERENCE PHENOMENA AND LEVEL-CROSSING SIGNALS

### 4.4.a. Background

The level-crossing phenomenon can be described in several ways. One standard way is to consider an initially excited superposition state

$$|\psi(t)\rangle = C_1 \exp(-i\omega_1 t)|\,1\rangle + C_2 \exp(-i\omega_2 t)|\,2\rangle \qquad (4.73)$$

which is coupled to a lower level $|0\rangle$ by an operator $\hat{V}$, as perhaps in spontaneous emission. The transition rate will be given by

$$W(t) = |\langle 0|V|\psi(t)\rangle|^2 = \text{const.} + 2\,\text{Re}\, C_2 C_1^* e^{i\omega_{12}t} V_{10} V_{02}. \qquad (4.74)$$

The quantum beat at frequency $\omega_{12} = \omega_1 - \omega_2$ gives for a steady state measurement the Fourier-transformed line shape

$$L(\omega_{12}) = \frac{\gamma^2}{\omega_{12}^2 + \gamma^2}, \qquad (4.75)$$

where $\gamma$ is the common decay rate of the states $|1\rangle$ and $|2\rangle$. When the levels cross, $\omega_{12} = 0$, and a maximum is observed in the signal.

An alternative way to treat the level crossing is to use scattering theory. Here we explicitly introduce the excitation process and consider the level configuration of Fig. 4.9a. Population transfer from level $|0\rangle$ back to level $|0\rangle$ can take place through the two distinct channels of Fig. 4.9b. If the excited states again decay at the rate $\gamma$, the theory of inelastic scattering gives for the scattering amplitude

$$A_1 \propto \frac{1}{\Delta\omega_{10} - i\gamma} \tag{4.76}$$

and similarly for $A_2$. The observed signal will be proportional to

$$W \propto |A_1 + A_2|^2, \tag{4.77}$$

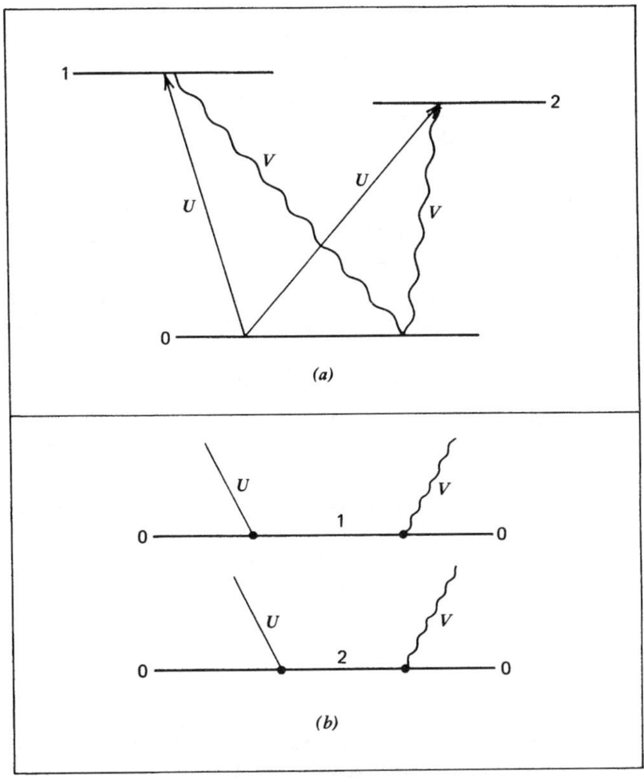


**Fig. 4.9** ($a$) An example showing how two channels of scattering can interfere when the state $|0\rangle$ is repopulated through the two intermediate levels $|1\rangle$ and $|2\rangle$. ($b$) Perturbation graphs giving rise to the interfering amplitudes $A_1$ and $A_2$ for the scattering.

which contains the interference term

$$W_{12} \simeq 2\,\mathrm{Re}\frac{1}{(\Delta\omega_{10} - i\gamma)(\Delta\omega_{20} + i\gamma)}. \tag{4.78}$$

When the levels are excited by the optical frequency $\Omega$ we have for the energy denominators

$$\Delta\omega_{i0} = \omega_i - \omega_0 - \Omega \qquad (i = 1,2). \tag{4.79}$$

If we have a broad band source, we must average the observable (4.78) over the frequency distribution of the source $D(\Omega)$ giving

$$\langle W_{12}\rangle = 2\,\mathrm{Re}\int D(\Omega)\frac{1}{(\omega_{10} - \Omega - i\gamma)(\omega_{20} - \Omega + i\gamma)}d\Omega$$

$$\simeq -4\pi\frac{2\gamma}{\omega_{12}^2 + 4\gamma^2}, \tag{4.80}$$

which again shows a Lorentzian line shape at the crossing point $\omega_{12} = 0$. The derivation is clearly different and rests on experimental assumptions other than that leading to Eq. (4.75).

The result (4.75) assumes that the superposition of states $|1\rangle$ and $|2\rangle$ can survive during the full lifetime $\gamma^{-1}$ of the excited levels. If the phases in (4.73) are mixed faster than $\gamma$, no signal will result. The second method explicitly needs the broad exciting radiation distributed according to $D(\Omega)$. Only after an average over this do we find a dependence on the difference $\omega_{21}$. At no point does it seem necessary to assume a linear superposition of the levels $|1\rangle$ and $|2\rangle$.

In this section we attempt a detailed discussion of the conditions when level crossings can be observed. We discuss the effects of the presence or absence of different types of coherence. The cases are illustrated with examples realizable with a laser source.

### 4.4.b. General Considerations

We consider a three-level system with the lowest level $|0\rangle$ coupled to both excited levels by the perturbation (see Fig. 4.10)

$$\hat{V} = 2\hbar[\;|1\rangle V_1\langle 0| + |2\rangle V_2\langle 0| \;+ \text{h.c.}]\cos\Omega t. \tag{4.81}$$

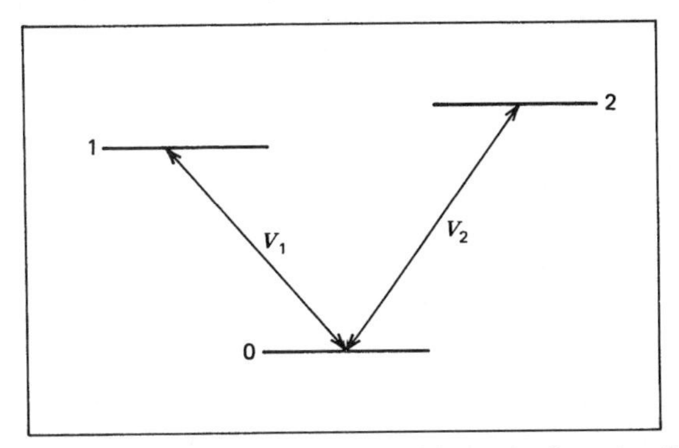


**Fig. 4.10** The three-level configuration to be treated in our level-crossing discussion. Each level $|1\rangle$, $|2\rangle$ is coupled to the ground state $|0\rangle$ through its own matrix element $V_1$ and $V_2$, respectively.

We assume only level $|0\rangle$ to be populated in equilibrium, and, after using the rotating wave approximation, we obtain the steady state equations

$$i\gamma_0\rho_{00} = i\lambda + V_1(\rho_{10} - \rho_{01}) + V_2(\rho_{20} - \rho_{02}) \tag{4.82a}$$

$$(i\gamma_{10} - \omega_{10})\rho_{10} = V_1(\rho_{00} - \rho_{11}) - V_2\rho_{12} \tag{4.82b}$$

$$(i\gamma_{20} - \omega_{20})\rho_{20} = V_2(\rho_{00} - \rho_{22}) - V_1\rho_{21} \tag{4.82c}$$

$$(i\gamma_{12} - \omega_{12})\rho_{12} = V_1\rho_{02} - V_2\rho_{10} \tag{4.82d}$$

$$i\gamma_1\rho_{11} = V_1(\rho_{01} - \rho_{10}) \tag{4.82e}$$

$$i\gamma_2\rho_{22} = V_2(\rho_{02} - \rho_{20}), \tag{4.82f}$$

where we have defined the rotating wave density matrix elements

$$\rho_{i0} = e^{i\Omega t}\langle C_i C_0^*\rangle \tag{4.83}$$

and

$$\omega_{i0} = \frac{1}{\hbar}(E_i - E_0) - \Omega \tag{4.84}$$

for $i = 1$ and 2. For later reference we illustrate the way in which the density matrix elements (4.82) are coupled by the diagram in Fig. 4.11.

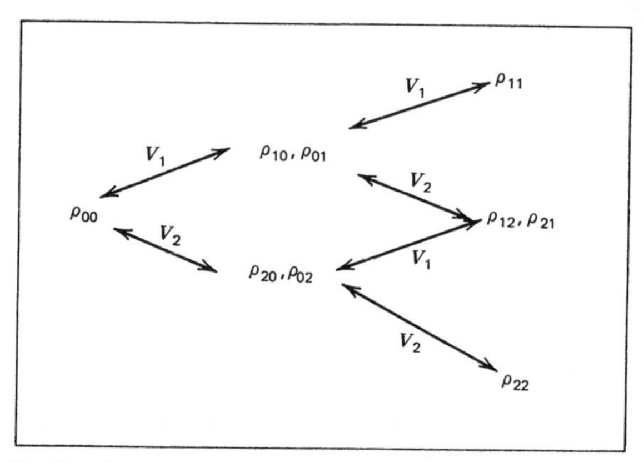


**Fig. 4.11** Coupling diagram showing all second-order processes starting from $\rho_{00}$.

If we assume the observable to be the total absorption of the transmitted radiation, it is proportional to

$$W = V_1 \text{Im}\, \rho_{10} + V_2 \text{Im}\, \rho_{20}. \tag{4.85}$$

We follow the calculation of Im $\rho_{10}$ in detail; the quantity Im $\rho_{20}$ is obtained in an analogous way.

From (4.82b) we find the result

$$\rho_{10} = \frac{1}{i\gamma_{10} - \omega_{10}} \left[ V_1(\rho_{00} - \rho_{11}) - V_2 \rho_{12} \right]. \tag{4.86}$$

When the fields are so weak that the ground state remains undepleted we insert only the equilibrium value $\rho_{00} = (\lambda/\gamma_0)$ into (4.86) and obtain the linear absorption term

$$W_0 = -V_1^2 \frac{\gamma_{10}}{\gamma_{10}^2 + \omega_{10}^2} \left( \frac{\lambda}{\gamma_0} \right). \tag{4.87}$$

Higher-order effects give corrections and modifications of this. From Eq. (4.86) we see that these are due to two different field-induced effects. The first term is given by the modified population difference $(\rho_{00} - \rho_{11})$ and the second by the field-induced coherence $\rho_{12}$ in the upper states.

From Fig. 4.11 we see that to reach $\rho_{11}$ the atom must use two interactions $V_1$, giving an absorption proportional to at least $V_1^4$. But the depletion of the ground state $\rho_{00}$ enters to the same order, and then we can lose population both to level 1 and level 2, giving terms proportional to $V_1^2$ and $V_2^2$, respectively. Thus the population-induced part of the dipole moment contains both a straightforward absorption proportional to at least $V_1^4$ and cross-terms proportional to at least $V_1^2V_2^2$. All energy denominators are, however, of the form $[\omega_{i0} - i\gamma_{i0}]^{-1}$, and $\omega_{12}$ does not appear.

The coherence-induced term $V_2\rho_{12}$ can be obtained from the ground state in two ways. From Fig. 4.11 we can see that both give a cross-term proportional to $V_1V_2$, and this gives in absorption at least the factor $V_1^2V_2^2$. The level difference $\omega_{12}$ occurs in the denominator, but so do also the dipole transition denominators. A more detailed form for this coherence term is presented in the next section.

In addition to the decay rates of the populations, we have the coherence decay parameters $\gamma_{10}$, $\gamma_{20}$, $\gamma_{12}$. These must satisfy the following conditions:

$$\gamma_{i0} \geq \tfrac{1}{2}(\gamma_i + \gamma_0) \qquad (i = 1,2) \tag{4.88}$$

$$\gamma_{12} \geq \tfrac{1}{2}(\gamma_1 + \gamma_2). \tag{4.89}$$

They are sensitive to phase perturbations and describe the loss of coherence due to randomizing mechanisms. If one $\gamma_{ij}$ grows to infinity, it implies that perturbations do not allow the growth of the corresponding coherence $\rho_{ij}$.

If we assume that the ground state population dominates the right-hand side of (4.86) and (4.82c), we find that we can directly solve for $\rho_{12}$, $\rho_{11}$, and $\rho_{22}$ from (4.82d–f) without any further approximations. Thus the populations and coherence elements in the excited states $\{1,2\}$ can be discussed directly with this assumption. Only when we need to evaluate their effect on the observable do we have to consider the correction terms to Eq. (4.86).

It turns out that the scheme just described can be carried out straightforwardly in perturbation theory, and we proceed to do this in the following section.

### 4.4.c. The Perturbation Calculation

To be able to carry out the program discussed in the preceding section we expand the density matrix in the form

$$\rho = \sum_{n=0}^{\infty} \rho^{(n)}, \tag{4.90}$$

where $\rho^{(n)}$ is proportional to the couplings $V_i$ to the power $n$. The first solution is

$$\rho_{00}^{(0)} = \frac{\lambda}{\gamma_0}, \tag{4.91}$$

and inserting this we obtain first

$$\rho_{i0}^{(1)} = \frac{V_i}{i\gamma_{i0} - \omega_{i0}}\left(\frac{\lambda}{\gamma_0}\right), \tag{4.92}$$

giving the linear absorption (4.87). Assuming (4.92), we obtain directly the results

$$\rho_{11}^{(2)} = \left(\frac{2V_1^2}{\gamma_1}\right)\frac{\gamma_{10}}{\omega_{10}^2 + \gamma_{10}^2}\left(\frac{\lambda}{\gamma_0}\right) \tag{4.93}$$

$$\rho_{22}^{(2)} = \left(\frac{2V_2^2}{\gamma_2}\right)\frac{\gamma_{20}}{\omega_{20}^2 + \gamma_{20}^2}\left(\frac{\lambda}{\gamma_0}\right) \tag{4.94}$$

$$\rho_{12}^{(2)} = \left(\frac{V_1V_2}{\omega_{12} - i\gamma_{12}}\right)\left[\frac{1}{\omega_{20} + i\gamma_{20}} - \frac{1}{\omega_{10} - i\gamma_{10}}\right]\left(\frac{\lambda}{\gamma_0}\right). \tag{4.95}$$

Here we see explicitly that the population transfer to $\rho_{ii}^{(2)}$ is a simple rate process, whereas the coherence (4.95) is more complicated. Incidentally, Eq. (4.93) shows that a slow decay to a totally passive level $|3\rangle$ from $|1\rangle$ will monitor the same quantity as the linear absorption (4.87). Owing to the coherence term (4.95), this does not hold true for decay back to level $|0\rangle$. In addition to (4.93–4.95), we have the depletion of the ground state

$$\rho_{00}^{(2)} = -\frac{2\lambda}{\gamma_0^2}\left[V_1^2\frac{\gamma_{10}}{\omega_{10}^2 + \gamma_{10}^2} + V_2^2\frac{\gamma_{20}}{\omega_{20}^2 + \gamma_{20}^2}\right], \tag{4.96}$$

which explicitly consists of the contributions lost to the levels 1 and 2 in (4.93–4.94).

In the final step we calculate the observable in the form

$$\rho_{10}^{(3)} = -\frac{1}{\omega_{10} - i\gamma_{10}}\left[V_1\left(\rho_{00}^{(2)} - \rho_{11}^{(2)}\right) - V_2\rho_{12}^{(2)}\right], \tag{4.97}$$

where the population and coherence-induced parts are still retained. Combining the results (4.93) and (4.96) we obtain

$$\rho_{00}^{(2)} - \rho_{11}^{(2)} = -2V_1^2 \frac{\gamma_{10}}{\omega_{10}^2 + \gamma_{10}^2}\left[\frac{1}{\gamma_0} + \frac{1}{\gamma_1}\right]\left(\frac{\lambda}{\gamma_0}\right) - \frac{2V_2^2}{\gamma_0}\frac{\gamma_{20}}{\omega_{20}^2 + \gamma_{20}^2}\left(\frac{\lambda}{\gamma_0}\right). \tag{4.98}$$

Inserting (4.98) and (4.95) into (4.97) and calculating the averaged observable, we obtain for the nonlinear part of the response

$$\overline{W} = V_1 \mathrm{Im}\langle \rho_{10} \rangle = [S_1 + S_2 + S_3 + S_4]\left(\frac{\lambda}{\gamma_0}\right) \tag{4.99}$$

where

$$S_1 = \frac{2V_1^4}{\gamma_1}\left\langle \frac{\gamma_{10}^2}{\left(\omega_{10}^2 + \gamma_{10}^2\right)^2} \right\rangle \tag{4.100a}$$

$$S_2 = \frac{2V_1^4}{\gamma_0}\left\langle \frac{\gamma_{10}^2}{\left(\omega_{10}^2 + \gamma_{10}^2\right)^2} \right\rangle \tag{4.100b}$$

$$S_3 = \frac{2V_1^2V_2^2}{\gamma_0}\left\langle \left(\frac{\gamma_{10}}{\left(\omega_{10}^2 + \gamma_{10}^2\right)}\right)\left(\frac{\gamma_{20}}{\left(\omega_{20}^2 + \gamma_{20}^2\right)}\right) \right\rangle \tag{4.100c}$$

$$S_4 = V_1^2V_2^2 \mathrm{Im}\left\{ \frac{1}{\omega_{12} - i\gamma_{12}} \left\langle \frac{1}{\omega_{10} - i\gamma_{10}}\left[\frac{1}{\omega_{20} + i\gamma_{20}} - \frac{1}{\omega_{10} - i\gamma_{10}}\right] \right\rangle \right\}. \tag{4.100d}$$

Here we have assumed that the denominator $(\omega_{12} - i\gamma_{12})$ can be taken outside the average (see below).

The various terms $S_1$–$S_4$ are easily interpreted as definite paths in the coupling scheme of Fig. 4.11. We have to start at $\rho_{00}$ (where we pump) and end up in the observable $\rho_{10}$ using three steps. Figure 4.12 shows how to get the terms in (4.99); the details are easily checked through following the iteration process. From Fig. 4.11 it is easy to see that Fig. 4.12 exhausts all the possibilities available.

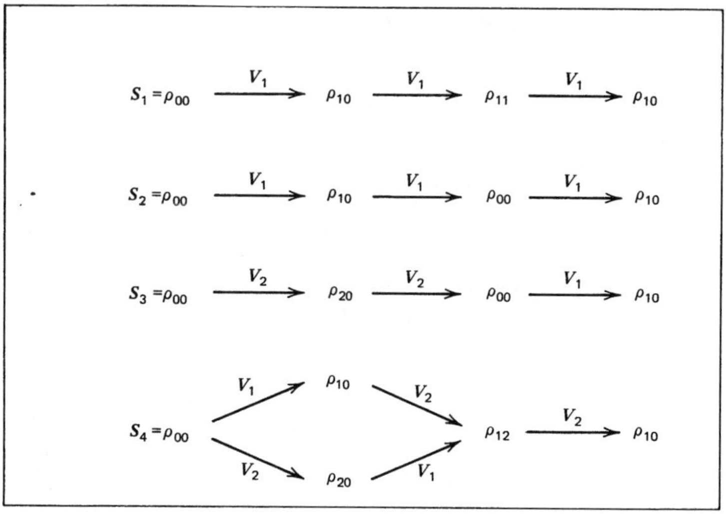

$$S_1 = \rho_{00} \xrightarrow{V_1} \rho_{10} \xrightarrow{V_1} \rho_{11} \xrightarrow{V_1} \rho_{10}$$

$$S_2 = \rho_{00} \xrightarrow{V_1} \rho_{10} \xrightarrow{V_1} \rho_{00} \xrightarrow{V_1} \rho_{10}$$

$$S_3 = \rho_{00} \xrightarrow{V_2} \rho_{20} \xrightarrow{V_2} \rho_{00} \xrightarrow{V_1} \rho_{10}$$

$$S_4 = \rho_{00} \begin{cases} \xrightarrow{V_1} \rho_{10} \xrightarrow{V_2} \\ \xrightarrow{V_2} \rho_{20} \xrightarrow{V_1} \end{cases} \rho_{12} \xrightarrow{V_2} \rho_{10}$$

Fig. 4.12 Illustration of the perturbation processes leading to $\rho_{10}$. The terms $S$ are obtained from these by multiplication with $V_1$.

We can see that $S_1$ is induced by the population $\rho_{11}$, $S_2$ and $S_3$ by the depletion of population $\rho_{00}$, and finally $S_4$ is induced by the coherence $\rho_{12}$.

The terms $S_1$ and $S_2$ are easy to interpret physically. Combining them with the result of Eq. (4.87) we have

$$\begin{aligned} \overline{W} &= S_0 + S_1 + S_2 \\ &= -V_1^2 \left\langle \frac{\gamma_{10}}{\omega_{10}^2 + \gamma_{10}^2} \left[ 1 - 2\left( \frac{1}{\gamma_1} + \frac{1}{\gamma_0} \right) \frac{V_1^2 \gamma_{10}}{\omega_{10}^2 + \gamma_{10}^2} \right] \right\rangle . \end{aligned} \tag{4.101}$$

The correction term is obviously due to a series expansion of

$$\begin{aligned} &\frac{\gamma_{10}}{\omega_{10}^2 + \gamma_{10}^2 + 2[(1/\gamma_1) + (1/\gamma_0)] V_1^2 \gamma_{10}} \\ &= \frac{\gamma_{10}}{\omega_{10}^2 + \gamma_{10}^2} \left[ \frac{1}{1 + 2\left( \frac{1}{\gamma_1} + \frac{1}{\gamma_0} \right) V_1^2 \frac{\gamma_{10}}{\omega_{10}^2 + \gamma_{10}^2}} \right] \end{aligned} \tag{4.102}$$

which corresponds to simple power broadening with the dimensionless saturation parameter

$$I_1 = \frac{2(\gamma_1 + \gamma_0)}{\gamma_1 \gamma_0 \gamma_{10}} V_1^2, \tag{4.103}$$

as is well known from our earlier cases. Thus the absorption caused by $S_1$ and $S_2$ only add to the constant part after the averaging process, and all physically interesting features are found in the two terms $S_3$ and $S_4$. In the following section we discuss how various physically realizable cases of coherence affect the shape and size of the observed signal.

#### 4.4.d. Some Cases of Physical Interest

Let us consider several special limits of the level-crossing situation described before. These do not exhaust all possibilities but display most phenomena likely to occur in physical situations.

I. Monochromatic Light, No Excited State Coherence

This corresponds to the case when we have only one frequency $\Omega$ and the excited state coherence element $\rho_{12}$ is never allowed to develop. This can be due to a strong perturbation randomizing the phase of $\rho_{12}$. Mathematically this limit occurs when

$$\gamma_{12} \gg \gamma_{10}, \gamma_{20}, \tag{4.104}$$

which by (4.100d) implies that $S_4$ is negligible. In addition to the power broadened resonance (4.102), we have the term $S_3$ in (4.100c), which, when no average is involved, is the product of two Lorentzians situated at the points $\omega_{10} = 0$ and $\omega_{20} = 0$. When $\omega_{10} = \omega_{20}$ they just fall at the top of each other, but no special level-crossing signal can be observed in this case.

II. Nonresonant Pumping, Excited State Coherence

In this case we assume that either the pumping light detuning or the dipole decay rates dominate the width of the excited state coherence,

$$\omega_{10}, \omega_{20} \gg \gamma_{12} \tag{4.105}$$

or

$$\gamma_{10}, \gamma_{20} \gg \gamma_{12}. \tag{4.106}$$

In each situation we expect a resonant behavior of $S_4$ when $\omega_{12} = 0$. In the former situation (4.105) we have nonresonant pumping, and the contribution $S_4$ gives a level-crossing signal in Raman scattering due to upper state coherence.

If we assume the detunings to totally dominate the line widths, we obtain from (4.100d)

$$S_4 = \frac{V_1^2 V_2^2}{\omega_{10}^2 \omega_{20}} \gamma_{12} \frac{\omega_{12}}{\omega_{12}^2 + \gamma_{12}^2}. \tag{4.107}$$

The signal, thus, is dispersion shaped and contains the large detunings $\omega_{10}, \omega_{20}$ in the denominators. This implies that the resulting signal is quite weak and requires extremely large fields.

When the observed system is perturbed by polarizing atomic encounters we have dipole selection rules for the collisional interaction, and $\rho_{10}, \rho_{20}$ are destroyed much more readily than the quadrupole coherence $\rho_{12}$. Then the situation (4.106) prevails, and we assume that we can tune the monochromatic light into near resonance. The signal $S_3$ gives a constant, but (4.100d) becomes

$$S_4 = \frac{V_1^2 V_2^2}{\gamma_{10}^2 \gamma_{20}} \left( \frac{\gamma_{12}}{\omega_{12}^2 + \gamma_{12}^2} \right) (\gamma_{10} + \gamma_{20}). \tag{4.108}$$

In this limit we observe a level-crossing Lorentzian of width $\gamma_{12}$.

In both cases (4.105–4.106) the phases of the dipole moments $\rho_{10}, \rho_{20}$ are rapidly averaged toward zero. This forces the levels $|1\rangle, |2\rangle$ to act as transient intermediate levels only, without any real population occurring. In (4.105) this is due to the rapid flipping of the phase by the detunings, and in (4.106) it derives from the external phase perturbations. It should be noted that an average over an inhomogeneous distribution of $\omega_{10}, \omega_{20}$ does not affect the results (4.107–4.108) as long as the range of the averaging is less than $|\omega_{10}|, |\omega_{20}|$ or $\gamma_{10}, \gamma_{20}$ respectively.

Next, let us consider the case of a broad source with the frequency distribution $D(\Omega)$. The power broadened Lorentzian (4.102) gives a constant absorption, whereas from (4.100c) we find

$$S_3 = \frac{2V_1^2 V_2^2}{\gamma_0} \int D(\Omega) \frac{\gamma_{10}\gamma_{20}}{\left((\omega_{10} - \Omega)^2 + \gamma_{10}^2\right)\left((\omega_{20} - \Omega)^2 + \gamma_{20}^2\right)} d\Omega$$

$$= 2\pi \frac{V_1^2 V_2^2}{\gamma_0} D(\bar{\omega}) \frac{(\gamma_{10} + \gamma_{20})}{\left(\omega_{12}^2 + (\gamma_{10} + \gamma_{20})^2\right)}, \tag{4.109}$$

where we have taken out the spectral density $D$ at an average frequency $\bar{\omega}$ and calculated the integral by the residue method.

From (4.100d) we find the result

$$S_4 = V_1^2 V_2^2 \operatorname{Im}\left\{ \frac{1}{\omega_{12} - i\gamma_{12}} \int D(\Omega) \frac{1}{(\omega_{10} - \Omega) - i\gamma_{10}} \right.$$

$$\left. \times \left[ \frac{1}{(\omega_{20} - \Omega + i\gamma_{20})} - \frac{1}{(\omega_{10} - \Omega - i\gamma_{10})} \right] d\Omega \right\}$$

$$= -2\pi V_1^2 V_2^2 D(\bar{\omega}) \operatorname{Im}\left\{ \frac{i}{(\omega_{12} - i\gamma_{12})[\omega_{12} - i(\gamma_{10} + \gamma_{20})]} \right\}$$

$$= -2\pi \frac{V_1^2 V_2^2}{(\gamma_{10} + \gamma_{20})} D(\bar{\omega}) \frac{(\gamma_{10} + \gamma_{20})}{\left[\omega_{12}^2 + (\gamma_{10} + \gamma_{20})^2\right]}$$

$$\times \left[ 1 - \frac{\gamma_{12}(\gamma_{10} + \gamma_{20} + \gamma_{12})}{\omega_{12}^2 + \gamma_{12}^2} \right]. \tag{4.110}$$

We can now consider the cases in the following sections.

## III. Broad Band Excitation, No Excited State Coherence

This case corresponds to the inequality (4.104) applied to the results (4.109) and (4.110). Near resonance we find the proportionality

$$S_3 \propto \frac{1}{\gamma_0(\gamma_{10} + \gamma_{20})} \tag{4.111}$$

and

$$S_4 \propto \frac{1}{\gamma_{12}(\gamma_{10} + \gamma_{20})}. \tag{4.112}$$

If the lower level decays very slowly

$$\gamma_0 \ll \gamma_{12}, \tag{4.113}$$

the latter signal is much weaker. Thus the signal $S_3$ dominates in this regime and a simple Lorentzian of width $(\gamma_{10} + \gamma_{20})$ is found.

The result derived here resembles the simple case (4.80) of the introduction. Both require an average over the spectrum of the source and give a line width of the order of $2\gamma$. To obtain a pure $S_3$ signal we must, however, require a large $\gamma_{12}$ giving from (4.110)

$$\left[1-\frac{\gamma_{12}(\gamma_{10}+\gamma_{20}+\gamma_{12})}{\omega_{12}^2+\gamma_{12}^2}\right]_{\gamma_{12}\to\infty}\to 0 \tag{4.114}$$

This implies that no coherence $\rho_{12}$ is allowed to build up, that is, the particle is never allowed to exist in a linear superposition of the states $|1\rangle$ and $|2\rangle$. That $\rho_{12}$ is necessary for the appearance of $S_4$ can be seen from Fig. 4.12.

## IV. Broad Band Excitation, Persistent Excited State Coherence

We assume the coherence $\rho_{12}$ to decay more slowly than the elements $\rho_{10}$, $\rho_{20}$, that is, the limit (4.106). In Case II we argued that such a situation occurs naturally in many cases of physical interest. If we consider the behavior in the region

$$|\omega_{12}| \ll (\gamma_{10}+\gamma_{20}), \tag{4.115}$$

all other terms of the signal become constant except the second term of (4.110) giving

$$S_4^1 = 2\pi\frac{V_1^2V_2^2}{\gamma_{12}(\gamma_{10}+\gamma_{20})}D(\bar{\omega})\left(1+\frac{\gamma_{12}}{\gamma_{10}+\gamma_{20}}\right)\frac{\gamma_{12}^2}{\omega_{12}^2+\gamma_{12}^2}. \tag{4.116}$$

This is a simple Lorentzian level crossing of width $\gamma_{12}$. Its strength at resonance is such that the approximate ratio of (4.116) and the resonance in Case II given by (4.109) is

$$\frac{S_4^1}{S_3} = \frac{\gamma_0}{\gamma_{12}}. \tag{4.117}$$

The magnitude of this ratio depends on the lower-level decay rate $\gamma_0$, which enters none of our resonance functions except in the combination $(\gamma_{10}+\gamma_{20})$.

In general, of course, we observe both signals $S_3$ and $S_4$. We combine them to obtain

$$S_3+S_4 = \frac{2\pi V_1^2V_2^2D(\bar{\omega})}{\gamma_0(\gamma_{10}+\gamma_{20})}\frac{(\gamma_{10}+\gamma_{20})^2}{\left[\omega_{12}^2+(\gamma_{10}+\gamma_{20})^2\right]}$$
$$\times\left[\left(1-\frac{\gamma_0}{\gamma_{10}+\gamma_{20}}\right)+\frac{\gamma_0}{\gamma_{12}}\left(1+\frac{\gamma_{12}}{\gamma_{10}+\gamma_{20}}\right)\frac{\gamma_{12}^2}{\omega_{12}^2+\gamma_{12}^2}\right]. \tag{4.118}$$

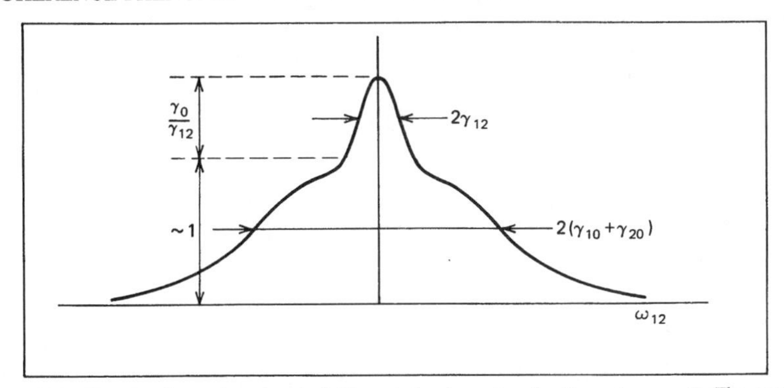


**Fig. 4.13** Two level-crossing signals falling at the top of each other at $\omega_{21} = 0$. The purely coherent one is of width $\gamma_{12}$, and the population-induced signal is of width $(\gamma_{10} + \gamma_{20})$.

Because mostly $\gamma_0, \gamma_{12} \ll \gamma_{10} + \gamma_{20}$, we see basically a broad Lorentzian of width $(\gamma_{10} + \gamma_{20})$, at the top of which we have another Lorentzian of the narrower width $\gamma_{12}$ and the strength $(\gamma_0/\gamma_{12})$ determined by (4.117). They are both level-crossing resonances, the broad one like Case III and Eq. (4.80) and the narrow one like Case IV. The general situation is illustrated in Fig. 4.13.

## V. Doppler-Broadened-Level Crossings

If we assume both transitions in Fig. 4.10 to be caused by one and the same traveling wave, we introduce instead of (4.84) the relation

$$\omega_{i0} \to \omega_{i0} - \Omega - kv \tag{4.119}$$

where $kv$ is the Doppler shift. With a broad velocity distribution (the Doppler limit) the calculations of (4.109)–(4.110) follow directly, and we can use the result (4.118). The distribution $D(\bar{\omega})$ is now the Maxwellian velocity distribution.

The fact that the cross-term $S_3$ occurs even without any coherence allowed within the $|1\rangle$, $|2\rangle$ system can be understood easily here. The terms $S_2$ and $S_3$ represent depletion of the lower state $|0\rangle$ due to the transitions $V_1$ and $V_2$ respectively; see Fig. 4.12. These create two Bennett holes in the population distribution on $|0\rangle$ when the two quantities (4.119) vanish. A subsequent transition $0 \to 1$ will suffer from saturation when it feels these holes; the hole caused by $V_1$ always being present, whereas the hole caused by $V_2$ meets the resonance condition of the transition only for

$$\omega_{10} - \Omega = kv = \omega_{20} - \Omega \tag{4.120}$$

giving the level-crossing requirement $\omega_{12} = 0$. Because of its origin in population processes this resonance feels both widths $(\gamma_{10} + \gamma_{20})$.

When the two transitions $0 \to 1$ and $0 \to 2$ are affected by counterpropagating fields the situation changes. We find the averaging over velocity to give

$$S_3 \propto \left\langle \frac{1}{\left[(\omega_{10} - \Omega - kv)^2 + \gamma_{10}^2\right]\left[(\omega_{20} - \Omega + kv)^2 + \gamma_{20}^2\right]} \right\rangle$$

$$\propto \left\{ \frac{1}{\left[\frac{1}{2}(\omega_{10} + \omega_{20}) - \Omega\right]^2 + \left[\frac{1}{2}(\gamma_{10} + \gamma_{20})\right]^2} \right\} \tag{4.121}$$

and

$$S_4 \propto \left\langle \frac{1}{[\omega_{10} - \Omega - kv - i\gamma_{10}][\omega_{20} - \Omega + kv + i\gamma_{20}]} \right\rangle$$

$$= 0. \tag{4.122}$$

In this case the only resonant signal observed is the crossover Lamb dip at

$$\Omega = \tfrac{1}{2}(\omega_{10} + \omega_{20}), \tag{4.123}$$

and no level-crossing signals occur.

### 4.4.e. Level Crossing in a Lower Level

In this section, so far, we have considered only the case where upper-state levels cross. This is the familiar situation when the final decay is spontaneous emission, but here we can assume all transitions to be induced. In that case the situation becomes quite symmetric, and it is possible to consider a level-crossing also in the lower state (the inverted V-configuration).

The treatment of this situation differs slightly because we must assume that levels $|1\rangle$ and $|2\rangle$ are populated initially. The same interconnection scheme as in Fig. 4.11 is valid; we have only to read it slightly differently. If the observable is again taken to be given by $\rho_{10}$, we obtain the scheme of Fig. 4.14 to be used in a perturbation approach. From the figure we can see that all the familiar processes occur again. We have the pure saturation processes proportional to $V_1^3$ and the interfering processes proportional to

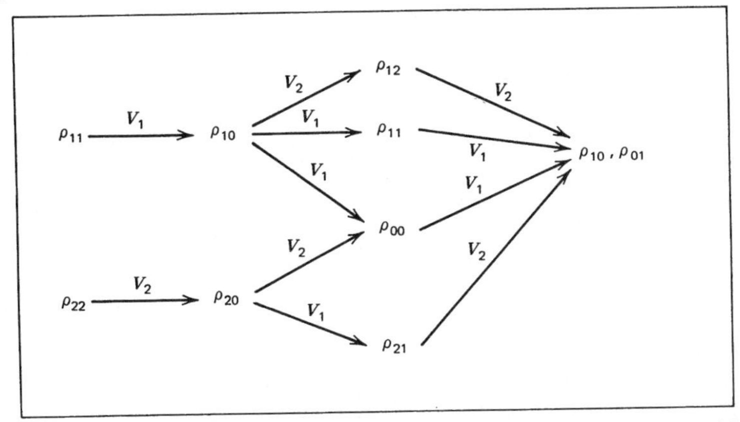


**Fig. 4.14** The processes involved when the two crossing levels $|1\rangle$ and $|2\rangle$ are situated below the common level $|0\rangle$. There are no level crossings in spontaneous emission. Induced level crossings occur in this $\Lambda$-configuration just as readily as in the V-configuration.

$V_1V_2^2$. Of these, two go through the coherence $\rho_{21}$ and show a direct level-crossing resonance near $\omega_{12} \simeq 0$, and one depends on the population accumulation on $\rho_{00}$ due to the level $|2\rangle$. The latter path passes both $\rho_{20}$ and $\rho_{10}$ and gives a product of two Lorentzians like $S_3$ in (4.100c). After averaging, this also gives a level-crossing behavior but with the width $(\gamma_{10} + \gamma_{20})$; see (4.109). It is, consequently, possible to identify all the processes of the V-configuration also in the inverted V case. This verifies that for induced processes both upper- and lower-state level crossings display the same behavior. The details of the calculation are identical with those already carried out and need not be given again.

In this case both the crossing levels can belong to the ground state, and their life times may be very long. Consequently the level-crossing signals can be extremely narrow, often of the order of one hertz.

### 4.4.f. Comments and References

The level-crossing effect was discovered by Hanle (1924), and it found wide applications in atomic physics; see Corney (1977) Chapter 15. The scattering theory approach is found in Breit (1933). In laser physics it was also seen as mode crossings in multimode measurements (Feld 1973). Many of the aspects discussed in this part of our work are also described in the article by Decomps et al. (1976). The level-crossing phenomenon and the related quantum beat phenomenon are reviewed by Haroche in Shimoda (1976). See also in this connection the article by Sargent (1978).

## 4.5. MULTIPHOTON PROCESSES

### 4.5.a. Background Information

In molecular systems the levels form a nearly dense set. The vibrational energies ideally give a harmonic oscillator ladder, but in real molecules anharmonicity plays an important role. For molecules with more than a few atoms, the level density rapidly becomes so dense that the individual levels merge and a quasi-continuum arises. In molecules, thus, a laser field couples many levels at resonance or very nearly at resonance.

The effect of coherent laser light on molecular systems has become an important issue because of its importance in isotope separation and laser chemistry research. The real situation is very complicated, and we can only discuss a highly simplified model. Within this we can, however, take up some questions of general interest and try to look into some of the physical features.

The model consists of a series of $N + 1$ nearly equally spaced levels from 0 to $N$, see Fig. 4.15. The levels are coupled by a laser field of frequency $\Omega$, however, in such a way that each level is coupled only to those immediately above and below itself. We can understand this so that only the levels near resonance are included as in the atomic case. The Hamiltonian includes no spatial effects and is given by

$$H = \hbar \sum_{n=0}^{N} |n\rangle \varepsilon_n \langle n| - \mu E \sum_{n=0}^{N-1} \cos \Omega t [|n\rangle\langle n+1| + |n+1\rangle\langle n|], \quad (4.124)$$

and we assume $\varepsilon_0 = 0$. Our model allows us to reach the level $n$ only by the absorption of $n$ photons from the field. It is then possible to perform a rotating wave approximation as before. In a model where no unique process can be utilized to reach level $n$, no unique rotating wave approximation is possible.

In this case we choose to perform the approximation on the amplitudes and not on the density matrix. We write the state

$$|\psi\rangle = \sum_{n=0}^{N} e^{-in\Omega t} C_n |n\rangle, \quad (4.125)$$

and the energy differences

$$\varepsilon_n - n\Omega = \Delta\omega_n \quad (4.126)$$

are the small detunings at the $n$th step of the ladder. Inserting (4.125) into the Schrödinger equation

$$i\hbar\frac{\partial}{\partial t}|\psi\rangle = H|\psi\rangle \tag{4.127}$$

and using the orthogonality of the states $|n\rangle$, we find with $\alpha = \mu E/2\hbar$

$$i\dot{C}_n = \Delta\omega_n C_n - \alpha(C_{n+1} + C_{n-1}). \tag{4.128}$$

Terms with oscillations different from $e^{-in\Omega t}$ are discarded in this result. This is the basic equation to start a discussion of the properties of multiphoton transitions in $N$-level systems. In this section its properties are discussed

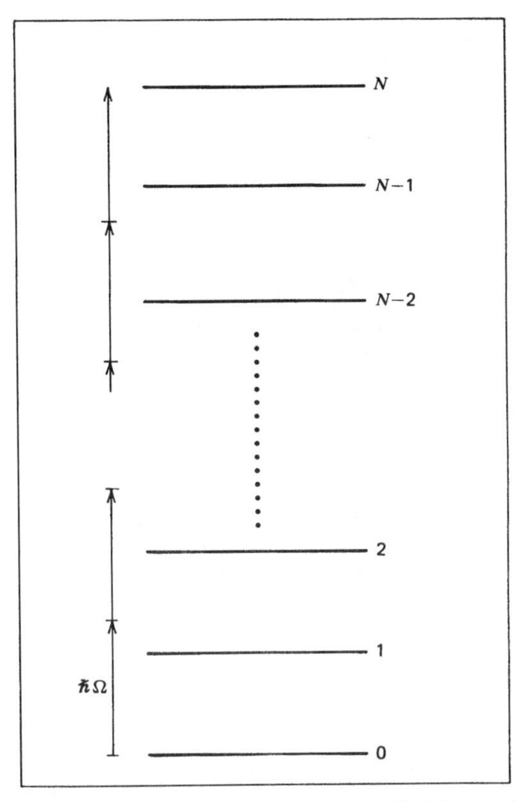


**Fig. 4.15** A multilevel system where $N$ photons of frequency $\Omega$ add up to give resonance with the $N$th level but all intermediate steps are off resonance.

without the use of a density matrix. Because $N$ can be any number ($> 1$) our results will also hold for two- and three-level systems. The density matrix treatment does, however, allow a much more detailed description of the relaxation processes. Dephasing of coherences and spontaneous decay especially, cannot be pictured without the use of a density matrix.

Equation (4.128) is of great interest in quantum electronics. It has been found to describe many systems; in addition to the multilevel case considered here, it appears in the theory of photon momentum effects and the theory of the free electron laser; see the discussion in Sec. 4.5.e.

The level anharmonicity is hidden in the detuning $\Delta\omega_n$. Its presence makes it difficult to treat time-dependent problems. These are, however, of great interest because the physical situations often involve large laser powers, and consequently the light intensity is available for a limited time only.

### 4.5.b. The Steady State Situation

Let us assume that each level in our ladder except the lowest one can decay exponentially into some unobserved states forming a quasicontinuum. Without the field coupling the levels each one would then decay exponentially like

$$|C_n(t)|^2 = e^{-\gamma_n t}|C_n(0)|^2. \tag{4.129}$$

This corresponds closely to the situation in a molecule where each optically active level is imbedded in a dense spectrum of other states causing nearly exponential nonradiative decay.

The ground state does not decay, $\gamma_0 = 0$, and the topmost level $|N\rangle$ is taken to decay mainly due to ionization. Then the rate

$$\left|\frac{d}{dt}|C_N|^2\right| = \gamma_N |C_N|^2 \tag{4.130}$$

can be used as a measure of success in our process. What leaks out of the ladder on the way up is losses, what leaks out at the top is the yield.

We assume that the molecule is initially in the ground state $C_n = \delta_{n0}$, and we Laplace transform Eq. (4.128) to

$$is\tilde{C}_n(s) - i\delta_{n0} = \left(\Delta\omega_n - \tfrac{1}{2}i\gamma_n\right)\tilde{C}_n(s) - \alpha\left[\tilde{C}_{n+1}(s) + \tilde{C}_{n-1}(s)\right] \tag{4.131}$$

which gives

$$\tilde{C}_n(s) = \frac{1}{s}\delta_{n0} + \alpha D(n)\left[\tilde{C}_{n+1}(s) + \tilde{C}_{n-1}(s)\right], \tag{4.132}$$

when $\Delta\omega_0 - i\gamma_0 = 0$ and

$$D(n) \equiv \frac{i}{s + \frac{1}{2}\gamma_n + i\,\Delta\omega_n}. \tag{4.133}$$

This equation is similar to the one we met earlier, Eq. (2.78) in Sec. 2.3. The solution is obtained in exactly the same way for $n > 0$, but with the largest nonvanishing $D$-function being $D_N$. The continued fractions are hence finite. We find

$$\frac{\tilde{C}_1(s)}{\tilde{C}_0(s)} = \cfrac{\alpha D(1)}{1 - \cfrac{\alpha^2 D(1) D(2)}{1 - \cdots \quad - \cfrac{\alpha^2 D(N-2) D(N-1)}{1 - \alpha^2 D(N-1) D(N)}}}. \tag{4.134}$$

For $n = 0$ we have $\tilde{C}_{-1} = 0$ and hence

$$\tilde{C}_0(s) = \frac{i}{\left[-is + \alpha\tilde{C}_1(s)/\tilde{C}_0(s)\right]}. \tag{4.135}$$

Thus we can know all the coefficients $\tilde{C}_n(s)$ in terms of finite continued fractions. The total yield from the upper level is given by the accumulated leakage from (4.130).

We can calculate this immediately from the Laplace transform $\tilde{C}(s)$ by considering

$$\begin{aligned}\frac{1}{2\pi}\int_{-\infty}^{\infty} d\xi |\tilde{C}(i\xi)|^2 &= \frac{1}{2\pi}\int_{-\infty}^{+\infty} d\xi \int_0^{\infty} e^{-i\xi t} C(t)\, dt \int_0^{\infty} e^{i\xi t'} C^*(t')\, dt' \\ &= \int_0^{\infty} dt \int_0^{\infty} dt'\, C(t) C^*(t') \delta(t - t') \\ &= \int_0^{\infty} dt\, |C(t)|^2. \end{aligned} \tag{4.136}$$

From (4.130) we then obtain

$$W = \gamma_N \int_0^{\infty} dt\, |C_N(t)|^2 = \frac{\gamma_N}{2\pi}\int_{-\infty}^{+\infty} d\xi\, |\tilde{C}_N(i\xi)|^2. \tag{4.137}$$

This gives the total yield from level $N$ and gives us an opportunity to investigate the losses into quasi-continua along the excitation road.

It is possible to obtain the time-dependent behavior by inverting the Laplace transform, but this is computationally involved, and it is often equally simple to integrate the original equations directly on a computer. Many such calculations have, in fact, been carried out.

It is also possible to solve the problem of adding particles at a steady rate into the level zero and following the flow through the various levels. The branching out from each level denotes losses, and the yield from level $N$ is again given by the rate (4.130).

Instead of (4.128) the equations become

$$i\dot{C}_n = i\sqrt{\lambda}\,\delta_{n0} + \left(\Delta\omega_n - \tfrac{1}{2}i\gamma_n\right)C_n - \alpha\left(C_{n+1} + C_{n-1}\right). \tag{4.138}$$

The steady state solution of these equations is determined by an equation very similar to (4.134). From the solution the leakage out of level $|n\rangle$ is given by

$$W_n = \gamma_n |C_n(t = \infty)|^2. \tag{4.139}$$

This is an alternative way of investigating the branching ratios in the multilevel excitation system.

### 4.5.c. The Effective Two-Level System

Assume that all intermediate levels are well off resonance, but the $N$-photon transition is nearly at resonance

$$\varepsilon_N \simeq N\Omega. \tag{4.140}$$

In this case only the initially populated level $|0\rangle$ and the $N$-photon resonant level $|N\rangle$ are appreciably populated

$$|C_0|^2 \simeq |C_N|^2 > 0. \tag{4.141}$$

All other populations $|C_n|^2$ remain very small.

The physical situation is similar to that of a set of coupled harmonic oscillators. If one point is made to vibrate, all oscillators must follow this one at its frequency, but only those that are in resonance can accumulate an appreciable amplitude.

Each amplitudes is driven by the stronger amplitude just below it; the one above has a much smaller effect. In this case we write Eq. (4.128) in the form

$$i\dot{C}_N = \Delta\omega_N C_N - \alpha C_{N-1} \tag{4.142}$$

$$i\dot{C}_n = \Delta\omega_n C_n - \alpha C_{n-1} \qquad (0 < n < N) \tag{4.143}$$

$$i\dot{C}_0 = -\alpha C_1. \tag{4.144}$$

In all Eqs. (4.143) we can neglect the time derivative because $\Delta\omega_n$ is large and solve

$$C_n \cong \frac{\alpha}{\Delta\omega_n} C_{n-1}. \tag{4.145}$$

When these are inserted into the equations and we assume that $C_N$ and $C_0$

are much larger than the other coefficient, we find the coupled equations

$$i\dot{C}_N = \Delta\omega_N C_N - A_N C_0 \tag{4.146a}$$

$$i\dot{C}_0 = -A_N C_N, \tag{4.146b}$$

where the $N$-photon matrix element is given by

$$A_N = \frac{\alpha^N}{\prod_{n=1}^{N-1} \Delta\omega_n}. \tag{4.147}$$

This is the transition matrix element obtained from $N$th-order time-dependent perturbation theory. We see that, when our assumptions hold, the $N$-photon resonance can, by Eqs. (4.146), be treated as an effective two-level system. The correction terms neglected in going from (4.142–4.144) to (4.146) give rise to power shifts similar to those discussed in Secs. 2.3 and 4.3.e.

The matrix element (4.147) describes a coherent process, which involves no accumulation of population on any of the intermediate levels in the excitation $0 \to 1 \to 2 \to \cdots \to N$. Thus it generalizes the treatment in part 4.3.c. The matrix element found there, in Eq. (4.53), is also a special case of (4.147) when the additional factor $\alpha_2$ from (4.48) is taken into account.

It is instructive to write down the present approximation in terms of a density matrix treatment. We have

$$i\dot{\tilde{\rho}}_{10} = (\Delta\omega_{10} - i\gamma_{10})\tilde{\rho}_{10} - \alpha(\rho_{00} - \rho_{11}) \tag{4.148a}$$

$$i\dot{\tilde{\rho}}_{j,0} = (\Delta\omega_{j0} - i\gamma_{j0})\tilde{\rho}_{j,0} - \alpha\tilde{\rho}_{j-1,0} + \alpha\tilde{\rho}_{j,1} \tag{4.148b}$$

$$i\dot{\tilde{\rho}}_{N,0} = (\Delta\omega_{N0} - i\gamma_{N0})\tilde{\rho}_{N,0} - \alpha\tilde{\rho}_{N-1,0}, \tag{4.148c}$$

where we have followed the coherent path up to the most extensive coherence $\tilde{\rho}_{N0}$. The path is indicated in Fig. 4.16a. When the lowest level $\rho_{00}$ is only slightly depleted we neglect all $\rho_{ii}$ with $i \neq 0$, and, because $|\Delta\omega_{j0}| \gg |\Delta\omega_{N0}|$, we assume $\dot{\tilde{\rho}}_{j0} = 0$ for all $j \neq 0, N$, and the solution of Eqs. (4.148) follows just as before in the form

$$i\dot{\tilde{\rho}}_{N0} = (\Delta\omega_{N0} - i\gamma_{N0})\tilde{\rho}_{N,0} - A'_N\rho_{00} \tag{4.149}$$

where we have

$$A'_N = \frac{\alpha^N}{\prod_{n=1}^{N-1} (\Delta\omega_{n0} - i\gamma_{n0})}. \tag{4.150}$$

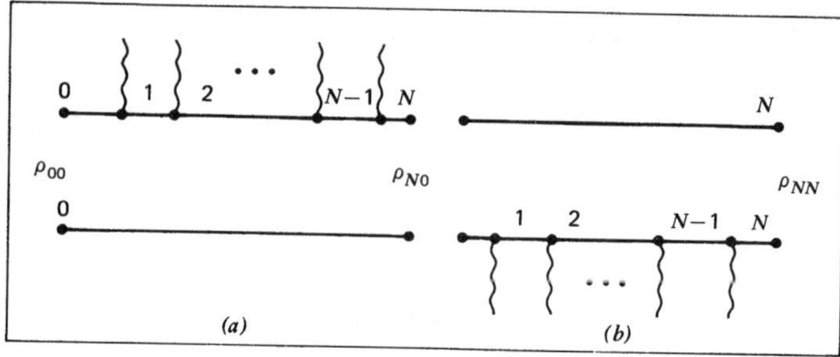


**Fig. 4.16** (*a*) Perturbation graph showing how we can absorb $N$ photons to build up the coherence $\rho_{N0}$. (*b*) We absorb $N$ more photons to obtain the excited state population $\rho_{NN}$. This is the only path contributing in ordinary time-dependent perturbation theory. It is also the most coherent path to reach $\rho_{NN}$. It corresponds to an effective two-level approximation of the states $|0\rangle$ and $|N\rangle$.

This is clearly a generalization of (4.147) to the case when each coherence decays according to its own dephasing rate $\gamma_{n0}$.

When the transition $|N\rangle \to |0\rangle$ is allowed, the coherence $\tilde{\rho}_{N0}$ acts as the polarization that drives a parametric up-conversion process where we generate radiation of the frequency $N\Omega$, which is $N$th harmonic generation from the original incident frequency $\Omega$. There is full coherence between the incident and the outgoing frequency. This type of parametric generation is one basic process in nonlinear optics.

If we want to transfer the population up to level $|N\rangle$ we must calculate the matrix element $\rho_{NN}$ using Fig. 4.16b. As before, we obtain

$$\begin{aligned}\dot{\rho}_{NN} &= i\alpha(\tilde{\rho}_{N-1,N} - \tilde{\rho}_{N,N-1}) \\ &= 2\alpha \operatorname{Im} \tilde{\rho}_{N,N-1} = 2 \operatorname{Im} B_N \rho_{00},\end{aligned} \tag{4.151}$$

where

$$\begin{aligned}&2 \operatorname{Im} B_N \\ &= 2 \operatorname{Im}\left\{\left(\frac{1}{\Delta\omega_{N0} - i\gamma_{N0}}\right)\left(\frac{\alpha^N}{\prod_{n=1}^{N-1}(\Delta\omega_{Nn} - i\gamma_{Nn})}\right)\left(\frac{\alpha^N}{\prod_{n=1}^{N-1}(\Delta\omega_{n0} - i\gamma_{n0})}\right)\right\} \\ &\approx \frac{2\gamma_{N0}}{\Delta\omega_{N0}^2 + \gamma_{N0}^2} \times \operatorname{Re}\left[\frac{\alpha^{2N}}{\prod_{n=1}^{N-1}(\Delta\omega_{Nn} - i\gamma_{Nn})(\Delta\omega_{n0} - i\gamma_{n0})}\right].\end{aligned} \tag{4.152}$$

The ignored contributions to the imaginary part are of order $(\Delta\omega_{N0}/\Delta\omega)$, where $\Delta\omega = \Delta\omega_{Nn}$ or $\Delta\omega_{n0}$. When all $\gamma$'s go to zero we find

$$2\,\mathrm{Im}\,B_N = 2\pi t_{N0}^2\delta(\Delta\omega_{N0}). \tag{4.153}$$

This is the result of $N$th-order time-dependent perturbation theory with $t_{N0}$ being the $N$th-order time-dependent perturbation theory matrix element coupling level $|0\rangle$ to level $|N\rangle$. This emerges naturally from our treatment, but the essentially new information is in the way the dephasing rates $\gamma_{ij}$ enter the result. There is a difference between the initial and final levels because all $\gamma_{Nn}$ and $\gamma_{n0}$ may be different. The real part in (4.152) cannot be written as the absolute value squared of a transition rate $T_{N0}$ between the two levels. The effects of collisions and decays can be included explicitly for each level in (4.152).

### 4.5.d. The Harmonic Oscillator

In most cases $\varepsilon_n$ is a nonlinear function of $n$. In a harmonic oscillator we have, however, an equal spacing between the levels. At resonance the excitation is then expected to be highly efficient. In the case of the harmonic oscillator we ought also to introduce the correct matrix elements $\sqrt{n}$. These give the equations of motion instead of (4.128)

$$i\dot{C}_n = \Delta n C_n - \alpha\left[\sqrt{n+1}\,C_{n+1} + \sqrt{n}\,C_{n-1}\right], \tag{4.154}$$

where the one-step detuning is given by

$$\Delta = \varepsilon - \Omega \tag{4.155}$$

when the oscillator energy levels are given by $\varepsilon n$.

To be able to solve Eq. (4.154) we introduce the new variable

$$x(n) = \frac{C_n}{\sqrt{n!}} \tag{4.156}$$

which transforms (4.154) into

$$i\dot{x}(n) = \Delta n x(n) - \alpha[(n+1)x(n+1) + x(n-1)]. \tag{4.157}$$

To solve this we can utilize a generating function method by introducing

$$G(z,t) = \sum_{n=0}^{\infty} x(n) z^n \tag{4.158}$$

and find

$$\sum_{n=0}^{\infty} nx(n)z^n = z\frac{\partial}{\partial z}\sum_{n=0}^{\infty} x(n)z^n = z\frac{\partial}{\partial z}G$$

$$\sum_{n=0}^{\infty}(n+1)x(n+1)z^n = \frac{1}{z}\sum_{n=0}^{\infty} nx(n)z^n = \frac{\partial}{\partial z}G$$

$$\sum_{n=0}^{\infty} x(n-1)z^n = z\sum_{n=0}^{\infty} x(n)z^n = zG. \tag{4.159}$$

Equation (4.157) then transforms into

$$i\frac{\partial G}{\partial t} + (\alpha - \Delta z)\frac{\partial G}{\partial z} = -\alpha zG. \tag{4.160}$$

This partial differential equation can be solved by determining its characteristic trajectories $z = z(t)$. For this purpose we set

$$\frac{dt}{i} = \frac{dz}{\alpha - \Delta z} = -\frac{dG}{\alpha zG} \tag{4.161}$$

giving the differential equations

$$\frac{dz}{dt} = -i(\alpha - \Delta z) \tag{4.162}$$

$$\frac{dG}{dz} = -\frac{\alpha z}{\alpha - \Delta z}G. \tag{4.163}$$

These can be integrated to give

$$\alpha - \Delta z = C_1 e^{+i\Delta t} \tag{4.164}$$

$$G = C_2 e^{\alpha z/\Delta}(\alpha - \Delta z)^{(\alpha/\Delta)^2}, \tag{4.165}$$

where $C_1$ and $C_2$ are constants of integration. The general solution is obtained by setting $C_2$ equal to a function of $C_1$, and hence we have

$$G(z,t) = e^{\alpha z/\Delta}(\alpha - \Delta z)^{(\alpha/\Delta)^2}\Phi\left[e^{-i\Delta t}(\alpha - \Delta z)\right]. \tag{4.166}$$

The unknown function $\Phi$ can be determined from the initial conditions. The easiest case occurs when we are in the ground state $|0\rangle$ at $t = 0$. Then

$G(z, t = 0) = 1$, and

$$G(z,0) = 1 = e^{\alpha z/\Delta}(\alpha - \Delta z)^{(\alpha/\Delta)^2}\Phi(\alpha - \Delta z). \tag{4.167}$$

Setting $\alpha - \Delta z = s$ we have

$$\Phi(s) = s^{-(\alpha/\Delta)^2}e^{-(\alpha/\Delta)^2}e^{\alpha s/\Delta^2}, \tag{4.168}$$

giving for (4.166) the result

$$G(z, t) = e^{+i\alpha^2 t/\Delta}\exp\left[\frac{\alpha^2}{\Delta^2}A(t)\right]\exp\left[-\frac{\alpha}{\Delta}A(t)z\right], \tag{4.169}$$

where

$$A(t) = e^{-i\Delta t} - 1. \tag{4.170}$$

The coefficients $x(n)$ are now easily extracted as

$$x(n) = \frac{C_n}{\sqrt{n!}} = \frac{e^{+i\alpha^2 t/\Delta}e^{\alpha^2 A/\Delta^2}}{n!}\left(-\frac{\alpha A}{\Delta}\right)^n, \tag{4.171}$$

giving the occupation probability

$$|C_n|^2 = \frac{1}{n!}\exp\left[\frac{\alpha^2}{\Delta^2}(A + A^*)\right]\left(\frac{\alpha^2}{\Delta^2}AA^*\right)^n. \tag{4.172}$$

We have

$$AA^* = 2(1 - \cos\Delta t) = 4\sin^2\frac{\Delta t}{2} = -(A + A^*) \tag{4.173}$$

giving

$$|C_n|^2 = \frac{1}{n!}e^{-B}B^n. \tag{4.174}$$

This is a Poisson distribution over the states with the average occupation level given by

$$\langle n\rangle = B = 4\left(\frac{\alpha}{\Delta}\right)^2\sin^2\left(\frac{\Delta t}{2}\right). \tag{4.175}$$

When we have a detuning $\Delta \neq 0$, the detunings accumulate as we climb the ladder and force the excitation to turn down eventually. At exact resonance $\Delta = 0$ there is no detuning at any stage, and from (4.175) we find the result

$$\langle n \rangle = \alpha^2 t^2; \tag{4.176}$$

thus no limit exists for the level of excitation.

**Example.** The driven classical oscillator behaves in very much the same way as the quantum mechanical one presented here. Its equation of motion is taken to be

$$\ddot{x} + \varepsilon^2 x = \frac{e\mathscr{E}_0}{m} \cos \Omega t. \tag{4.177}$$

If the oscillator initially is in the lowest state of excitation (the ground state!), the initial conditions are $x(0) = 0$ and $\dot{x}(0) = 0$. The solution is

$$x(t) = -\frac{e\mathscr{E}_0}{m} \frac{\cos \varepsilon t - \cos \Omega t}{\varepsilon^2 - \Omega^2}, \tag{4.178}$$

which for exact resonance $\Omega \to \varepsilon$ gives

$$x(t) \to \frac{e\mathscr{E}_0}{2m\Omega} t \sin \Omega t. \tag{4.179}$$

The unbounded part of the total energy becomes

$$E(t) = \frac{m}{2}(\dot{x}^2 + \varepsilon^2 x^2) \to \frac{m}{2}\left(\frac{e\mathscr{E}_0}{2m}\right)^2 t^2. \tag{4.180}$$

This corresponds exactly to the result (4.176).

It is also possible to solve the problem for initial conditions differing from those of Eq. (4.167). We can, for instance, require an initial Poisson distribution

$$C_n = \frac{e^{-N/2} N^{n/2}}{\sqrt{n!}}; \tag{4.181}$$

the phase is of no importance. We obtain

$$G(z, t = 0) = \exp\left[-\tfrac{1}{2}N + z\sqrt{N}\right], \tag{4.182}$$

and, using the same method as before, we find

$$G(z, t) = e^{+i\alpha^2 t/\Delta} e^{-N/2} \exp\left[\frac{\alpha}{\Delta}\left(\frac{\alpha}{\Delta} - \sqrt{N}\right) A\right]$$
$$\times \exp\left[z\left(e^{i\Delta t}\sqrt{N} - A\frac{\alpha}{\Delta}\right)\right]. \tag{4.183}$$

When we put $t = 0$ we regain (4.182), and when we put $N = 0$ we regain (4.169), as we expect.

Again, it is easy to expand (4.183) in a power series in $z$ and extract the occupation probabilities. For $\Delta \neq 0$ we obtain an oscillating behavior, but for $\Delta = 0$ we have again the Poisson distribution (4.174) but now with

$$B = \langle n \rangle = N + \alpha^2 t^2. \tag{4.184}$$

Thus, at resonance, the excitation process takes over from the average level $N$ just as if it started from the ground state $N = 0$ given by (4.176).

These calculations give an idea of the efficiency of multistep excitation. In real molecules only a few steps remain resonant before anharmonicity detunes the transitions. The selective molecular excitation in isotope separation and laser chemistry does, however, take place in the first few steps.

### 4.5.e. Comments and References

The multiphoton research area is huge, and the amount of literature is immense. We here refer only to the proceedings of two recent conferences edited by Eberly and Lambropoulos (1978) and Janossy and Varro (1980). Further references can be found from these. Molecules in strong fields have attracted much interest because of potential applications to laser chemistry and isotope separation. We refer the readers to a recent review article by Letokhov and Makarov (1981).

Equation (4.128) with $\Delta\omega_n$ a quadratic function of $n$ occurs in many fields of physics. In addition to the anharmonic molecular level scheme, it can be used to describe free electrons in a laser beam, light diffraction by ultrasound, and atomic beam deflection by a standing light wave. The literature in these areas is too vast to be included; I only give some recent references to enable the reader to trace the earlier work.

The use of Eq. (4.128) in free-electron–laser theories is reviewed in Stenholm and Bambini (1981). In acoustic deflection experiments the equation is called the Raman–Nath equation, and its properties are summarized in Berry (1966). The beam deflection work can be studied from the recent references, Bernhardt and Shore (1981) and Arimondo et al. (1981). Many earlier references can be found in these.

The classical example in Sec. 4.5.d is discussed in Eberly (1980). The relationship between quantum systems and the classical oscillator is discussed by Feld (1977).

Many authors have replaced a multilevel system by an effective two-level system. A recent work is Allen and Stroud (1982). Here the calculation is carried out in detail, the validity of its various assumptions is discussed, and some correction terms to our simple treatment are given.

CHAPTER 5

# Effects of Field Fluctuations on Spectroscopy

## 5.1. STOCHASTIC BEHAVIOR OF PHYSICAL PARAMETERS

The first lasers were based on resonant transitions, and inside a high-quality optical cavity their intrinsic line width was very narrow. Drifts and mechanical instabilities increased the observed bandwidth of the output, but these could be overcome technically. Then the laser oscillator was made monochromatic enough to serve as a time standard, an atomic clock.

The use of dye lasers and more recently color center lasers changed the situation. Here the intrinsic oscillational band is broad, and it requires special efforts to obtain narrow frequency operation. The tunability of these devices derives just from these very broad bandwidths. The output of the device will, however, show large fluctuations. A professionally stabilized dye laser has a bandwidth of about 10 MHz, which is of the same order of magnitude as the atomic line-width parameters. Only with highly refined techniques can one improve the performance below this value.

Thus the noise in a tunable laser is an essential feature of its operation. Even a well-stabilized single-mode laser contains its unavoidable quantum noise. Except close to the onset of oscillations, near threshold, these fluctuations are of less importance than the technical perturbations. In addition to those deriving from the multimode structure, we have acoustic vibrations, thermal drifts, pump-laser fluctuations, and inhomogeneities in the dye liquid and its flow. Away from threshold these limit the performance of laser spectroscopy in most practical cases.

There are two essentially different limits of the random perturbation process. One assumes the random variables to remain constant most of the time, but at random moments they undergo sudden jumps within a very short time. This is the basis of much mathematical work on random processes. Binary collisions in a dilute gas can often be represented by this type of model; see the discussion in Sec. 1.6 Eqs. (1.118)–(1.125).

The opposite limit assumes the random variables to change smoothly but in an erratic way. Processes described by Fokker–Planck equations belong to this class. They can be understood as the limit of a jump process when the jumps occur very frequently but each one contributes only a minute change to the stochastic variable. This way the Brownian motion model is justified. In Sec. 1.7 we discussed such processes in Eqs. (1.132)–(1.136) and Eqs. (1.139)–(1.157). Laser field fluctuations, especially, can be modeled by this type of approach. Hence we choose to use a Fokker–Planck approach to the line-width problem in this chapter.

The field amplitude that effects the transitions between the energy levels of matter is taken to be a stochastic quantity $\mathscr{E}(t)$. In this chapter we introduce the basic physical implications of this for some of the nonlinear effects introduced earlier in this book. Many works have appeared in this area over the last few years, and we introduce only the basic formulation of the stochastic field problem in laser spectroscopy. For details and further developments we refer to the literature given in Sec. 5.5.

Much of the work on stochastic processes is based on the theory of Brownian motion, in which the velocity of a heavy particle suspended in a medium obeys the equation of motion

$$\dot{v}(t) = -\Gamma v(t) + F(t), \tag{5.1}$$

where the velocity suffers viscous damping but sees a fluctuating force $F(t)$ because of the thermal motion of the medium. The average value of $F$, the average force $\overline{F(t)}$, is assumed to be zero. This equation is called a *Langevin equation*. The fluctuations of the medium are taken to be so fast that the values of the random force $F(t)$ are essentially uncorrelated for different times. In this way we introduce the essential assumption that the medium acts as a heat bath that is unable to sustain correlations.

Mathematically, this can be written as the assumption

$$\overline{F(t)F(t')} = 2\Gamma^2 D\delta(t - t'), \tag{5.2}$$

where $D$ is the diffusion coefficient measuring the strength of the fluctuations. All higher-order correlation functions are assumed to factor into products of pairs and single-force averages. This means that all odd correlation functions are zero because the average of the random force is zero, $\overline{F} = 0$. These properties describe a Gaussian random process.

If we introduce a Fourier transform of the random force by setting

$$\tilde{F}(\omega) = \int_{-\infty}^{+\infty} e^{i\omega t} F(t)\, dt, \tag{5.3}$$

we can define a quantity $\Phi(\omega)$ by

$$\overline{\tilde{F}(\omega)\tilde{F}(\omega')} = 2\Gamma^2 D \int_{-\infty}^{+\infty} e^{i(\omega+\omega')t}\,dt$$

$$= 2\pi\Phi(\omega)\delta(\omega+\omega'). \tag{5.4}$$

The quantity

$$\Phi(\omega) = 2\Gamma^2 D \tag{5.5}$$

is called the spectrum of the noise, and (5.5) shows that it is here frequency independent. The noise is then called *white*.

The formal solution of (5.1) is

$$v(t) = \int_{-\infty}^{t} e^{-\Gamma(t-\tau)} F(\tau)\,d\tau. \tag{5.6}$$

The correlation function of the solution (5.6) is easily seen to be

$$\overline{v(t)v(t')} = \int_{-\infty}^{t} e^{-\Gamma(t-\tau)} \int_{-\infty}^{t'} e^{-\Gamma(t'-\tau')} \overline{F(\tau)F(\tau')}\,d\tau\,d\tau'$$

$$= 2\Gamma^2 D e^{-\Gamma(t+t')} \int_{-\infty}^{t'} e^{2\Gamma\tau}\,d\tau$$

$$= D\Gamma e^{-\Gamma(t-t')} \tag{5.7}$$

when we assume $t > t'$. Taking the opposite case, $t < t'$, we can combine

$$\overline{v(t)v(t')} = \overline{v^2} e^{-\Gamma|t-t'|}, \tag{5.8}$$

which gives the relationship

$$\overline{v^2} = D\Gamma. \tag{5.9}$$

The ensuing spectrum of $v(t)$ is easily evaluated just as in Eq. (5.4) to be given by

$$\overline{\tilde{v}(\omega)\tilde{v}(\omega')} = \int\int e^{i\omega(t+\tau)} e^{i\omega' t}\,\overline{v(t+\tau)v(t)}\,dt\,d\tau$$

$$= 2\pi\delta(\omega+\omega') \int_{-\infty}^{+\infty} e^{i\omega\tau}\,\overline{v(t+\tau)v(t)}\,d\tau, \tag{5.10}$$

where we have used the fact that the process is stationary and hence

$\overline{v(t+\tau)v(t)}$ does not depend on $t$. The spectrum $\Phi(\omega)$ can now be evaluated from (5.8) to be

$$\Phi_v(\omega) = \int_{-\infty}^{+\infty} e^{i\omega\tau}\overline{v(t+\tau)v(t)}d\tau$$

$$= \frac{2\Gamma^2}{\omega^2 + \Gamma^2}D. \tag{5.11}$$

The Brownian motion variable $v(t)$ thus shows a Lorentzian spectrum.

We have assumed that the only nontrivial correlation function of the random force $F(t)$ is (5.2), that is, all higher-order correlation functions can be decoupled into products of the second-order correlation function. Then it is shown in textbooks on statistical mechanics that the probability distribution $P(v, t)$ for the random variable $v$ at time $t$ obeys the partial differential equation

$$\frac{\partial}{\partial t}P(v,t) = \Gamma\frac{\partial}{\partial v}[vP(v,t)] + \Gamma^2 D\frac{\partial^2}{\partial v^2}P(v,t). \tag{5.12}$$

This is called the Fokker–Planck equation. Its steady state solution is given by

$$\Gamma D\frac{dP}{dv} = -vP, \tag{5.13}$$

which integrates to

$$P \propto \exp\left[-\frac{v^2}{2\Gamma D}\right]. \tag{5.14}$$

This agrees well with the result (5.9). If we assume a Maxwellian thermal distribution we obtain

$$D = \frac{kT}{\Gamma m}, \tag{5.15}$$

which is called *Einstein's relation*.

From Eq. (5.11) we see that the Brownian motion spectrum becomes white when the fluctuations become very rapid $\Gamma \to \infty$. This is the *fast fluctuation limit*; the correlation time $\Gamma^{-1}$ goes to zero, and we cannot use Eq.(5.1). The limiting process can, however, be carried out through defining the new stochastic variable $x(t)$ by setting

$$v(t) = \dot{x}(t) \tag{5.16}$$

and defining the new random force

$$g(t) = \lim_{\Gamma \to \infty} \left[ \frac{F(t)}{\Gamma} \right] \tag{5.17}$$

with the correlation function

$$\overline{g(t)g(t')} = 2D\delta(t - t'), \tag{5.18}$$

as follows from (5.2). From Eq. (5.1) we have

$$\ddot{x} = -\Gamma(\dot{x} - g) \tag{5.19}$$

and only by setting

$$\dot{x}(t) = g(t) \tag{5.20}$$

can we let $\Gamma$ go to infinity. From the solution

$$x(t) = \int_0^t g(\tau)\, d\tau \tag{5.21}$$

we obtain the result

$$\overline{x(t)x(t')} = \int_0^t \int_0^{t'} \overline{g(\tau)g(\tau')} d\tau\, d\tau' = 2Dt' \qquad \text{if } t' < t. \tag{5.22}$$

The general result is

$$\overline{x(t)x(t')} = 2D\,\mathrm{Min}(t, t'). \tag{5.23}$$

This shows that we have a pure diffusion process $\overline{x^2} = 2Dt$, and the Fokker–Planck equation becomes

$$\frac{\partial P}{\partial t} = D\frac{\partial^2 P}{\partial x^2}. \tag{5.24}$$

This process can be regarded as the limiting form of the one described by (5.12) when it is viewed over time spans such that $\Gamma t \gg 1$. In many cases its use considerably simplifies the discussion. The solution of (5.24), which is initially ($t = 0$) centered at $x = 0$, is known to be

$$P(x,t) = \frac{1}{\sqrt{4\pi Dt}} \exp\left[-\frac{x^2}{4Dt}\right]. \tag{5.25}$$

It is easy to see that this is in agreement with (5.23) for $t = t'$.

## 5.2. THEORY OF PHASE NOISE

### 5.2.a. Phase Noise in Linear Spectroscopy

We introduce a field with a fluctuating phase parameter

$$\mathscr{E}(t) = \tfrac{1}{2} E e^{i\varphi} e^{i\Omega t} + \text{c.c.} \,. \tag{5.26}$$

We consider the simple two-level system of Fig. 5.1 with the Hamiltonian

$$H = \hbar\omega|2\rangle\langle 2| + \left(\frac{\mu E}{2} e^{i[\Omega t + \varphi(t)]}|1\rangle\langle 2| + \text{h.c.}\right), \tag{5.27}$$

where we have already introduced the rotating wave approximation. The induced dipole moment obeys the equation

$$i\dot{\rho}_{21} = (\omega - i\gamma_{21})\rho_{21} + \frac{\mu E}{2\hbar} e^{-i(\varphi + \Omega t)}(\rho_{11} - \rho_{22}). \tag{5.28}$$

In the linear approximation we can neglect the saturation and set $\rho_{11} - \rho_{22} = \rho_{11}^0$. Then we can introduce a new variable $\tilde{\rho}_{21}$ by writing

$$\tilde{\rho}_{21} = e^{i(\varphi + \Omega t)}\rho_{21}, \tag{5.29}$$

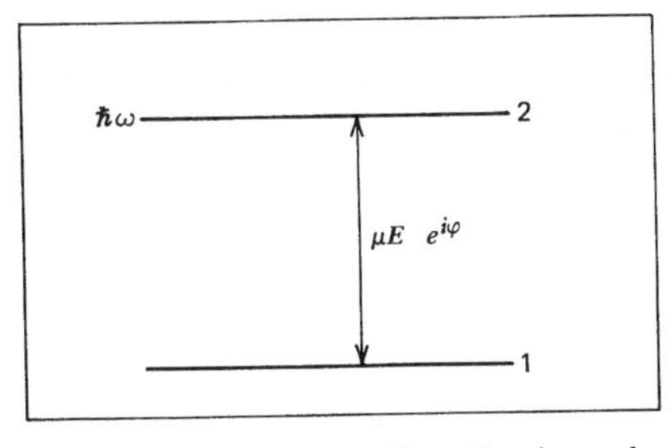


**Fig. 5.1** The two-level system with a dipole coupling $\mu E$ and a random phase $\varphi$.

which gives instead of (5.28) the equation

$$i\dot{\tilde{\rho}}_{21} = (\omega - \Omega - \dot{\varphi} - i\gamma_{21})\tilde{\rho}_{21} + \frac{\mu E}{2\hbar}\rho_{11}^0. \tag{5.30}$$

This transformation was introduced in Eq. (1.142) in Sec. 1.7. Now the new variable $\dot{\varphi}$ is also a random function, which we denote by $\nu$. In an operating laser, the amplitude is fixed in a self-consistent manner, but the phase of the oscillator is not determined. It is free to drift under the influence of many small but random perturbations due to the environment. Also, quantum noise causes dephasing of the laser field. The situation makes it plausible to describe the time evolution of the phase $\varphi$ by a Brownian motion model, and hence we write

$$\ddot{\varphi} = -\Gamma\dot{\varphi} + F(t) \tag{5.31}$$

just as in Eq. (5.1). The Langevin force $F(t)$ is assumed to satisfy a relation like Eq. (5.2). How do we combine the discussion of the preceding section with our density matrix treatment?

According to our discussion in Sec. 1.4, the density operator can be defined by Eq. (1.74)

$$\rho = \sum_{\alpha} |\alpha\rangle P_{\alpha} \langle\alpha|. \tag{5.32}$$

For orthogonal states $\{|\alpha\rangle\}$, $P_{\alpha}$ is the probability of occurrence of the state $|\alpha\rangle$. For each $\alpha$, we now assume that $P_{\alpha}$ is the probability distribution for the random variable $\nu$, and then it obeys a Fokker–Planck equation like (5.12)

$$\frac{\partial P_{\alpha}}{\partial t} = \Gamma\left[\frac{\partial}{\partial \nu}(\nu P_{\alpha}) + \Gamma D \frac{\partial^2}{\partial \nu^2} P_{\alpha}\right]. \tag{5.33}$$

For external field fluctuations the parameters $\Gamma$ and $D$ do not depend on which level $\alpha$ we discuss, and hence the density matrix (5.32) is seen to obey the equation of motion

$$i\hbar\frac{\partial \rho}{\partial t} = [H, \rho] + i\hbar\left(\Gamma\frac{\partial}{\partial \nu}\nu + \Gamma^2 D \frac{\partial^2}{\partial \nu^2}\right)\rho. \tag{5.34}$$

To this equation we can add phenomenological damping and pumping terms, as before.

The density matrix $\rho$ is the quantum mechanical generalization of a statistical distribution function; now we have made it contain also the probability distribution for the stochastic parameter $\nu$. The expectation value of any operator $\hat{O}(\nu)$ that depends on $\nu$ is given by

$$\langle \hat{O} \rangle = \int \mathrm{Tr}\left[\hat{O}(\nu)\rho(\nu)\right] d\nu. \tag{5.35}$$

This approach can be generalized directly to any stochastic parameter in the Hamiltonian of a quantum system.

Wanting to treat the effect of $\dot{\varphi} = \nu$ in Eq. (5.30), we must add the Fokker–Planck operator to its right-hand side.

To solve the problem we need to know the eigenfunctions and eigenvalues of the operator

$$Lu_n \equiv \frac{d}{d\nu}\left[\nu u_n(\nu)\right] + \Gamma D \frac{d^2}{d\nu^2} u_n(\nu) = \lambda_n u_n(\nu). \tag{5.36}$$

For the eigenvalue $\lambda_0 = 0$ we find the solution

$$u_0(\nu) \propto e^{-\nu^2/2\Gamma D}, \tag{5.37}$$

as we expect from (5.14). We try to solve (5.36) by the ansatz

$$u_n(\nu) = e^{-\nu^2/2\Gamma D} v_n(\nu). \tag{5.38}$$

When this is inserted into (5.36) we find the equation

$$\Gamma D v_n'' - \nu v_n' - \lambda_n v_n = 0. \tag{5.39}$$

Introducing

$$\xi = \frac{\nu}{\sqrt{2\Gamma D}} \tag{5.40}$$

we obtain the equation

$$\frac{d^2 v_n}{d\xi^2} - 2\xi \frac{dv_n}{d\xi} - 2\lambda_n v_n = 0. \tag{5.41}$$

This is the differential equation for the Hermite polynomials when we set

$\lambda_n = -n$, and its solution is

$$v_n(\xi) = H_n(\xi) = H_n\left(\frac{\nu}{\sqrt{2\Gamma D}}\right). \tag{5.42}$$

The solution of the original equation (5.36) is hence

$$u_n = C_n e^{-\xi^2} H_n(\xi). \tag{5.43}$$

The problem is not self-adjoint, and hence the functions (5.43) are not orthonormal. We normalize them by setting

$$\int H_m(\xi) u_n(\xi)\, d\xi = C_n \int H_m(\xi) H_n(\xi) e^{-\xi^2}\, d\xi$$

$$= C_n \sqrt{\pi}\, 2^n n! \delta_{nm} \equiv \delta_{nm} \tag{5.44}$$

which gives

$$C_n = \frac{1}{\sqrt{\pi}\, 2^n n!}. \tag{5.45}$$

The normalization differs, of course, from the one used for the quantum mechanical harmonic oscillator, which is a hermitean problem.

Later we will also need the "matrix elements" of the random variable $\nu$ in the following form

$$(\nu)_{nn'} = \int H_n(\xi) \nu u_{n'}(\xi)\, d\xi$$

$$= \sqrt{2\Gamma D} \int H_n(\xi) \xi H_{n'}(\xi) e^{-\xi^2}\, d\xi$$

$$= \sqrt{\Gamma D}\left[\sqrt{n+1}\, \delta_{n', n+1} + \sqrt{n}\, \delta_{n', n-1}\right]. \tag{5.46}$$

In steady state the dipole satisfies the equation obtained by combining (5.30) and (5.34)

$$(\omega - \Omega - \nu - i\gamma_{21})\tilde{\rho}_{21} = -\frac{\mu E}{2\hbar}\rho_{11}^0 - i\Gamma\left(\frac{\partial}{\partial \nu}\nu + \Gamma D\frac{\partial^2}{\partial \nu^2}\right)\tilde{\rho}_{21}. \tag{5.47}$$

Introducing the expansion

$$\tilde{\rho}_{21}(\nu) = \sum_{n=0}^{\infty} r_n u_n(\nu) \tag{5.48}$$

we find

$$\sum_n r_n(\omega - \Omega - \nu - i\gamma_{21})u_n(\nu) = -\frac{\mu E}{2\hbar}\rho_{11}^0(\nu) - i\sum_{n=0}^{\infty}(-\Gamma n)r_n u_n(\nu). \tag{5.49}$$

If we let $\Gamma$ go to zero, the fluctuations occur very slowly. The steady state develops according to the instantaneous value of the phase change $\nu$; that is, the slow changes in $\nu$ are followed adiabatically. Observed over a long time the system must, however, have time to go through all possible values of the rate of change of the phase, and hence the observed quantity must appear as an average over the instantaneous value with the distribution $u_0(\nu)$, as given in (5.37). We thus find that the zeroth-order distribution must be given by $u_0$ according to (5.37), and hence

$$\rho_{11}^0(\nu) = r^0 u_0(\nu). \tag{5.50}$$

We multiply Eq. (5.49) by $H_m(\xi)$ and integrate over $\xi$ to obtain

$$(\omega - \Omega - i\gamma_{21})r_n - \sqrt{\Gamma D}\left[\sqrt{n+1}\,r_{n+1} + \sqrt{n}\,r_{n-1}\right] = -\frac{\mu E}{2\hbar}r^0\delta_{n0} + i\Gamma n r_n. \tag{5.51}$$

We can write this in the form

$$x_n = D(n)\left[\sqrt{n+1}\,x_{n+1} + \sqrt{n}\,x_{n-1}\right] + \delta_{n0} \tag{5.52}$$

where

$$x_n = -\frac{2r_n\hbar}{\mu E r^0} \tag{5.53}$$

and

$$D(n) = \frac{\sqrt{\Gamma D}}{\omega - \Omega - i(\gamma_{21} + n\Gamma)}. \tag{5.54}$$

This type of recurrence relation was solved earlier in terms of a continued fraction in Chapter 2, Secs. 2.3 and 2.7. The difference here is that there now appear only positive integers $n$.

We set for $n > 0$

$$\frac{x_n}{x_{n-1}} = \frac{D(n)\sqrt{n}}{1 - D(n)\sqrt{n+1}\,(x_{n+1}/x_n)}. \tag{5.55}$$

Setting $n = 1$ we iterate

$$\frac{r_1}{r_0} = \cfrac{D(1)}{1 - \cfrac{2D(1)D(2)}{1 - \cfrac{3D(2)D(3)}{1 - \cfrac{4D(3)D(4)}{1 - \cdots}}}}. \tag{5.56}$$

Equation (5.51) for $n = 0$ gives

$$\left(\omega - \Omega - i\gamma_{21} - \sqrt{\Gamma D}\,\frac{r_1}{r_0}\right) r_0 = -\frac{\mu E}{2\hbar} r^0 \tag{5.57}$$

and

$$r_0 = -\cfrac{(\mu E r^0/2\hbar)}{\omega - \Omega - i\gamma_{21} - \cfrac{\sqrt{\Gamma D}\,D(1)}{1 - \cfrac{2D(1)D(2)}{1 - \cdots}}}$$

$$= -\frac{\mu E r^0}{2\hbar}\,\cfrac{1}{\omega - \Omega - i\gamma_{21} - \cfrac{\Gamma D}{\cfrac{\omega - \Omega - i(\gamma_{21} + \Gamma)}{1 - \cdots}}}. \tag{5.58}$$

The observable quantity is the averaged value

$$\overline{\rho_{21}} = \int_{-\infty}^{+\infty} \rho_{21}(\nu)\, d\nu$$

$$= r_0. \tag{5.59}$$

When the amplitude of the fluctuations goes to zero $D \to 0$, we find the result without fluctuations having the resonance denominator

$$\rho_{21} \propto \frac{1}{\omega - \Omega - i\gamma_{21}}. \tag{5.60}$$

To see the effect of fluctuations we introduce the second approximation of (5.58) and find

$$r_0 = -\frac{\mu E r^0}{2\hbar} \frac{\omega - \Omega - i(\gamma_{21} + \Gamma)}{(\omega - \Omega - i\gamma_{21})[\omega - \Omega - i(\gamma_{21} + \Gamma)] - \Gamma D}. \tag{5.61}$$

This holds in the limit of a small $D$, and hence the roots are given by

$$\omega - \Omega = i\gamma_{21} + \frac{i\Gamma}{2}\left(1 \pm \sqrt{1 - \frac{4D}{\Gamma}}\right) = i\gamma_{21} + \frac{i\Gamma}{2}\left(1 \pm 1 \mp \frac{2D}{\Gamma}\right). \tag{5.62}$$

Because $D \ll \Gamma$, by assumption, the pole of width $\gamma_{21} + \Gamma$ cancels nearly exactly, and we find the response

$$r_0 = -\frac{\mu E r^0}{2\hbar} \frac{1}{\omega - \Omega - i(\gamma_{21} + D)}. \tag{5.63}$$

The fluctuations are thus contributing a line width $D$. This is easy to understand because we expect from time-dependent perturbation theory a transition rate caused by the fluctuating field of the order of magnitude

$$(\text{fluctuating field})^2 \times \text{density of states} \propto \overline{\nu\nu} \times \Gamma^{-1} \tag{5.64}$$

which gives $D$ according to the assumptions about the fluctuations, as seen, for example, in Eq. (5.37).

Next we consider the limit of fast fluctuations, that is, the limit when $\Gamma \to \infty$. Because, from (5.54), $D(n) \propto \Gamma^{-1/2}$ each $D$ factor decreases the magnitude of the following contribution. Letting $\Gamma$ grow in Eq. (5.51), with $n = 0$, we find

$$r_0 \propto \Gamma^0, \qquad r_1 \propto \Gamma^{-1/2}. \tag{5.65}$$

For $n = 2$ we find for large $\Gamma$ the result

$$\sqrt{\Gamma D}\left[\sqrt{3}\, r_3 + \sqrt{2}\, r_1\right] + i2\Gamma r_2 = 0. \tag{5.66}$$

From here we find that

$$r_1 \propto \Gamma^{-1/2}, \qquad r_2 \propto \Gamma^{-1}. \tag{5.67}$$

Continuing this way we find that the higher-order terms go down with increasing powers of $\Gamma^{-1/2}$. The lowest nontrivial approximation consists of $r_0$ and $r_1$, which is just the result of (5.61)–(5.63). This is thus valid when the fluctuation rate $\Gamma$ grows to infinity independently of the magnitude measured by $D$. This is the fast fluctuation limit, when the correlation time of the randomly fluctuating quantity $\Gamma^{-1}$ becomes exceedingly short. The same result can be obtained immediately from the continued fraction representation (5.58) where we estimate the factors

$$D(n)D(n+1) \sim \frac{D}{\Gamma}. \tag{5.68}$$

If we define the fast fluctuation limit in such a way that $\Gamma \to \infty$ but keep $\Gamma D$ constant, only the first stage contributes

$$\sqrt{\Gamma D}\, D(1) \propto D. \tag{5.69}$$

Thus we need to keep only the lowest-order term, which gives the result (5.63). For fixed $\overline{\nu^2} = D\Gamma$, the width $\gamma_{21} + D$ decreases when the fluctuation rate $\Gamma$ increases, hence the phenomenon is called *motional narrowing.*

The limit corresponds to the pure diffusion limit of Eqs. (5.16)–(5.22), and the result agrees with our conclusions in Sec. 1.7; the diffusion rate directly adds to the other relaxation rates.

It is also of interest to study the opposite limit, $\Gamma \to 0$. According to our foregoing argument, this limit should give the instantaneous response $\tilde{\rho}_{21}(\nu)$ for the given value of the random parameter $\nu$ averaged with the distribution of $\nu$.

We set

$$z = \omega - \Omega - i\gamma_{21} \tag{5.70}$$

and let $\Gamma \ll |z|$ when we find from (5.58)

$$r_0 \propto \cfrac{1}{z - \cfrac{\Gamma D}{z - \cfrac{2\Gamma D}{z - \cfrac{3\Gamma D}{z - \dots}}}}. \tag{5.71}$$

Using the well-known property† of the Gaussian function

$$\frac{1}{\sqrt{2\pi a^2}}\int_{-\infty}^{+\infty}\frac{e^{-t^2/2a^2}\,dt}{z-t}=\cfrac{1}{z-\cfrac{a^2}{z-\cfrac{2a^2}{z-\cfrac{3a^2}{z-\cfrac{4a^2}{z-\dots}}}}} \tag{5.72}$$

we can write Eq. (5.71) in the form

$$\begin{aligned} r_0 &\propto \frac{1}{\sqrt{2\pi\Gamma D}}\int_{-\infty}^{+\infty}\frac{e^{-t^2/2\Gamma D}}{z-t}\,dt \\ &= \frac{1}{\sqrt{2\pi\Gamma D}}\int_{-\infty}^{+\infty}\frac{e^{-\nu^2/2\Gamma D}\,d\nu}{\omega-\Omega-i\gamma_{21}-\nu}. \end{aligned} \tag{5.73}$$

This result corresponds exactly to the result of (5.47) when $\Gamma = 0$, but we retain the average over the distribution function (5.37). This proves that, in the slow fluctuation limit, our intuitive description can be reproduced from our formal treatment valid for all values of $\Gamma$.

This concludes our description of phase fluctuation effects on linear spectroscopy. We next show how these results appear when saturation is included.

### 5.2.b. Phase Noise on the Two-Level System

We look again at the model in Sec. 5.2.a with the two-level Hamiltonian (5.27). In addition to Eqs. (5.28) and (5.29), we have the corresponding equations for $\rho_{11}$ and $\rho_{22}$, as in Chapter 2. Following the notation of that chapter we introduce

$$\begin{aligned} C &= \tilde{\rho}_{12} + \tilde{\rho}_{21} \\ S &= i(\tilde{\rho}_{12} - \tilde{\rho}_{21}) \\ N &= \rho_{11} - \rho_{22}. \end{aligned} \tag{5.74}$$

†See, for example, Abramovitz and Stegun (1970) page 298, Eq. 7.1.15.

These variables correspond to those used in Chapter 2; in Sec. 2.3 they are seen to be equivalent with a Bloch equation description if we have only one decay rate for the diagonal elements $\gamma_1$, and we use $\gamma_2$ to denote the off-diagonal decay rate $\gamma_{21}$. We introduce the coupling frequency

$$\alpha = \frac{\mu E}{2\hbar}. \tag{5.75}$$

We assume that $\rho_{11} = \lambda/\gamma_1$ when no fields act. The equations of motion are easily found to be

$$\dot{N} = \lambda - \gamma_1 N + 2\alpha S + LN \tag{5.76}$$

$$\dot{S} = -\gamma_2 S - (\Delta\omega + \nu)C - 2\alpha N + LS \tag{5.77}$$

$$\dot{C} = -\gamma_2 C + (\Delta\omega + \nu)S + LC, \tag{5.78}$$

where $\Delta\omega = \omega - \Omega$ and $L$ is the Fokker–Planck operator defined by (5.36). All density matrix elements, of course, see the same fluctuations, which leads directly to (5.76)–(5.78).

We use an expansion just like that in (5.48) and write

$$N = \sum_{n=0}^{\infty} n_n u_n \tag{5.79}$$

$$S = \sum_{n=0}^{\infty} s_n u_n \tag{5.80}$$

$$C = \sum_{n=0}^{\infty} c_n u_n. \tag{5.81}$$

Inserted into Eqs. (5.76)–(5.78) they give

$$\dot{n}_n + (\gamma_1 + n\Gamma) n_n = \bar{\lambda}\delta_{n0} + 2\alpha s_n \tag{5.82}$$

$$\dot{s}_n + (\gamma_2 + n\Gamma) s_n = -\Delta\omega c_n - \sqrt{\Gamma D}\left(\sqrt{n+1}\, c_{n+1} + \sqrt{n}\, c_{n-1}\right) - 2\alpha n_n \tag{5.83}$$

$$\dot{c}_n + (\gamma_2 + n\Gamma) c_n = \Delta\omega s_n + \sqrt{\Gamma D}\left(\sqrt{n+1}\, s_{n+1} + \sqrt{n}\, s_{n-1}\right). \tag{5.84}$$

In steady state we can eliminate $n_n$ and set

$$n_n = \frac{\bar{\lambda}}{\gamma_1}\delta_{n0} + \frac{2\alpha}{\gamma_1 + \Gamma n}s_n \tag{5.85}$$

to find

$$\frac{(\gamma_2 + n\Gamma)(\gamma_1 + n\Gamma) + 4\alpha^2}{\gamma_1 + n\Gamma}s_n = -\Delta\omega c_n - \sqrt{\Gamma D}\left(\sqrt{n+1}\,c_{n+1} + \sqrt{n}\,c_{n-1}\right) - \frac{2\alpha\bar{\lambda}}{\gamma_1}\delta_{n0}, \tag{5.86}$$

which couples to

$$(\gamma_2 + n\Gamma)c_n = \Delta\omega\, s_n + \sqrt{\Gamma D}\left(\sqrt{n+1}\,s_{n+1} + \sqrt{n}\,s_{n-1}\right). \tag{5.87}$$

The structure of these equations is just like the one encountered in Eq. (5.51). They can be solved by the use of matrix continued fractions, but the expressions rapidly become opaque, and no physical interpretation can be seen. The details are most easily found in the references listed in Sec. 5.5. Already the lowest approximation having $s_0$, $s_1$, $c_0$, and $c_1$ nonzero becomes rather complicated. The observables are, as in Eq. (5.59), the quantities $n_0$, $s_0$, and $c_0$.

It is of some interest to see how the fast fluctuation limit $\Gamma \to \infty$ emerges in this case. From Eqs. (5.83–5.84) we see that

$$c_0 \propto s_0 \propto \Gamma^0 \tag{5.88}$$

and then again

$$c_1 \propto s_1 \propto \Gamma^{-1/2}. \tag{5.89}$$

If we assume $\Gamma$ to greatly exceed the two line widths $\gamma_1$ and $\gamma_2$ and also the coupling frequency $\alpha$, we find the equations

$$\begin{aligned}
\left(\gamma_2 + \frac{4\alpha^2}{\gamma_1}\right)s_0 &= -\Delta\omega\, c_0 - \sqrt{\Gamma D}\,c_1 - \frac{2\alpha\bar{\lambda}}{\gamma_1} \\
\gamma_2 c_0 &= \Delta\omega\, s_0 + \sqrt{\Gamma D}\,s_1 \\
\Gamma s_1 &= -\sqrt{\Gamma D}\,c_0 \\
\Gamma c_1 &= \sqrt{\Gamma D}\,s_0.
\end{aligned} \tag{5.90}$$

We solve these equations to find

$$s_0 = -\frac{2\alpha\bar{\lambda}}{\gamma_1}\frac{\gamma_2 + D}{\Delta\omega^2 + (\gamma_2 + D)^2\left(1 + \dfrac{4\alpha^2}{(\gamma_2 + D)\gamma_1}\right)}. \tag{5.91}$$

The result clearly corresponds to the phase diffusion limit. The off-diagonal element $\gamma_2$ obtains an additional contribution from the diffusion rate $D$. In the limit of $\Gamma$ going to infinity, the result (5.91) is the exact one. The terms $c_n$, $s_n$ with $n > 1$ go to zero faster than $\Gamma^{-1/2}$ in this limit.

To see the effects of higher-order terms we look at the position of exact resonance $\Delta\omega = 0$. Then there are only even terms in $s_n$ and only odd in $c_n$, and the problem simplifies. We define

$$x_n = -\frac{\gamma_1}{2\alpha\bar{\lambda}}\begin{cases} s_n & n \text{ even} \\ c_n & n \text{ odd} \end{cases} \tag{5.92}$$

and

$$\begin{aligned} D(n) &= -\frac{\sqrt{\Gamma D}\,(\gamma_1 + n\Gamma)}{(\gamma_2 + n\Gamma)(\gamma_1 + n\Gamma) + 4\alpha^2} \qquad n \text{ even} \\ &= \frac{\sqrt{\Gamma D}}{\gamma_2 + N\Gamma} \qquad n \text{ odd}. \end{aligned} \tag{5.93}$$

The equations then become equal to Eq. (5.52) with the solution (5.56)

$$\frac{c_1}{s_0} = \cfrac{\cfrac{\sqrt{\Gamma D}}{\gamma_2 + \Gamma}}{1 + \cfrac{\cfrac{2\Gamma D(\gamma_1 + 2\Gamma)}{[(\gamma_2 + 2\Gamma)(\gamma_1 + 2\Gamma) + 4\alpha^2](\gamma_2 + \Gamma)}}{1 + \cfrac{\cfrac{3\Gamma D(\gamma_1 + 2\Gamma)}{[(\gamma_2 + 2\Gamma)(\gamma_1 + 2\Gamma) + 4\alpha^2](\gamma_2 + 3\Gamma)}}{1 + \cdots}}}. \tag{5.94}$$

The observable is then obtained from

$$\left[\frac{\gamma_1\gamma_2 + 4\alpha^2}{\gamma_1} + \sqrt{\Gamma D}\,\frac{c_1}{s_0}\right]s_0 = -\frac{2\alpha\bar{\lambda}}{\gamma_1}. \tag{5.95}$$

In the limit $\Gamma \to \infty$ we regain the result of Eq. (5.91). Other limits are easily evaluated. For numerical work, a continued fraction like (5.94) is found to converge very rapidly, usually needing no more than 10 steps.

## 5.3. AMPLITUDE FLUCTUATIONS IN A SINGLE-MODE LASER

### 5.3.a. The Two-Level System

In single-mode operation a laser has a self-stabilized amplitude. The phase is free to drift and hence gives the most important contribution to the line width. But it is also useful to consider the effect of amplitude fluctuations about the mean value $E_0$. We consequently write the laser field

$$E(t) = \left[E_0 + \hbar\varepsilon(t)\right] e^{i\varphi(t)}, \tag{5.96}$$

where $E_0$ describes the coherent part of the amplitude. We assume that the phase noise can be treated in the fast fluctuation limit, adding a diffusion constant to the off-diagonal density matrix elements. For the random amplitude we assume the Fokker–Planck operator

$$L = \Gamma \frac{\partial}{\partial \varepsilon} \varepsilon + \Gamma^2 \kappa \frac{\partial^2}{\partial \varepsilon^2}. \tag{5.97}$$

The equations for the two-level system treated previously now become, instead of (5.76–5.78),

$$\dot{N} = \lambda - \gamma_1 N + 2\alpha S + \mu\varepsilon S + LN \tag{5.98}$$

$$\dot{S} = -\gamma_2 S - \Delta\omega\, C - 2\alpha N - \mu\varepsilon N + LS \tag{5.99}$$

$$\dot{C} = -\gamma_2 C + \Delta\omega\, S + LC. \tag{5.100}$$

Again, using an expansion like that in (5.79)–(5.81), we obtain

$$\dot{n}_n + (\gamma_1 + n\Gamma) n_n = \lambda\delta_{n0} + 2\alpha s_n + \mu\sqrt{\Gamma\kappa}\left(\sqrt{n+1}\, s_{n+1} + \sqrt{n}\, s_{n-1}\right) \tag{5.101}$$

$$\dot{s}_n + (\gamma_2 + n\Gamma) s_n = -\Delta\omega\, c_n - 2\alpha n_n - \mu\sqrt{\Gamma\kappa}\left(\sqrt{n+1}\, n_{n+1} + \sqrt{n}\, n_{n-1}\right) \tag{5.102}$$

$$\dot{c}_n + (\gamma_2 + n\Gamma) c_n = \Delta\omega\, s_n. \tag{5.103}$$

We turn to consider the steady state case and eliminate $c_n$ by the use of

(5.103). We find

$$s_n = -2\alpha D_2(n) n_n - \mu\sqrt{\Gamma\kappa}\, D_2(n)\left(\sqrt{n+1}\, n_{n+1} + \sqrt{n}\, n_{n-1}\right) \quad (5.104)$$

and

$$n_n = \frac{\bar{\lambda}}{\gamma_1}\delta_{n0} + 2\alpha D_1(n) s_n + \mu\sqrt{\Gamma\kappa}\, D_1(n)\left(\sqrt{n+1}\, s_{n+1} + \sqrt{n}\, s_{n-1}\right). \quad (5.105)$$

The functions $D(n)$ are defined by

$$D_2(n) = \frac{1}{2}\left[\frac{1}{\gamma_2 + n\Gamma + i\,\Delta\omega} + \frac{1}{\gamma_2 + n\Gamma - i\,\Delta\omega}\right] \quad (5.106a)$$

and

$$D_1(n) = \frac{1}{\gamma_1 + n\Gamma}. \quad (5.106b)$$

The structure of the Eqs. (5.104)–(5.105) is similar to the structure of Eqs. (5.86) and (5.87). For $\alpha \propto E_0 \to 0$ we can obtain one single continued fraction, but this limit is not acceptable because the representation (5.96) is physically meaningful only when $E_0 > \hbar\sqrt{\overline{\varepsilon^2}}$.

As in all previous cases, the observable is given by

$$\bar{N} = \int \frac{d\varepsilon}{\sqrt{2\Gamma\kappa}} N(\varepsilon) = n_0 \quad (5.107)$$

because of (5.44).

Just as before, we can obtain the result in the limit of large $\Gamma$ by noting from Eqs. (5.104)–(5.105) that we must have

$$s_n, n_n \propto \Gamma^{-n/2} \quad (5.108)$$

because $D(n) \sim \Gamma^{-1}$ for all $n \neq 0$. For the first few values we find the equations

$$\begin{aligned} n_0 &= \frac{\bar{\lambda}}{\gamma_1} + \frac{2\alpha}{\gamma_1} s_0 + \frac{\mu\sqrt{\Gamma\kappa}}{\gamma_1} s_1 \\ s_0 &= -2\alpha D_2(0) n_0 - \mu\sqrt{\Gamma\kappa}\, D_2(0) n_1 \\ n_1 &= \mu\sqrt{\frac{\kappa}{\Gamma}}\, s_0 \\ s_1 &= -\mu\sqrt{\Gamma\kappa}\, D_2(1) n_0, \end{aligned} \quad (5.109)$$

where, when $\Gamma \gg \gamma_2$, we have

$$D_2(0) = \frac{\gamma_2}{\gamma_2^2 + \Delta\omega^2} \tag{5.110}$$

$$D_2(1) \simeq \frac{\Gamma}{\Gamma^2 + \Delta\omega^2}. \tag{5.111}$$

The solution is

$$n_0 = \frac{\bar{\lambda}}{\gamma_1}\left[1 + \frac{\mu^2\kappa}{\gamma_1}\frac{\Gamma^2}{\Delta\omega^2 + \Gamma^2} + \frac{4\alpha^2\gamma_2}{\gamma_1\left(\Delta\omega^2 + \gamma_2^2 + \mu^2\kappa\gamma_2\right)}\right]^{-1}. \tag{5.112}$$

At resonance $\Delta\omega \cong 0$, we have two dips, one of width $\Gamma$, the other of width $\gamma_2(1 + \mu^2\kappa/\gamma_2)^{1/2}$. The effective power due to the noise is given by $\mu^2\kappa \propto \overline{(\mu\varepsilon)^2}$. When $|\Delta\omega| \ll \Gamma$, we can see only one dip, but there is a saturation by the factor $(1 + \mu^2\kappa/\gamma_1)^{-1}$ due to the fluctuating field transferring population betwen the two levels.

In the mathematical limit $\Gamma \to \infty$ we obtain the resonance curve

$$n_0 = \frac{\bar{\lambda}}{\gamma_1 + \mu^2\kappa}\left\{1 - \frac{\dfrac{4\alpha^2\gamma_2}{(\gamma_1 + \mu^2\kappa)}}{\Delta\omega^2 + \gamma_2^2\left[1 + \dfrac{\mu^2\kappa}{\gamma_2} + \dfrac{4\alpha^2}{\gamma_2(\gamma_1 + \mu^2\kappa)}\right]}\right\}. \tag{5.113}$$

Comparing this to the result without fluctuations, we notice that the diagonal decay rate has changed into

$$\gamma_1' = \gamma_1 + \mu^2\kappa, \tag{5.114}$$

which acts to decrease the saturating power of the coherent part of the field $\alpha$. Because the fluctuations tend to decrease the field amplitude occasionally, they affect the higher order terms' efficiency to act. Relaxations get a chance to restore the original populations. As a result saturation decreases. The effective line width has been increased to

$$\gamma_2' = \gamma_2\left[1 + \frac{\mu^2\kappa}{\gamma_2} + \frac{4\alpha^2}{\gamma_2(\gamma_1 + \mu^2\kappa)}\right]^{1/2}. \tag{5.115}$$

There appears an effective line width due to the power broadening by the fluctuating light. This adds quadratically with that due to the coherent light ($\propto \alpha^2$) because the cross-terms clearly average to zero. Including the lowest order in $\kappa$, only, we have

$$\gamma_2' = \gamma_2 + \frac{\mu^2\kappa}{2}, \tag{5.116}$$

which can be explained in terms of time-dependent perturbation theory rates just like the result in (5.64).

### 5.3.b. The Three-Level System with a Probe

We showed in Sec. 4.3 that many questions of interest to nonlinear spectroscopy can be investigated in a three-level system where one level of a strongly saturated system is probed by a weak field, coupling it to a third level. We now investigate how the randomness of the field coupling levels 1 and 2 is seen when we couple level 2 to another level 3 by a weak and nonfluctuating field. The level structure is shown in Fig. 5.2.

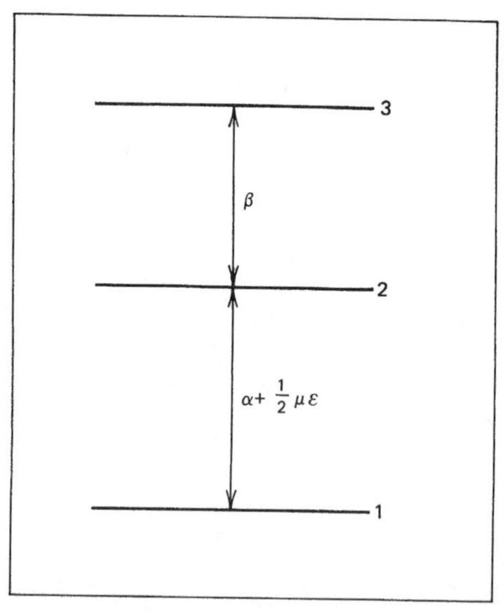


Fig. 5.2 The three-level system. The lower pair of levels is coupled by a field containing a random amplitude fluctuation $\varepsilon$ added to the transition frequency $\alpha$. The upper transition is probed by the field $\beta$ without fluctuations.

If the coupling frequency of the second laser is given by $\beta$ and its frequency by $\Omega_\beta$, we introduce

$$\Delta\omega_{32} = \omega_{32} - \Omega_\beta \tag{5.117}$$

$$\Delta\omega_{31} = \omega_{31} - \Omega - \Omega_\beta, \tag{5.118}$$

and the steady state density matrix elements from Sec. 4.3.b are seen to obey

$$(\Delta\omega_{32} - i\gamma_2)\tilde{\rho}_{32} - \alpha\tilde{\rho}_{31} - \tfrac{1}{2}\mu\varepsilon\tilde{\rho}_{31} = -\beta\rho_{22} - iL\tilde{\rho}_{32} \tag{5.119}$$

$$(\Delta\omega_{31} - i\gamma_3)\tilde{\rho}_{31} - \alpha\tilde{\rho}_{32} - \tfrac{1}{2}\mu\varepsilon\tilde{\rho}_{32} = -\beta\tilde{\rho}_{21} - iL\tilde{\rho}_{31}, \tag{5.120}$$

where the relaxation rate for the coherence $\tilde{\rho}_{31}$ is denoted by $\gamma_3$. The operator $L$ is the same as in Eq. (5.97).

To treat the field $\beta$ as a probe, we calculate $\rho_{22}$ and $\rho_{21}$ with $\beta = 0$, that is, just like the previous section.

Because we have in this case

$$M = \rho_{11} + \rho_{22} = \frac{\bar{\lambda}}{\gamma_1} \tag{5.121}$$

we can write

$$\rho_{22} = \frac{1}{2}(M - \bar{N}) = \sum_{n=0}^{\infty} \frac{1}{2}\left(\frac{\bar{\lambda}}{\gamma_1}\delta_{n0} - n_n\right)u_n(\varepsilon) \tag{5.122}$$

$$\tilde{\rho}_{21} = \frac{1}{2}(C + iS) = \sum_{n=0}^{\infty} \frac{1}{2}(c_n + is_n)u_n(\varepsilon). \tag{5.123}$$

If we now, in addition, expand

$$\tilde{\rho}_{32} = \sum_{n=0}^{\infty} r_n u_n(\varepsilon) \tag{5.124}$$

$$\tilde{\rho}_{31} = \sum_{n=0}^{\infty} q_n u_n(\varepsilon) \tag{5.125}$$

we find

$$[\Delta\omega_{32} - i(\gamma_2 + n\Gamma)]r_n - \alpha q_n - \frac{\mu}{2}\sqrt{\Gamma\kappa}\left(\sqrt{n+1}\,q_{n+1} + \sqrt{n}\,q_{n-1}\right)$$
$$= -\beta\frac{1}{2}\left(\frac{\bar{\lambda}}{\gamma_1}\delta_{n0} - n_n\right) \tag{5.126}$$

$$[\Delta\omega_{31} - i(\gamma_3 + n\Gamma)]q_n - \alpha r_n - \frac{\mu}{2}\sqrt{\Gamma\kappa}\left(\sqrt{n+1}\,r_{n+1} + \sqrt{n}\,r_{n-1}\right)$$
$$= -\frac{\beta}{2}(c_n + is_n). \tag{5.127}$$

Again the problem becomes rather complicated. In addition to the matrix structure of the equations, we also have inhomogeneous terms for each $n$. We need to construct the continued fraction equivalent of a Green function for the difference operator. This can be done, but the algebraic complexity is apt to obscure the physics of the ensuing expressions. We look instead at the limit of large $\Gamma$ again, the fast fluctuation limit.

In addition to the result (5.113) we need the coefficients $s_0$, $s_1$, and $n_1$. From Eqs. (5.109) we obtain

$$s_0 = -\frac{2\alpha\gamma_2}{\Delta\Omega^2 + \gamma_2^2 + \mu^2\kappa\gamma_2} n_0$$

$$s_1 = -\mu\sqrt{\frac{\kappa}{\Gamma}}\, n_0$$

$$n_1 = \mu\sqrt{\frac{\kappa}{\Gamma}}\, s_0 \tag{5.128}$$

when $|\Delta\omega| \ll \Gamma$. We also need from (5.103)

$$c_0 = \frac{\Delta\omega}{\gamma_2} s_0$$

$$c_1 = \frac{\Delta\omega}{\Gamma} s_1 \simeq 0. \tag{5.129}$$

The equations to be solved are thus

$$(\Delta\omega_{32} - i\gamma_2) r_0 - \alpha q_0 - \frac{\mu}{2}\sqrt{\Gamma\kappa}\, q_1 = -\frac{\beta}{2}\left(\frac{\bar{\lambda}}{\gamma_1} - n_0\right) \tag{5.130}$$

$$(\Delta\omega_{31} - i\gamma_3) q_0 - \alpha r_0 - \frac{\mu}{2}\sqrt{\Gamma\kappa}\, r_1 = -\frac{\beta}{2}\left(\frac{\Delta\omega + i\gamma_2}{\gamma_2}\right) s_0 \tag{5.131}$$

$$-i\Gamma r_1 - \alpha q_1 - \frac{\mu}{2}\sqrt{\Gamma\kappa}\, q_0 = +\frac{\beta}{2} n_1 \tag{5.132}$$

$$-i\Gamma q_1 - \alpha r_1 - \frac{\mu}{2}\sqrt{\Gamma\kappa}\, r_0 = -\frac{\beta}{2}\left(\frac{\Delta\omega}{\Gamma} + i\right) s_1. \tag{5.133}$$

If we, in the last two equations, keep only the terms of order $\Gamma^{1/2}$, we obtain

$$r_1 = \frac{i\mu}{2}\sqrt{\frac{\kappa}{\Gamma}}\, q_0, \qquad q_1 = \frac{\mu}{2}\sqrt{\frac{\kappa}{\Gamma}}\, r_0. \tag{5.134}$$

Inserting back into (5.130) and (5.131) we find the result

$$\left[\Delta\omega_{32} - i\left(\gamma_2 + \frac{\mu^2\kappa}{4}\right)\right] r_0 - \alpha q_0 = -\frac{\beta}{2}\left(\frac{\bar{\lambda}}{\gamma_1} - n_0\right) \tag{5.135}$$

$$\left[\Delta\omega_{31} - i\left(\gamma_3 + \frac{\mu^2\kappa}{4}\right)\right] q_0 - \alpha r_0 = -\frac{\beta}{2}\left(\frac{\Delta\omega + i\gamma_2}{\gamma_2}\right) s_0$$

$$= +\alpha\beta \frac{(\Delta\omega + i\gamma_2)}{(\Delta\omega^2 + \gamma_2^2 + \mu^2\kappa\gamma_2)} n_0. \tag{5.136}$$

These equations are exactly like those for the noiseless case, see Eqs. (4.44–4.45), as far as the mathematical structure is concerned. Their solution is thus similar to (4.50). They are exact in the motional narrowing limit $\Gamma \to \infty$. The off-diagonal decay rates $\gamma_2$ and $\gamma_3$ have been replaced by the effective widths

$$\gamma_i' = \gamma_i + \frac{\mu^2\kappa}{4} \qquad (i = 2, 3). \tag{5.137}$$

The broadening observed here is less by a factor of $\frac{1}{4}$ when compared with that of the two-level system, Eq. (5.114). The reason for this is that only one of the levels in $\rho_{32}$ and $\rho_{31}$ is coupled to the random field. Hence only half the amplitude is seen, and consequently the fluctuation intensity $\kappa \propto \overline{\varepsilon\varepsilon}$ is divided by 4.

In both Eqs. (5.135–5.136) the inhomogeneous terms are expressed in terms of $n_0$. The coupling in the second equation, which gives rise to the coherent contributions (see Sec. 4.3), is, however, multiplied by the function

$$\frac{\Delta\omega + i\gamma_2}{(\Delta\omega^2 + \gamma_2^2 + \mu^2\kappa\gamma_2)} \underset{\Delta\omega \to 0}{\Rightarrow} \frac{i}{\gamma_2 + \mu^2\kappa}. \tag{5.138}$$

This shows that the influence of coherence-induced processes as compared with the population-induced processes goes down by the increased line width $\gamma_2 + \mu^2\kappa$. Thus field fluctuations tend to decrease the role of coherent transitions relative to the incoherent rate contributions. In this aspect, amplitude fluctuations behave just like phase-destructing processes; see the discussion in Sec. 4.3.d.

The solution to Eqs. (5.135)–(5.136) is

$$r_0 = \frac{-\dfrac{\bar{\lambda}\beta\gamma_2}{\gamma_1^2}}{\left(\Delta\omega^2 + \gamma_2^2 + \mu^2\kappa\gamma_2\right)\left(1 + \dfrac{\mu^2\kappa}{\gamma_1}\right) + \dfrac{4\alpha^2\gamma_2}{\gamma_1}}$$
$$\times \left\{ \frac{2\alpha^2\left[\Delta\omega_{31} - i\left(\gamma_3 + \dfrac{\mu^2\kappa}{4}\right)\right] - (\alpha^2\gamma_1/\gamma_2)(\Delta\omega + i\gamma_2)}{\left[\Delta\omega_{32} - i\left(\gamma_2 + \dfrac{\mu^2\kappa}{4}\right)\right]\left[\Delta\omega_{31} - i\left(\gamma_3 + \dfrac{\mu^2\kappa}{4}\right)\right] - \alpha^2} \right.$$
$$\left. + \frac{\dfrac{\mu^2\kappa}{\gamma_2}\left[\Delta\omega_{31} - i\left(\gamma_3 + \dfrac{\mu^2\kappa}{4}\right)\right]\left(\Delta\omega^2 + \gamma_2^2 + \mu^2\kappa\gamma_2\right)}{\left[\Delta\omega_{32} - i\left(\gamma_2 + \dfrac{\mu^2\kappa}{4}\right)\right]\left[\Delta\omega_{31} - i\left(\gamma_3 + \dfrac{\mu^2\kappa}{4}\right)\right] - \alpha^2} \right\}. \tag{5.139}$$

When we put $\kappa = 0$ we regain the expression from Eq. (4.50) with the velocity $v = 0$. The first two terms proportional to $\alpha^2$ are the transitions induced by the coherent field with the appropriately broadened line width parameters; the third term is proportional to $\beta\kappa \propto \beta\overline{\varepsilon^2}$ and describes transitions due to the random part of the field between levels 1 and 2. The resonance $\Delta\omega_{31} \simeq 0$ tends to cancel in this term, and hence it combines with the first term proportional to $2\alpha^2$ to give the incoherent rate, as we might expect. The random field can only enhance the incoherent two-step process at the expense of the coherent, two-quantum process.

Our treatment of the amplitude fluctuations is unsatisfactory because we have separated the random part in (5.96). The separation of random contributions to the rate is rather artificial, and actually the factor $\alpha_1^2\alpha_2^2$ should be averaged with the $\alpha$'s containing the total amplitudes. The method of the next section is applicable but rapidly becomes very complicated.

## 5.4. THE FREE-RUNNING MULTIMODE LASER

### 5.4.a. Statistics of Multimode Light

When a laser is operating in many modes two extreme situations are possible: either the modes are phase locked and the output consists of a regular pulse train which repeats itself, or the modes oscillate independently of each other as if the others did not exist. In reality one often encounters intermediate situations, but they are very complicated to describe theoretically. The locked regime does not concern us here, but the free-running laser produces an output that is strongly fluctuating. Many solid state and dye lasers correspond to the model considered here.

The free-running modes have phases that drift nearly freely with respect to each other. If we can neglect the interaction between the modes, their relative intensities do not fluctuate too much. Hence the mathematical model writes the electric field as

$$E(t) = \hbar \operatorname{Re}[\varepsilon(t)] \tag{5.140}$$

where

$$\varepsilon(t) = e^{i\Omega t} \sum_{j=1}^{M} \varepsilon_j e^{-i\Delta\Omega_j t - i\varphi_j}. \tag{5.141}$$

Here $\Omega$ is the average frequency and $\{\varepsilon_j\}$ are the amplitudes of the $M$ modes; $\varphi_j$ are the random phases. The average intensity is

$$\overline{|\varepsilon|^2} = \sum_{j=1}^{M} |\varepsilon_j|^2. \tag{5.142}$$

If we define a characteristic function for the field amplitude by setting

$$\chi_M(\lambda, \lambda^*) = \overline{e^{-i\lambda\varepsilon(t) - i\lambda^*\varepsilon^*(t)}}, \tag{5.143}$$

this function generates the moments

$$\overline{\varepsilon(t)^n \varepsilon^*(t)^m} = \left[\left(i\frac{\partial}{\partial\lambda}\right)^n \left(i\frac{\partial}{\partial\lambda^*}\right)^m \chi_M(\lambda, \lambda^*)\right]_{\lambda=0}. \tag{5.144}$$

As we assume the phases $\varphi_j$ to be random in the interval $(-\pi, \pi)$, we can evaluate the integrals in terms of the Bessel function $J_0$

$$\frac{1}{2\pi}\int_{-\pi}^{\pi} \exp\left[-i\lambda\varepsilon_i e^{-i(\varphi_i + \Delta\Omega t)} - i\lambda\varepsilon_i e^{i(\varphi_i + \Delta\Omega_i)t}\right] d\varphi$$

$$= \frac{1}{\pi}\int_0^{\pi} \exp(-i2|\lambda|\varepsilon_i \cos x)\, dx = J_0(2|\lambda|\varepsilon_i), \tag{5.145}$$

because the phases $\Omega t$ and $\arg\lambda$ can be incorporated into the integration variables. For the total function we can write, when $M$ is large enough,

$$\chi_M(\lambda, \lambda^*) = \prod_{i=1}^{M} J_0(2|\lambda|\varepsilon_i) \cong \prod_{i=1}^{M} (1 - \lambda^2|\varepsilon_i|^2)$$

$$\cong \prod_{i=1}^{M} e^{-\lambda^2|\varepsilon_i|^2} = \exp\left(-\lambda^2\overline{|\varepsilon|^2}\right), \tag{5.146}$$

where (5.142) has been used.

To obtain the probability distribution of the sum variable $\varepsilon(t)$ we must undo the Fourier transform contained in the characteristic function $\chi_M(\lambda, \lambda^*)$

$$P(\varepsilon, \varepsilon^*) = \int \frac{d\lambda\, d\lambda^*}{\pi^2} e^{i\lambda\varepsilon + i\lambda^*\varepsilon^*} \chi_M(\lambda, \lambda^*), \tag{5.147}$$

which inverts (5.143). Separating the integration variables into $\lambda = \lambda' + i\lambda''$ we find

$$\frac{1}{\pi^2} \int d\lambda' \int d\lambda''\, e^{iA\lambda'} e^{iB\lambda'} \exp\left(-\lambda'^2\overline{|\varepsilon|^2} - \lambda''^2\overline{|\varepsilon|^2}\right)$$

$$= \frac{1}{\pi\overline{|\varepsilon|^2}} \exp\left[-\frac{(A^2 + B^2)}{4\overline{|\varepsilon|^2}}\right], \tag{5.148}$$

with

$$A = \varepsilon + \varepsilon^*, \qquad B = i(\varepsilon - \varepsilon^*). \tag{5.149}$$

When this is inserted we find from (5.147)–(5.148) that the probability distribution becomes

$$P(\varepsilon, \varepsilon^*) = \frac{1}{\pi\overline{|\varepsilon|^2}} \exp\left(-\frac{\varepsilon\varepsilon^*}{\overline{|\varepsilon|^2}}\right). \tag{5.150}$$

This result is, of course, just a special case of the central limit theorem; if many uncorrelated random variables are combined, the result is described by a Gaussian distribution.

In this section we take the attitude that any multimode laser can be described by a Gaussian amplitude distribution in the complex $\varepsilon$ plane. The distribution function is taken to be (5.150). A convenient model for this is the stochastic process described by the Fokker–Planck operator

$$L = \Gamma\left(\frac{\partial}{\partial\varepsilon}\varepsilon + \frac{\partial}{\partial\varepsilon^*}\varepsilon^* + 4\Gamma\kappa\frac{\partial^2}{\partial\varepsilon\,\partial\varepsilon^*}\right), \tag{5.151}$$

where we have introduced as the diffusion parameter

$$\kappa = \frac{\overline{|\varepsilon|^2}}{2\Gamma}. \tag{5.152}$$

When this is combined with the density matrix equation in the manner presented in the previous sections we can formulate any problem in laser spectroscopy with a complex Gaussian random field. Such a field is often called *chaotic*.

By writing the operator in the form

$$L = \Gamma\frac{\partial}{\partial\varepsilon}\left[\varepsilon + 2\Gamma\kappa\frac{\partial}{\partial\varepsilon^*}\right] + \Gamma\frac{\partial}{\partial\varepsilon^*}\left[\varepsilon^* + 2\Gamma\kappa\frac{\partial}{\partial\varepsilon}\right], \tag{5.153}$$

we can immediately see that it gives zero when applied to the distribution function (5.150), which hence constitutes the steady state solution to the stochastic problem.

Introducing the real and imaginary parts of the field

$$\varepsilon = x + iy \tag{5.154}$$

we can write

$$\frac{\partial}{\partial\varepsilon} = \frac{1}{2}\left(\frac{\partial}{\partial x} - i\frac{\partial}{\partial y}\right), \qquad \frac{\partial}{\partial\varepsilon^*} = \frac{1}{2}\left(\frac{\partial}{\partial x} + i\frac{\partial}{\partial y}\right) \tag{5.155}$$

giving

$$\frac{\partial}{\partial\varepsilon}\varepsilon + \frac{\partial}{\partial\varepsilon^*}\varepsilon^* = \frac{\partial}{\partial x}x + \frac{\partial}{\partial y}y \tag{5.156}$$

and

$$\frac{\partial^2}{\partial\varepsilon\,\partial\varepsilon^*} = \frac{1}{4}\left(\frac{\partial^2}{\partial x^2} + \frac{\partial^2}{\partial y^2}\right). \tag{5.157}$$

The total operator (5.151) then becomes

$$L = \Gamma\left[\frac{\partial}{\partial x}x + \Gamma\kappa\frac{\partial^2}{\partial x^2} + \frac{\partial}{\partial y}y + \Gamma\kappa\frac{\partial^2}{\partial y^2}\right]. \tag{5.158}$$

The operator thus separates into two operators just like the one used in Eq. (5.36). The eigenvalue problem

$$L\varphi(x, y) = \lambda\varphi(x, y) \tag{5.159}$$

has the factorized solution

$$\varphi(x, y) = C_{nm}\exp\left[-\frac{x^2+y^2}{2\Gamma\kappa}\right]H_n\left[\frac{x}{\sqrt{2\Gamma\kappa}}\right]H_m\left[\frac{y}{\sqrt{2\Gamma\kappa}}\right] \tag{5.160}$$

as given in Sec. 5.2. The eigenvalues are

$$\lambda_{nm} = -\Gamma(n+m). \tag{5.161}$$

In this way we can reduce the problem to the one we have treated before.

It is also possible to introduce intensity and phase variables by setting

$$\varepsilon = \sqrt{I}\,e^{i\varphi}. \tag{5.162}$$

Then the problem can be solved in terms of the associated Laguerre polynomials. For this treatment we refer to the literature.

### 5.4.b. The Two-Level System

The equations for a two-level system in a chaotic field are, after the rotating wave approximation has been made,

$$i\dot{\rho}_{11} = i(\lambda - \gamma_1\rho_{11}) - \frac{\mu}{2}\varepsilon\tilde{\rho}_{12} + \frac{\mu}{2}\varepsilon^*\tilde{\rho}_{21} + iL\rho_{11} \tag{5.163}$$

$$i\dot{\rho}_{22} = -i\gamma_1\rho_{22} + \frac{\mu}{2}\varepsilon\tilde{\rho}_{12} - \frac{\mu}{2}\varepsilon^*\tilde{\rho}_{21} + iL\rho_{22} \tag{5.164}$$

$$i\dot{\tilde{\rho}}_{21} = (\Delta\omega - i\gamma_2)\tilde{\rho}_{21} + \frac{\mu}{2}\varepsilon(\rho_{11} - \rho_{22}) + iL\rho_{21}, \tag{5.165}$$

where $\varepsilon$ is the complex random field and $L$ is the Fokker–Planck operator (5.158). The detuning is assumed to be defined with respect to the center of the frequency spectrum of $\varepsilon(t)$.

Introducing real parameters, as before, by setting

$$\varepsilon = x + iy, \qquad N = \rho_{11} - \rho_{22} \tag{5.166}$$

$$\tilde{\rho}_{21} = \tfrac{1}{2}(C + iS) \tag{5.167}$$

and inserting into Eqs. (5.163)–(5.165), we obtain

$$\dot{N} = \lambda - \gamma_1 N + \mu(xS - yC) + LN \tag{5.168}$$

$$\dot{S} = -\gamma_2 S - \Delta\omega\, C - \mu x N + LS \tag{5.169}$$

$$\dot{C} = -\gamma_2 C + \Delta\omega\, S + \mu y N + LC. \tag{5.170}$$

These equations can now be treated by an expansion in a double series in the functions (5.43) as

$$N = \sum_{n,m} n(n,m) u_n(x) u_m(y) \tag{5.171}$$

$$S = \sum_{n,m} s(n,m) u_n(x) u_m(y) \tag{5.172}$$

$$C = \sum_{n,m} c(n,m) u_n(x) u_m(y). \tag{5.173}$$

Inserting these into (5.168)–(5.170) and retaining the possibility that the coefficients may depend on time, we obtain

$$\begin{aligned} \dot{n}(n,m) &+ [\gamma_1 + \Gamma(n+m)] n(n,m) \\ &= \bar{\lambda}\delta_{n0}\delta_{m0} \\ &\quad + \mu\sqrt{\Gamma\kappa}\left[\sqrt{n+1}\, s(n+1,m) + \sqrt{n}\, s(n-1,m)\right. \\ &\quad \left. - \sqrt{m+1}\, c(n,m+1) - \sqrt{m}\, c(n,m-1)\right] \end{aligned} \tag{5.174}$$

$$\begin{aligned} \dot{s}(n,m) &+ [\gamma_2 + \Gamma(n+m)] s(n,m) + \Delta\omega c(n,m) \\ &= -\mu\sqrt{\Gamma\kappa}\left[\sqrt{n+1}\, n(n+1,m) + \sqrt{n}\, n(n-1,m)\right] \end{aligned} \tag{5.175}$$

$$\begin{aligned} \dot{c}(n,m) &+ [\gamma_2 + \Gamma(n+m)] c(n,m) - \Delta\omega\, s(n,m) \\ &= \mu\sqrt{\Gamma\kappa}\left[\sqrt{m+1}\, n(n,m+1) + \sqrt{m}\, n(n,m-1)\right]. \end{aligned} \tag{5.176}$$

These equations form a formidable coupled problem, which has such a complicated structure that no simple treatment is possible in the general case.

As before, the problem simplifies considerably in the limit of a very large $\Gamma$. Then the main contribution comes from $n(0,0)$, $s(0,0)$, and $c(0,0)$.

Through the factor $(\Gamma\kappa)^{1/2}$ these couple to the terms with $n = 1$, $m = 0$ or $n = 0$, $m = 1$. These are found to be of order $\Gamma^{-1/2}$ and contribute. In their equations we can ignore all terms except those of order $\Gamma^{-1/2}$ and obtain

$$s(1,0) = -\mu\sqrt{\frac{\kappa}{\Gamma}}\, n(0,0) \tag{5.177}$$

$$c(0,1) = \mu\sqrt{\frac{\kappa}{\Gamma}}\, n(0,0) \tag{5.178}$$

$$n(1,0) = \mu\sqrt{\frac{\kappa}{\Gamma}}\, s(0,0), \qquad n(0,1) = -\mu\sqrt{\frac{\kappa}{\Gamma}}\, c(0,0). \tag{5.179}$$

Inserting into the equations for $m = n = 0$ we find

$$\dot{n}(0,0) + \gamma_1 n(0,0) = \bar{\lambda} - 2\mu^2\kappa n(0,0) \tag{5.180}$$

$$\dot{s}(0,0) + \gamma_2 s(0,0) + \Delta\omega c(0,0) = -\mu^2\kappa s(0,0) \tag{5.181}$$

$$\dot{c}(0,0) + \gamma_2 c(0,0) - \Delta\omega s(0,0) = -\mu^2\kappa c(0,0). \tag{5.182}$$

The first is a typical rate equation for the population difference. The total population disappears at the rate $\gamma_1$, hence also the difference. The lower level is pumped at the rate $\bar{\lambda}$, and the random field $\kappa \propto \overline{\varepsilon^2}$ tends to make the populations equal. In steady state we obtain

$$n(0,0) = \frac{\bar{\lambda}/\gamma_1}{1 + 2\mu^2\kappa/\gamma_1}. \tag{5.183}$$

The saturation term should be compared with the random field saturation term in Eq. (5.112). Although we argued that we could not let the coherent field amplitude $\alpha$ go to zero there, the result obtained in the limit $\Gamma \to \infty$ was given correctly. The factor 2 here comes from the two independent components of the random field.

The dipole moment Eqs. (5.181–5.182) decouple from the population equations in this approximation. The eigenfrequencies of the two coupled equations are

$$\lambda_{\pm} = \left(\gamma_2 + \mu^2\kappa\right) \pm i\,\Delta\omega. \tag{5.184}$$

Thus the system oscillates with the detuning $\Delta\omega$ and decays with the increased decay rate

$$\gamma_2' = \gamma_2\left(1 + \frac{\mu^2\kappa}{\gamma_2}\right). \tag{5.185}$$

This agrees with the phase fluctuation result from Eq. (5.116) and has a similar interpretation. The complex fluctuating field, of course, contains fully developed phase fluctuations as well as amplitude fluctuations.

We can see that the chaotic field is more efficient in saturating the transition, Eq. (5.183), than in broadening the line shape.

In conclusion we note that in the fast fluctuation limit we obtain a pure rate description for the population difference. This corresponds to what we might expect from a highly random field. The averaged dipole moment is not zero, but it decouples totally from the time variation of the population difference. The fluctuations can, in this limit, be seen only as a broadening of the resonance curve. This is the response we would see if we coupled a weak coherent resonant light to the two-level system.

## 5.5. COMMENTS AND REFERENCES

The theory of random processes is very rich. Many classical articles are reprinted in Wax (1954), which is still a most valuable reference. A modern review of stochastic processes in physics is presented by van Kampen (1982).

The relaxation effects caused by a randomly fluctuating parameter were treated in the theory of magnetic resonance, see Abragam (1961) and Slichter (1978). Applications of these ideas to spectroscopy were early introduced by Burshtein (1965) and Burshtein and Oseledchik (1967). Burshtein also realized the similarity between random change of phase and an erratic level shift due to atomic collisions (Burshtein 1966b), see the discussion in Sec. 1.6. The method of Burshtein was mainly developed from a field model with random jumps as discussed in Sec. 1.7.

The fluctuations due to quantum noise were early treated by stochastic methods; the Fokker–Planck equation was introduced and solved. For references and discussions of this approach we refer the reader to Louisell (1973).

Recently many papers have investigated the effects of noise on various processes in quantum electronics. It is useless to try to list them all, and new applications will, no doubt, appear continuously. Only the main papers of interest to our approach are mentioned here.

The addition of noise to a quantum mechanical process can always be treated as a linear problem, but it gives a partial differential equation. For the basic theory of this approach see van Kampen (1976), Sec. 23. To add a noise operator to the density matrix equation is well known in NMR; perhaps the classic reference is Anderson (1954). The method is implicit in the early work by Burshtein, and fully developed in Zusman and Burshtein (1972).

The problem of stochastically varying optical fields was taken up by Eberly (1976) and Agarwal (1976). It was immediately followed by a large number of works. Zoller (1977) formulated the problem by adding a relaxation operator to the density matrix equation of motion in the fashion of Anderson and Burshtein. Together with his collaborators he has developed this method into a powerful technique for treating laser noise problems. We mainly follow their presentations. The Hermite polynomial approach to phase noise is treated in Dixit et al. (1980), where they show that the method leads to a continued fraction. The motional narrowing limit is well known for magnetic resonance. The laser line width problem is discussed by Avan and Cohen-Tannoudji (1977). The reduction of the chaotic field problem to Hermite poynomials is classic (Louisell 1973), the Laguerre polynomial solution is given by Zoller (1979a) and applied to a two-level system in Zoller (1979b). The three-level system with phase noise is discussed in Dixit et al. (1980). The general formulation of a problem with a chaotic multimode field is treated in Zoller et al. (1981); here we have followed their treatment quite closely.

A treatment in terms of Gaussian functional integrals has been presented by Wodkiewicz (1979a, b). He derives for the averaged atomic quantities effective equations of motion, which are equivalent with those derived from the Fokker–Planck treatment, see Zoller et al. (1981).

The influence of noise processes on the coherent contributions to three-level resonances is discussed by Mostowski and Stenholm (1982) along the lines of this book. The Green function method for difference equations is described in Appendix 3 of Aminoff and Stenholm (1976).

CHAPTER 6

# Elements of Electromagnetic Field Quantization

## 6.1. INTRODUCTION

In laser spectroscopy there is usually little need to quantize those fields that cause the induced transitions between the atomic levels. They emerge from coherent sources which provide strong fields with well-defined amplitudes and phases. Such fields are well described within the semiclassical approximation. There are, however, phenomena that cannot be treated entirely classically. For the situations of interest to us the most important one is the spontaneous emission. In the semiclassical approximation there is no emission from an excited state into vacuum. This follows only from a quantum theory of the radiation field. Such a spontaneous decay is also accompanied by a level shift, the Lamb shift, which has been measured with very great accuracy.

Experiments directly observing quantum properties of the field do, of course, require a quantized treatment. One example of this kind is a photon-counting experiment which measures the quantum fluctuations in the field quantities. These are not treated here.

There are fundamental arguments why the electromagnetic field should be quantized if the atomic matter is. These derive from the fact that quantized matter contains inherent fluctuations manifested in the uncertainty relations. If the excitation energy can be transferred totally to a classical field, the quantum uncertainty of the combined system would disappear, contrary to our understanding of quantum mechanics. Purely logically, the quantization of the field variables is, however, an additional postulate that cannot be derived directly from the quantum properties of matter. It requires us to make the field variables into noncommuting operators. To be able to do this we need a Hamiltonian formulation of the classical theory, because only for such a theory do we have a simple prescription for quantization.

The electromagnetic theory turns out to be equivalent with a set of harmonic oscillators. The corresponding integer quantum numbers are often interpreted directly as a number indicating the amount of independent particles residing in the field, the number of photons. This interpretation is in agreement with the expressions for the energy and momentum of the field, thus satisfying their conservation laws, but it is not an unavoidable consequence of the theory. The particle nature of the photon is rather elusive, it cannot be localized and it can only be observed in the act of absorption. Too literal an interpretation of the photon as a particle may lead to confusion. On the other hand, the concept of a multiphoton transition retains its validity also in the semiclassical description of resonant quantum processes. We have seen many examples of this in earlier parts of this book. The heuristic value of the photon concept is great, but it is best to think of the photon simply as the unit of radiation energy needed to accomplish a transition at the given frequency. The quantization procedure of the field does not imply the existence of a specific particle to be named a "photon."

We can quantize the field in an optical cavity; to obtain a quantum theory of the laser this step is necessary. To obtain spontaneous emission we need to use traveling waves in free space. The quantization procedure does require a set of eigenmodes for the radiation, but any choice of a complete basis will be equally good. A transformation of the basis directly generates a transformation of the quantized operators.

## 6.2. DECOMPOSITION OF THE CLASSICAL FIELDS

All treatments of the radiation field must start from Maxwell's equation [Eqs. (1.1)–(1.4)] of Sec. 1.2. For quantization it is useful to introduce the vector potential **A** and scalar potential $\varphi$ so that

$$\mathbf{B} = \nabla \times \mathbf{A}, \qquad \mathbf{E} = -\frac{\partial}{\partial t}\mathbf{A} - \nabla\varphi. \tag{6.1}$$

In empty space $\mathbf{B} = \mu_0 \mathbf{H}$ and $\mathbf{D} = \varepsilon_0 \mathbf{E}$, and then Maxwell's equations give

$$\nabla^2 \mathbf{A} - \mu_0 \varepsilon_0 \frac{\partial^2 \mathbf{A}}{\partial t^2} - \nabla(\nabla \cdot \mathbf{A}) = -\mu_0 \mathbf{j} + \mu_0 \varepsilon_0 \nabla \frac{\partial \varphi}{\partial t}. \tag{6.2}$$

The potentials are not unique but can be transformed by the arbitrary gauge function $\chi(r, t)$ so that the new potentials are

$$\mathbf{A}' = \mathbf{A} + \nabla\chi$$

$$\varphi' = \varphi - \frac{\partial}{\partial t}\chi. \tag{6.3}$$

We can choose $\chi$ so that

$$\nabla \cdot \mathbf{A} = 0, \qquad \frac{\partial \varphi}{\partial t} = 0. \tag{6.4}$$

This is called the Coulomb gauge. Then we find

$$\left(\nabla^2 - \frac{1}{c^2}\frac{\partial^2}{\partial t^2}\right)\mathbf{A} = -\mu_0 \mathbf{j} \tag{6.5}$$

$$-\nabla \cdot \mathbf{E} = \nabla^2 \varphi = -\frac{\rho}{\varepsilon_0}. \tag{6.6}$$

Thus the vector potential becomes a transverse vector ($\nabla \cdot \mathbf{A} = 0$), and the potential $\varphi$ is determined by the instantaneous distribution of charges. Because $\nabla \cdot \mathbf{B} = 0$ the magnetic field is also transverse, but the electric field can be divided into two parts

$$\mathbf{E} = \mathbf{E}_\perp + \mathbf{E}_\parallel$$

$$\nabla \cdot \mathbf{E}_\perp = 0, \qquad \nabla \times \mathbf{E}_\parallel = 0. \tag{6.7}$$

From Eqs. (6.1) we can see directly that

$$\mathbf{E}_\perp = -\frac{\partial}{\partial t}\mathbf{A}, \qquad \mathbf{E}_\parallel = -\nabla\varphi. \tag{6.8}$$

The longitudinal electric field $\mathbf{E}_\parallel$ contains the Coulomb interaction between the charges as given by the potential (1.12). This determines the static structure of the atoms and molecules that are investigated by the radiation. The quantities $\mathbf{A}$, $\mathbf{E}_\perp$, and $\mathbf{B}$ contain the radiation fields.

A charged particle situated in the field has the Hamiltonian

$$H = \frac{1}{2m}(\mathbf{p} - q\mathbf{A})^2 + q\varphi. \tag{6.9}$$

From here we find the equations of motion for the particle

$$\dot{\mathbf{r}} = \frac{\partial H}{\partial \mathbf{p}} = \frac{\mathbf{p} - q\mathbf{A}}{m} = \mathbf{v} \tag{6.10a}$$

$$\dot{\mathbf{p}} = -\nabla H \tag{6.10b}$$

and consequently

$$\dot{\mathbf{v}} = \frac{1}{m}\left[\dot{\mathbf{p}} - q\frac{\partial}{\partial t}\mathbf{A} - q(\mathbf{v}\cdot\nabla)\mathbf{A}\right]$$

$$= -\frac{q}{m}\left[\nabla\varphi + \frac{\partial}{\partial t}\mathbf{A} - \nabla(\mathbf{v}\cdot\mathbf{A}) + (\mathbf{v}\cdot\nabla)\mathbf{A}\right]$$

$$= \frac{q}{m}[\mathbf{E} + \mathbf{v}\times\mathbf{B}]. \tag{6.10c}$$

This is the Lorentz force law. It is straightforward to quantize the particle variables **r** and **p** by the usual prescription.

For free fields we neglect the current density **j** in Eq. (6.5) and expand the vector potential in the free space modes of a cavity of volume $V$:

$$\mathbf{A}(\mathbf{r}, t) = \frac{1}{\sqrt{\varepsilon_0 V}} \sum_{k,\sigma} \left[\mathbf{C}_{k\sigma}(t)e^{i\mathbf{k}\cdot\mathbf{r}} + \mathbf{C}^*_{k\sigma}e^{-i\mathbf{k}\cdot\mathbf{r}}\right], \tag{6.11}$$

where **k** are the wave vectors and $\sigma$ the polarization index of the field. The transversality condition gives

$$\mathbf{k}\cdot\mathbf{C}_{k\sigma} = 0,$$

and hence there are two independent polarizations for each **k**. From Eq. (6.5) we find

$$\frac{d}{dt}\mathbf{C}_{k\sigma} = -i\omega_k\mathbf{C}_k \tag{6.12}$$

with

$$\omega_k = c|\mathbf{k}|.$$

The energy density of the electromagnetic field is given by

$$\mathscr{E} = \tfrac{1}{2}\left(\varepsilon_0\mathbf{E}^2 + \frac{1}{\mu_0}\mathbf{B}^2\right)$$

$$= \tfrac{1}{2}\left(\varepsilon_0\mathbf{E}_\perp^2 + \frac{1}{\mu_0}\mathbf{B}^2\right) + \tfrac{1}{2}\varepsilon_0\mathbf{E}_\parallel^2. \tag{6.13}$$

The transverse contributions to the energy can be calculated from

$$\mathbf{E}_\perp = \frac{i}{\sqrt{\varepsilon_0 V}}\sum_{k\sigma}\omega_k\left[\mathbf{C}_{k\sigma}e^{i\mathbf{k}\cdot\mathbf{r}} - \mathbf{C}^*_{k\sigma}e^{-i\mathbf{k}\cdot\mathbf{r}}\right] \tag{6.14a}$$

$$\mathbf{B} = \frac{i}{\sqrt{\varepsilon_0 V}}\sum_{k\sigma}\left[\mathbf{k}\times\mathbf{C}_{k\sigma}e^{i\mathbf{k}\cdot\mathbf{r}} - \mathbf{k}\times\mathbf{C}^*_{k\sigma}e^{-i\mathbf{k}\cdot\mathbf{r}}\right] \tag{6.14b}$$

using the equations

$$\frac{1}{V}\int d^3r e^{i(\mathbf{k}-\mathbf{k}')\cdot\mathbf{r}} = \delta_{\mathbf{k},\mathbf{k}'} \tag{6.15}$$

and

$$\begin{aligned}(\mathbf{k}\times\mathbf{C}_{k\sigma})\cdot(\mathbf{k}\times\mathbf{C}^*_{k\sigma'}) &= k^2\mathbf{C}_{k\sigma}\cdot\mathbf{C}^*_{k\sigma'} - (\mathbf{k}\cdot\mathbf{C}_{k\sigma})(\mathbf{k}\cdot\mathbf{C}^*_{k\sigma'})\\ &= k^2|\mathbf{C}_{k\sigma}|^2\delta_{\sigma\sigma'}.\end{aligned} \tag{6.16}$$

For the radiation part of the field the volume integral of the energy density becomes

$$E = \frac{1}{2}\int d^3r\left(\varepsilon_0\mathbf{E}_\perp^2 + \frac{1}{\mu_0}\mathbf{B}^2\right) = 2\sum_{k\sigma}\omega_k^2\mathbf{C}_{k\sigma}\cdot\mathbf{C}^*_{k\sigma}. \tag{6.17}$$

We now introduce the polarization eigenvectors $\boldsymbol{\varepsilon}_{k\sigma}$ and define the scalar variables $Q_{k\sigma}$ and $P_{k\sigma}$ by

$$\mathbf{C}_{k\sigma} = \frac{1}{2}\left(Q_{k\sigma} + \frac{iP_{k\sigma}}{\omega_k}\right)\boldsymbol{\varepsilon}_{k\sigma} \tag{6.18}$$

and we have

$$\boldsymbol{\varepsilon}_{k\sigma}\cdot\boldsymbol{\varepsilon}_{k\sigma'} = \delta_{\sigma\sigma'}. \tag{6.19}$$

Equation (6.17) now becomes

$$E = \sum_{k\sigma}\frac{1}{2}\left[P_{k\sigma}^2 + \omega_k^2Q_{k\sigma}^2\right] = \sum_{k\sigma}H_{k\sigma}. \tag{6.20}$$

From Eq. (6.12) we easily obtain

$$\dot{Q}_{k\sigma} = P_{k\sigma} = \frac{\partial H_{k\sigma}}{\partial P_{k\sigma}} \tag{6.21a}$$

$$\dot{P}_{k\sigma} = -\omega_k^2Q_k = -\frac{\partial H_{k\sigma}}{\partial Q_{k\sigma}}. \tag{6.21b}$$

If we interpret (6.20) as a sum of harmonic oscillator Hamiltonians, Eqs. (6.21) show that $P_{k\sigma}$ and $Q_{k\sigma}$ are the canonical coordinates for such oscillators. They are the natural objects for the quantization of the theory; see Sec. 6.3.

*Remark.* We must also evaluate the energy density due to the longitudinal part of (6.13). From the solution to (6.6) we find

$$\mathbf{E}_{\parallel} = -\nabla\varphi = -\nabla\frac{1}{4\pi\varepsilon_0}\int\frac{\rho(\mathbf{r}')}{|\mathbf{r}-\mathbf{r}'|}d^3r'. \tag{6.22}$$

Its energy now becomes

$$U_{\parallel} = \frac{1}{2}\varepsilon_0\int d^3r\nabla\varphi\cdot\nabla\varphi = -\frac{\varepsilon_0}{2}\int d^3r\varphi\nabla^2\varphi$$

$$= \frac{1}{2}\int d^3r\varphi(\mathbf{r})\rho(\mathbf{r}) = \frac{1}{2}\int\frac{d^3rd^3r'}{4\pi\varepsilon_0}\frac{\rho(\mathbf{r}')\rho(\mathbf{r})}{|\mathbf{r}-\mathbf{r}'|}. \tag{6.23}$$

This is the Coulomb interaction energy between the charges of the system as we may have expected. Except for the infinite self-energy this expression is of an obvious physical significance as the static energy forming our bound neutral systems.

We now must add the interaction between matter and radiation just as in Eq. (6.9). The full Hamiltonian becomes

$$H = \frac{1}{2m}(\mathbf{p}-q\mathbf{A})^2 + U_{\parallel}(\mathbf{r}) + \frac{1}{2}\sum_{k\sigma}\left(P_{k\sigma}^2 + \omega_k^2Q_{k\sigma}^2\right). \tag{6.24}$$

The potential $U_{\parallel}$ combines with the kinetic part to give the atomic bound states; for simplicity we have assumed only one charged particle, an electron ($q = -e$).

The Hamiltonian equations of motion are now modified to

$$\dot{Q}_{k\sigma} = P_{k\sigma} - \frac{iq}{2\omega_k\sqrt{\varepsilon_0 V}}\mathbf{v}\cdot\boldsymbol{\varepsilon}_{k\sigma}\left(e^{i\mathbf{k}\cdot\mathbf{r}} - e^{-i\mathbf{k}\cdot\mathbf{r}}\right) \tag{6.25a}$$

$$\dot{P}_{k\sigma} = -\omega_k^2Q_{k\sigma} + \frac{q}{2\sqrt{\varepsilon_0 V}}\mathbf{v}\cdot\boldsymbol{\varepsilon}_{k\sigma}\left(e^{i\mathbf{k}\cdot\mathbf{r}} + e^{-i\mathbf{k}\cdot\mathbf{r}}\right). \tag{6.25b}$$

In terms of the field amplitudes $\mathbf{C}_{k\sigma}$ this becomes

$$\dot{\mathbf{C}}_{k\sigma} = -i\omega_k\mathbf{C}_{k\sigma} + \frac{iq\boldsymbol{\varepsilon}_{k\sigma}}{2\omega_k\sqrt{\varepsilon_0 V}}\mathbf{v}\cdot\boldsymbol{\varepsilon}_{k\sigma}e^{-i\mathbf{k}\cdot\mathbf{r}}, \tag{6.26}$$

which gives for the field the equation of motion

$$\begin{aligned}\frac{\partial}{\partial t}\mathbf{E}(\mathbf{x}) &= \frac{i}{\sqrt{\varepsilon_0 V}}\sum_{k\sigma}\omega_k\left(\dot{\mathbf{C}}_{k\sigma}e^{i\mathbf{k}\cdot\mathbf{x}} - \dot{\mathbf{C}}^*_{k\sigma}e^{-i\mathbf{k}\cdot\mathbf{x}}\right)\\
&= \frac{1}{\sqrt{\varepsilon_0 V}}\sum_{k\sigma}\omega_k^2\left(\mathbf{C}_{k\sigma}e^{i\mathbf{k}\cdot\mathbf{x}} + \mathbf{C}^*_{k\sigma}e^{-i\mathbf{k}\cdot\mathbf{x}}\right)\\
&\quad - \frac{q}{2\varepsilon_0 V}\mathbf{v}\cdot\sum_{k\sigma}\boldsymbol{\varepsilon}_{k\sigma}\boldsymbol{\varepsilon}_{k\sigma}\left(e^{i\mathbf{k}\cdot(\mathbf{x}-\mathbf{r})} + e^{-i\mathbf{k}\cdot(\mathbf{x}-\mathbf{r})}\right)\\
&= c^2\nabla\times\mathbf{B}(\mathbf{x}) - \frac{q}{\varepsilon_0}\mathbf{v}\delta(\mathbf{x}-\mathbf{r}),\end{aligned} \tag{6.27}$$

because $\mathbf{v}$ is a transverse vector proportional to $\mathbf{E}_\perp$, and the completeness relation for the field eigenstates

$$\mathbf{U}_{k\sigma} = \frac{\boldsymbol{\varepsilon}_{k\sigma}}{\sqrt{V}}e^{i\mathbf{k}\cdot\mathbf{r}} \tag{6.28}$$

has been used. For one particle the current density is

$$\mathbf{j}(\mathbf{x}) = q\mathbf{v}\delta(\mathbf{x}-\mathbf{r}). \tag{6.29}$$

Here $\mathbf{r}$ is the quantum operator of the particle's position and $\mathbf{x}$ is the $c$-number variable giving the point where the field is to be evaluated. To obtain the magnetic field we have used

$$\begin{aligned}\omega_k^2\mathbf{C}_{k\sigma}e^{i\mathbf{k}\cdot\mathbf{x}} &= -c^2\nabla^2\left(\mathbf{C}_{k\sigma}e^{i\mathbf{k}\cdot\mathbf{x}}\right)\\
&= c^2\nabla\times\left(\nabla\times\left(\mathbf{C}_{k\sigma}e^{i\mathbf{k}\cdot\mathbf{x}}\right)\right),\end{aligned} \tag{6.30}$$

which directly gives the magnetic field. Thus the Hamiltonian (6.24) with the dynamic interpretation of $P_{k\sigma}$ and $Q_{k\sigma}$ gives the correct Maxwell's equation.

When the extension of the atomic system is much less than the wavelength

$$|\mathbf{x}| \ll \lambda, \tag{6.31}$$

we can replace the exponents by unity

$$e^{i\mathbf{k}\cdot\mathbf{x}} \simeq 1, \tag{6.32}$$

and the vector potential can be evaluated at the position of the atom, the origin say, and $A = A(0)$; this is called the *dipole approximation*. We now perform a gauge transformation of the vector potential. The phase of the wave function changes

$$\Psi' = e^{iq\chi/\hbar}\Psi \tag{6.33}$$

together with (6.3). The form of the Schrödinger equation is easily seen to remain invariant under the gauge transformation; see Section 1.3.

Because $\chi$ is real, the transformation is also a canonical transformation. In terms of the new state $\Psi'$ the Hamiltonian becomes

$$H' = H(\mathbf{p} - q\mathbf{A} - q\nabla\chi) - q\frac{\partial\chi}{\partial t}. \tag{6.34}$$

The last term can be considered as the gauge transformation of the scalar potential. Now choosing in the dipole approximation

$$\chi = -\mathbf{A}\cdot\mathbf{r}, \tag{6.35}$$

the new Hamiltonian becomes

$$H' = H(\mathbf{p}) + q\mathbf{r}\cdot\dot{\mathbf{A}} = H(\mathbf{p}) - q\mathbf{r}\cdot\mathbf{E}_\perp . \tag{6.36}$$

This is the dipole coupling of the particle to the electric field by the dipole operator $\mathbf{d} = q\mathbf{r}$. In lowest-order perturbation theory the two couplings are identical because for any states

$$\begin{aligned}\left\langle n\left|\frac{Ap}{m}\right|n'\right\rangle &= A\langle n|\dot{r}|n'\rangle = -i\omega_{nn'}A\langle n|r|n'\rangle \\ &\simeq -i\omega A\langle n|r|n'\rangle = \dot{A}\langle n|r|n'\rangle \\ &= -\langle n|rE|n'\rangle,\end{aligned} \tag{6.37}$$

where the resonance condition $\omega \simeq \omega_{nn'}$ has been used. For higher-order processes the correctly transformed states must be used. It is, however, expedient to use the dipole form of the interaction when treating bound atomic systems. All our earlier calculations have used this.

We have carried out the eigenmode decomposition of the electromagnetic field completely for classical variables. The introduction of the Hamiltonian variables $P_{k\sigma}$ and $Q_{k\sigma}$ in no way assumes quantization, it only forms the necessary preliminaries to it.

We have used the free space eigenmodes (6.28). As already mentioned, we could have expanded in the cavity eigenmodes of Eq. (1.21). The process of introducing harmonic oscillator coordinates remains the same, but now the amplitudes relate to different eigenconfigurations of the field. After quantization the photons sit in different modes of the field; their states are linear transformations of those obtained by the method chosen here.

## 6.3. QUANTIZATION OF THE FIELD

When we have a Hamiltonian formulation, we have a well-defined procedure for quantization. All we need to do to transform the results of the preceding section into a quantum theory of light is to postulate the commutators

$$[Q_{k\sigma}, P_{k'\sigma'}] = i\hbar\delta_{kk'}\delta_{\sigma\sigma'}. \tag{6.38}$$

After this all results of Sec. 6.2 can be interpreted in terms of quantized fields if some care is exercised with respect to the order of the operators. Complex conjugation is replaced with hermitean conjugation.

According to the practice in quantum theory, we introduce the operators

$$a_{k,\sigma} = \sqrt{\frac{\omega_k}{2\hbar}}\left(Q_{k\sigma} + i\frac{P_{k\sigma}}{\omega_k}\right)$$

$$a_{k,\sigma}^{\dagger} = \sqrt{\frac{\omega_k}{2\hbar}}\left(Q_{k\sigma} - i\frac{P_{k\sigma}}{\omega_k}\right). \tag{6.39}$$

Direct evaluation shows that these satisfy the commutation relation

$$\left[a_{k\sigma}, a_{k'\sigma'}^{\dagger}\right] = \delta_{kk'}\delta_{\sigma\sigma'}. \tag{6.40}$$

Thus we have obtained Boson operators where $a_{k\sigma}^{\dagger}$ creates an excitation of the harmonic oscillator mode ($\mathbf{k}\sigma$).

Insertion of $P_{k\sigma}$ and $Q_{k\sigma}$ into the definition (6.18) shows that $a_{k\sigma}$ is a scaled quantum version of $\mathbf{C}_{k\sigma}$, and the vector potential (6.11) now becomes

$$\mathbf{A}(\mathbf{r}, t) = \sum_{k\sigma}\sqrt{\frac{\hbar}{2\varepsilon_0\omega_k V}}\left[a_{k\sigma}\boldsymbol{\varepsilon}_{k\sigma}e^{i\mathbf{k}\cdot\mathbf{r}} + a_{k\sigma}^{\dagger}\boldsymbol{\varepsilon}_{k\sigma}e^{-i\mathbf{k}\cdot\mathbf{r}}\right]. \tag{6.41}$$

Inserting the inverse of (6.39) into the expression (6.20) for the total energy gives the expression

$$\begin{aligned} H &= \frac{1}{2}\sum_{k\sigma}\hbar\omega_k\left(a_{k\sigma}a_{k\sigma}^{\dagger} + a_{k\sigma}^{\dagger}a_{k\sigma}\right) \\ &= \sum_{k\sigma}\hbar\omega_k\left(a_{k\sigma}^{\dagger}a_{k\sigma} + \tfrac{1}{2}\right), \end{aligned} \tag{6.42}$$

where $a_{k\sigma}^{\dagger}a_{k\sigma}$ is the occupation number operator of the mode $(\mathbf{k}, \sigma)$. The term $\Sigma\omega_k/2$ is the (infinite) zero-point energy which is henceforth omitted.

The momentum of the field is calculated from the momentum density

$$\Pi = \varepsilon_0 \mathbf{E} \times \mathbf{B} \tag{6.43}$$

which is proportional to the energy flux vector (Poynting vector) $\mathbf{S} = \mathbf{E} \times \mathbf{H}$. Calculating the total flux of momentum we find

$$\mathbf{P} = \int \Pi \, d^3r = \sum_{k\sigma} \omega_k \mathbf{k}\left[\mathbf{C}_{k\sigma} \cdot \mathbf{C}_{k\sigma}^{\dagger} + \mathbf{C}_{k\sigma}^{\dagger} \cdot \mathbf{C}_{k\sigma}\right] \tag{6.44}$$

as terms like

$$\sum \mathbf{k}\omega_k \mathbf{C}_k \cdot \mathbf{C}_{-k}^{\dagger} = 0 \tag{6.45}$$

are omitted because the terms to be summed are odd with respect to $k$. Also

$$\begin{aligned} \mathbf{C}_k \times \left(\mathbf{k} \times \mathbf{C}_k^{\dagger}\right) &= \mathbf{k}\mathbf{C}_k \cdot \mathbf{C}_k^{\dagger} - (\mathbf{k} \cdot \mathbf{C}_k)\mathbf{C}_k^{\dagger} \\ &= \mathbf{k}\mathbf{C}_k \cdot \mathbf{C}_k^{\dagger}. \end{aligned} \tag{6.46}$$

In terms of the operators $a_{k\sigma}$ we find that

$$\mathbf{P} = \sum_{k\sigma} \hbar \mathbf{k} a_{k\sigma}^{\dagger} a_{k\sigma}. \tag{6.47}$$

There is no zero-point flux of momentum. Equation (6.47) shows the each quantum of the field mode $(\mathbf{k}, \sigma)$ carries the momentum $\hbar\mathbf{k}$. This agrees with a naive interpretation in terms of photons.

When introducing the dipole approximation the gauge transformation (6.35) is no longer explicitly time dependent, but $\mathbf{A}$ has now become an operator on the field variables, too. We can write

$$\chi = -\mathbf{A} \cdot \mathbf{r} = -\sum_{k\sigma} \left( A_k x_{k\sigma} a_{k\sigma} + A_k^* x_{k\sigma} a_{k\sigma}^{\dagger} \right) \tag{6.48}$$

where

$$x_{k\sigma} = \mathbf{r} \cdot \boldsymbol{\varepsilon}_{k\sigma}. \tag{6.49}$$

The action on the particle coordinates remains the same

$$\begin{aligned} \mathbf{p}' &= e^{-iq\chi/\hbar} \mathbf{p} e^{iq\chi/\hbar} = \mathbf{p} + q\nabla\chi = \mathbf{p} - q\mathbf{A} = m\mathbf{v} \\ \mathbf{r}' &= \mathbf{r} \end{aligned} \tag{6.50}$$

but the fields are transformed

$$a'_{k\sigma} = e^{-iq\chi/\hbar} a_{k\sigma} e^{iq\chi/\hbar} = a_{k\sigma} - i\frac{q}{\hbar}[\chi, a_{k\sigma}]$$

$$= a_{k\sigma} - \frac{iq}{\hbar} x_{k\sigma} \sqrt{\frac{\hbar}{2\varepsilon_0 \omega_k V}} \,. \tag{6.51}$$

This transforms the field energy according to

$$H'_f = \sum_{k\sigma} \hbar\omega_k \left( a'^{\dagger}_{k\sigma} - \frac{iqx_{k\sigma}A_k}{\hbar} \right) \left( a'_{k\sigma} + \frac{iqx_{k\sigma}A^*_k}{\hbar} \right)$$

$$= \sum_{k\sigma} \hbar\omega_k a'^{\dagger}_{k\sigma} a'_{k\sigma} + iq\mathbf{r} \cdot \sum_{k\sigma} \sqrt{\frac{\hbar\omega_k}{2\varepsilon_0 V}}\, \boldsymbol{\varepsilon}_{k\sigma} \left( a'^{\dagger}_{k\sigma} - a'_{k\sigma} \right)$$

$$+ q^2 \sum_{k\sigma} (\mathbf{r} \cdot \boldsymbol{\varepsilon}_{k\sigma})^2 \frac{1}{2\varepsilon_0 V}$$

$$= H_f - q\mathbf{r} \cdot \mathbf{E} + \frac{q^2 \mathbf{r}^2}{2V\varepsilon_0} \sum_{k\sigma} . \tag{6.52}$$

The interaction term emerges correctly, and the last term is an infinite dipole–dipole self-energy for the charged particle. This can again be omitted; it is no operator for the fields.

## 6.4. SOME PERTURBATION CALCULATIONS

First we want to use the result of time-dependent perturbation theory to calculate the decay from a level $|2>$ to a lower level $|1>$ by spontaneous emission. The rate is given by

$$\Gamma_{k\sigma} = \frac{2\pi}{\hbar^2} |V_{if}|^2 \delta(\omega_2 - \omega_k - \omega_1) \tag{6.53}$$

where $\hbar\omega_k$ is the energy of the emerging photon.

Initially the atomic state is $|2>$ and the field is vacuum, finally the state is $|1>$ and the mode $(\mathbf{k}, \sigma)$ has $n_{k\sigma} = \langle a^{\dagger}_{k\sigma} a_{k\sigma} \rangle = 1$. Thus in the dipole approximation (6.32)

$$\begin{aligned}\langle f|V|i\rangle &= \langle 1, n_{k\sigma} = 1|\frac{e}{m}\mathbf{A}\cdot\mathbf{p}|2\rangle \\ &= \frac{e}{m}\sum_{k'\sigma'}\langle 1, n_{k\sigma} = 1|\big(A_{k'}\boldsymbol{\varepsilon}_{k'\sigma'}\cdot\mathbf{p}a_{k'\sigma'} + A_{k'}\boldsymbol{\varepsilon}_{k'\sigma'}\cdot\mathbf{p}a^{\dagger}_{k'\sigma'}\big)|2\rangle \\ &= \frac{e}{m}A_k\langle 1|\boldsymbol{\varepsilon}_{k\sigma}\cdot\mathbf{p}|2\rangle = eA_k\langle 1|\boldsymbol{\varepsilon}_{k\sigma}\cdot\dot{\mathbf{r}}|2\rangle = -ie\omega_{21}A_k x_{12}, \quad (6.54)\end{aligned}$$

as in Eq. (6.37). From (6.41) we find that

$$A_k = \sqrt{\frac{\hbar}{2\varepsilon_0\omega_k V}}, \tag{6.55}$$

and hence, from (6.53), the total rate of decay becomes

$$\begin{aligned}\Gamma &= \sum_{k\sigma}\Gamma_{k\sigma} = \frac{\pi}{V}\sum_{k\sigma}\frac{e^2x_{12}^2}{\hbar\omega_k\varepsilon_0}\omega_{21}^2\delta(\omega_{21} - \omega_k) \\ &= \frac{\pi e^2}{(2\pi)^3\hbar\varepsilon_0}\int d^3k\sum_{\sigma}\omega_k|\boldsymbol{\varepsilon}_{k\sigma}\cdot\mathbf{r}_{12}|^2\frac{\delta\left(k - \frac{\omega_{21}}{c}\right)}{c} \\ &= \frac{e^2}{8\pi^2\hbar\varepsilon_0}\left(\frac{\omega_{21}}{c}\right)^3\int d\hat{\Omega}\sum_{\sigma}|\boldsymbol{\varepsilon}_{k\sigma}\cdot\mathbf{r}_{12}|^2. \quad (6.56)\end{aligned}$$

Because $\boldsymbol{\varepsilon}_{k\sigma}\cdot\hat{\mathbf{k}} = 0$ we have

$$\mathbf{r} = \sum_{\sigma}(\boldsymbol{\varepsilon}_{k\sigma}\cdot\mathbf{r})\boldsymbol{\varepsilon}_{k\sigma} + (\hat{\mathbf{k}}\cdot\mathbf{r})\hat{\mathbf{k}}, \tag{6.57}$$

and consequently

$$\sum_{\sigma}(\boldsymbol{\varepsilon}_{\sigma}\cdot\mathbf{r})^2 = \mathbf{r}\cdot\mathbf{r} - (\hat{\mathbf{k}}\cdot\mathbf{r})^2 = r^2(1 - \cos^2\theta). \tag{6.58}$$

The integral now becomes

$$\int d\hat{\Omega}(1 - \cos^2\theta) = 2\pi\int(1 - \cos^2\theta)\,d(\cos\theta) = \frac{8\pi}{3}. \tag{6.59}$$

With these results the decay rate (6.56) becomes

$$\Gamma = \frac{e^2 r_{12}^2 \omega_{21}^3}{3\pi\varepsilon_0 \hbar c^3}, \tag{6.60}$$

as given already in Eq. (1.113). Because a factor $\delta(\omega_{21} - \omega_k)$ occurs, the result is also obtained from the operator $-e\mathbf{r}\cdot\mathbf{E}$, as is easily verified with the aid of Eq. (6.37).

The spontaneous emission result (6.60) can be obtained only from the quantized theory. The semiclassical approach does not give it.

There is also a level shift connected with the quantized radiation field. This is obtained in second-order time-independent perturbation theory. If we want to know the shift of the level $|b>$ we must evaluate

$$\Delta E_b = \sum_s \frac{\langle b|V|s\rangle\langle s|V|b\rangle}{E_b - E_s}. \tag{6.61}$$

Introducing into (6.61) the virtual states with one photon excited, we find

$$\begin{aligned}\Delta E_b &= \sum_{a,k,\sigma} \frac{\left\langle b\left|\frac{e\mathbf{A}\cdot\mathbf{p}}{m}\right|a, n_{k\sigma} = 1\right\rangle\left\langle n_{k\sigma} = 1, a\left|\frac{e\mathbf{A}\cdot\mathbf{p}}{m}\right|b\right\rangle}{\hbar\omega_b - \hbar\omega_a - \hbar\omega_k} \\ &= \frac{e^2}{2\varepsilon_0(2\pi)^3}\sum_{a\sigma}\int d^3k \frac{\omega_{ba}^2}{\omega_k}\frac{|\langle b|\boldsymbol{\varepsilon}_{k\sigma}\cdot\mathbf{r}|a\rangle|^2}{\omega_{ba} - \omega_k}\end{aligned} \tag{6.62}$$

where we have used the results derived in this section.

At the upper limit this integral diverges linearly $\int dk \to \infty$. Bethe found that we have to subtract the energy of a free electron in the same approximation namely,

$$\begin{aligned}\Delta E_0 &= \sum_{ak\sigma} \frac{\langle b|\frac{e}{m}\mathbf{A}\cdot\mathbf{p}|a, k\sigma\rangle\langle k\sigma, a|\frac{e}{m}\mathbf{A}\cdot\mathbf{r}|b\rangle}{-\hbar\omega_k} \\ &= -\frac{e^2}{2\varepsilon_0}\sum_{a\sigma}\int \frac{d^3k}{(2\pi)^3}\omega_{ba}^2\frac{|\langle b|\boldsymbol{\varepsilon}_{k\sigma}\cdot\mathbf{r}|a\rangle|^2}{\omega_k^2}.\end{aligned} \tag{6.63}$$

The energy difference becomes, from (6.62) and (6.63)

$$\Delta E = \Delta E_b - \Delta E_0 = \frac{e^2}{2\varepsilon_0(2\pi)^3} \sum_{a\sigma} \int d^3k \frac{\omega_{ba}^2}{\omega_k} |\langle b|\boldsymbol{\varepsilon}_{k\sigma} \cdot \mathbf{r}|a\rangle|^2$$

$$\times \left[ \frac{1}{\omega_{ba} - \omega_k} + \frac{1}{\omega_k} \right]$$

$$= \frac{e^2}{2\varepsilon_0(2\pi)^3} \sum_a \omega_{ba}^3 \int \frac{k^2 dk}{\omega_k^2} \left( \frac{1}{\omega_{ba} - \omega_k} \right) \sum_\sigma \int d\hat{\Omega} |\langle b|\boldsymbol{\varepsilon}_{k\sigma} \cdot \mathbf{r}|a\rangle|^2. \tag{6.64}$$

The term to the right is recognized from (6.56) and can be evaluated the same way. We finally have

$$\Delta E = \frac{e^2}{6\pi^2 \varepsilon_0 c^3} \sum_a \omega_{ba}^3 r_{ab}^2 \int_0^K \frac{dk}{\frac{\omega_{ba}}{c} - k}$$

$$= \frac{e^2}{6\pi^2 \varepsilon_0 c^3} \sum_a \omega_{ab}^3 r_{ab}^2 \log \left| \frac{cK}{\omega_{ba}} \right|. \tag{6.65}$$

Here $K$ is some cutoff in $k$-space taken to satisfy $K < mc/\hbar$ because our approach excludes relativistic effects.

To proceed we evaluate

$$\sum_a (\omega_{ab} r_{ab})^2 (\omega_a - \omega_b) = \frac{1}{m^2 \hbar} \sum_a p_{ab}^2 (E_a - E_b)$$

$$= \frac{1}{2m^2 \hbar} \sum_{i=1}^{3} \langle b | \sum p_i [H, p_i] | b \rangle$$

$$= \frac{\hbar e}{2m^2} \sum_{i=1}^{3} \left\langle b \left| \frac{\partial^2 \varphi}{\partial x_i^2} \right| b \right\rangle$$

$$= \frac{\hbar e}{2m^2} \int \psi_b^*(x) \nabla^2 \varphi \psi_b(x) d^3x \tag{6.66}$$

where $\varphi$ is the Coulomb potential due to the central charge of strength $Ze$. But the Laplacian is given by the charge density, and hence

$$\sum_a (\omega_{ab} r_{ab})^2 \omega_{ab} = \frac{Z\hbar e^2}{2m^2 \varepsilon_0} |\psi_b(0)|^2, \tag{6.67}$$

where $\psi_b(0)$ is the wave function of the state $\psi_b$ at the position of the nucleus. We collect the results into

$$\Delta E = \frac{Z\hbar e^4}{12\pi^2 \varepsilon_0^2 c^3 m^2} |\psi_b(0)|^2 \log\left|\frac{cK}{\overline{\omega_{ba}}}\right|, \tag{6.68}$$

where $\overline{\omega_{ba}}$ is an average excitation energy. This is the Bethe result for the quantum shift of the levels.

For the hydrogen levels $2S_{1/2}$ and $2P_{1/2}$, which are degenerate according to Dirac's theory, this shift is of interest. Because the $p$-wave (nonrelativistically) does not reach the origin, only the $s$-state is shifted. Here we have

$$|\psi_s(0)|^2 = \frac{1}{\pi}\left(\frac{Z}{an}\right)^3, \tag{6.69}$$

and with $Z = 1$ and $n = 2$ we find

$$\Delta E = \frac{\alpha^5 mc^2}{6\pi} \log\left|\frac{cK}{\overline{\omega_{ba}}}\right| = 135.82 \log\left|\frac{cK}{\overline{\omega_{ba}}}\right| (\text{MHz}), \tag{6.70}$$

where $\alpha$ is the fine structure constant $(e^2/4\pi\varepsilon_0 \hbar c)$.

If we use the upper limit $mc/\hbar$ for $K$ and the experimentally obtained value

$$\hbar\overline{\omega_{n0}} = \overline{E_n - E_{2s}} = 242.2 \text{ eV}, \tag{6.71}$$

we find from (6.69) the result $\Delta E = 1040$ MHz.

This nonrelativistic estimate turns out to be in astonishing agreement with the measured Lamb shift between the levels $2S_{1/2}$ and $2P_{1/2}$. A more precise relativistic calculation leads to perfect agreement between the theory and experiments. It is a series of successful calculations like this one that have established quantum electrodynamics as a valid theory in spite of the numerous singularities that appear at first sight. The full renormalization theory teaches us how to eliminate these in a systematic manner. What is left then is a good description of experimental reality.

## 6.5. SPONTANEOUS DECAY TERMS FOR THE DENSITY MATRIX

As in the perturbation calculation of the preceding section, an excited atomic state can decay spontaneously with the emission of a photon into empty vacuum. Because all transitions must be induced by a preexisting field in a semiclassical description, this allows no spontaneous decay. It is, consequently, a genuine characteristic of the quantized field theory.

In spectroscopic applications we often do not record the spontaneously emitted photons. We notice spontaneous decay only because of its effect on the state of the atom. Thus we restrict our attention to the variables characterizing the atomic state. The escape of excitation energy into the radiation field is noted only as a dissipative mechanism. It was introduced without a derivation in Sec. 1.6. Here we want to derive the spontaneous emission terms as they appear in the density matrix equation of motion. Then the dissipation appears in a way that is analogous to the emergence of relaxation due to a heat bath. In this case the heat bath can be understood as the quantum fluctuations of the photon vacuum.

The atomic density matrix is a reduced one where we have traced out all the states of the field

$$\rho_{\text{at}} = \text{Tr}_{\text{field}}\,\rho. \tag{6.72}$$

To be specific, we choose to treat a two-level system and include only the states with one photon present in the Hilbert space of the quantized field states. Hence we write

$$\rho_{\text{at}} = \langle 0|\rho|0\rangle + \sum_q \langle q|\rho|q\rangle \tag{6.73}$$

where $|0\rangle$ is the field vacuum and $|q\rangle$ denotes a state with one photon in the field mode $\mathbf{q}$. There are, of course, terms with more photons in the definition (6.72), but they turn out to be unimportant for our case.

The Hamiltonian we want to treat is given by

$$H = \hbar\omega_{21}|2\rangle\langle 2| + \hbar\sum_q \Omega_q a_q^\dagger a_q$$
$$+ i\hbar\sum_q \lambda(q)\big(|2\rangle a_q\langle 1| - |1\rangle a_q^\dagger\langle 2|\big). \tag{6.74}$$

The processes included are easily recognized as those allowed in the rotating wave approximation. From Eqs. (6.14a), (6.39), (6.18), and (6.36) we can derive for the coupling constant

$$\lambda(q) = \frac{ex_{12}}{\hbar}\left[\frac{\hbar\omega_q}{2\varepsilon_0 V}\right]^{1/2}. \tag{6.75}$$

The states of interest to us are $|0,1\rangle$, $|0,2\rangle$, $|q,1\rangle$, and $|q,2\rangle$. The only off-diagonal element of the Hamiltonian is

$$\langle 0,2|H|q,1\rangle = i\hbar\lambda(q). \tag{6.76}$$

The equation of motion for the density matrix gives

$$\frac{d}{dt}\langle 0,2|\rho|0,2\rangle = \sum_q \lambda(q)(\langle q,1|\rho|0,2\rangle + \langle 0,2|\rho|q,1\rangle) \tag{6.77a}$$

$$\frac{d}{dt}\langle 0,2|\rho|0,1\rangle = -i\omega_{21}\langle 0,2|\rho|0,1\rangle + \sum_q \lambda(q)\langle q,1|\rho|0,1\rangle \tag{6.77b}$$

$$\frac{d}{dt}\langle q,1|\rho|q,1\rangle = -\lambda(q)(\langle 0,2|\rho|q,1\rangle + \langle q,1|\rho|0,2\rangle) \tag{6.77c}$$

$$\begin{aligned}\frac{d}{dt}\langle q,1|\rho|0,2\rangle &= i(\omega_{21} - \Omega_q)\langle q,1|\rho|0,2\rangle \\ &\quad -\lambda(q)\langle 0,2|\rho|0,2\rangle + \sum_{q'} \lambda(q')\langle q,1|\rho|q',1\rangle\end{aligned} \tag{6.77d}$$

and the additional terms needed. Our assumption that the radiation field acts as a random bath now suggests that we neglect the elements $\langle q|\rho|q'\rangle$ for $q \neq q'$. These designate coherences between different photon states. If each photon is allowed to escape the system irreversibly, no such coherence is able to take effect.

From (6.77d) we then obtain

$$\begin{aligned}\langle q,1|\rho|0,2\rangle = -\lambda(q)\int_0^t \exp\left[i(\omega_{21} - \Omega_q)(t - t')\right] \\ \times\langle 0,2|\rho(t')|0,2\rangle dt'.\end{aligned} \tag{6.78}$$

Inside the integral we need further approximations. The rapidly fluctuating exponent makes the integral vanish unless $t \simeq t'$, and hence the density matrix can be taken outside of the integral. We then calculate

$$\int_0^t dt' e^{i(\omega_{21} - \Omega_q)(t-t')} = \int_0^t d\tau e^{i(\omega_{21} - \Omega_q)\tau}. \tag{6.79}$$

The integral must be carried out in such a way that its value at infinite times is well defined. In scattering theory this is accomplished by the addition of a small imaginary part $\eta$ to the frequencies; at the end of the calculations it goes to zero. We evaluate

$$\int_0^t e^{i(\omega_{21}-\Omega_q+i\eta)\tau}d\tau = -\frac{1-e^{-\eta t}e^{i(\omega_{21}-\Omega_q)t}}{i(\omega_{21}-\Omega_q+i\eta)}$$

$$\underset{\substack{t\to\infty\\ \eta\to 0}}{\to} i\left[\frac{\mathscr{P}}{\omega_{21}-\Omega_q} - i\pi\delta(\omega_{21}-\Omega_q)\right], \tag{6.80}$$

where a well-known property of the $\delta$-function is used. The symbol $\mathscr{P}$ denotes the Cauchy principal value. In (6.78) we now have

$$\langle q,1|\rho|0,2\rangle = -i\lambda(q)\left[\frac{\mathscr{P}}{\omega_{21}-\Omega_q} - i\pi\delta(\omega_{21}-\Omega_q)\right]\langle 0,2|\rho|0,2\rangle. \tag{6.81}$$

This expression and its complex conjugate can be introduced into (6.77a). To calculate $\rho_{22}$ from (6.73) we also need the equation of motion for $\langle q,2|\rho|q,2\rangle$, but this can only couple to terms like $\langle q,2|\rho|q,q',1\rangle$, which contains a two-photon state $|q,q'\rangle$ and its correlation to the state $|q\rangle$. This must be ignored for consistency. We find from (6.73), (6.77a) and (6.81) the result

$$\dot{\rho}_{22} = -\Gamma\rho_{22} \tag{6.82}$$

where

$$\Gamma = 2\pi\sum_q \lambda(q)^2\delta(\omega_{21}-\Omega_q). \tag{6.83}$$

Introducing (6.75) we can easily see that this agrees with the result in (6.56). For simplicity we have not treated the polarization in this section; it is implicitly understood. Thus the present $\Gamma$ is given by (6.60) too.

We want to treat the matrix element $\rho_{21}$, which according to (6.77b) couples to $\langle q,1|\rho|0,1\rangle$. Just like in deriving (6.78) we obtain

$$\langle q,1|\rho|0,1\rangle = -\lambda(q)\int_0^t e^{-i\Omega_q(t-t')}\langle 0,2|\rho(t')|0,1\rangle\,dt'. \tag{6.84}$$

Now the matrix $\langle 0,2|\rho(t')|0,1\rangle$ cannot be taken outside the integral. It has a rapid time variation due to a factor $e^{-i\omega_{21}t'}$. The quantities

$$\langle 0,2|\rho(t)|0,1\rangle e^{i\omega_{21}t} \simeq e^{i\omega_{21}t'}\langle 0,2|\rho(t')|0,1\rangle \tag{6.85}$$

are, however, slowly varying, and hence

$$\begin{aligned}\langle q,1|\rho|0,1\rangle &= -\lambda(q)\langle 0,2|\rho(t)|0,1\rangle \\ &\quad\times \int_0^t e^{i(\omega_{21}-\Omega_q)(t-t')}dt'. \end{aligned} \tag{6.86}$$

From here we obtain, as before,

$$\langle q,1|\rho|0,1\rangle = -i\lambda(q)\left[\frac{\mathscr{P}}{\omega_{21}-\Omega_q} - i\pi\delta(\omega_{21}-\Omega_q)\right]\langle 0,2|\rho|0,1\rangle. \tag{6.87}$$

The element $\langle q,2|\rho|q,1\rangle$ can be treated in an analogous way in our basis set. Hence we insert (6.87) into (6.77b) and obtain

$$\frac{d}{dt}\rho_{21} = -i(\omega_{21}+\Delta\omega)\rho_{21} - \tfrac{1}{2}\Gamma\rho_{21}, \tag{6.88}$$

where the decay constant $\Gamma$ is the same as in (6.83). The frequency shift

$$\Delta\omega = -\mathscr{P}\sum_q \frac{\lambda(q)^2}{\Omega_q - \omega_{21}} \tag{6.89}$$

is the equivalent of a Lamb shift for our two-level system. When we transform it to an integral it behaves at the upper limit like $\int q^2 dq$ and is badly divergent. To regularize this shift is not of interest here, when considering a highly truncated level scheme. In the following discussion we neglect this frequency shift altogether.

There is one more term of interest, the lower level. The density matrix element $\langle 0,1|\rho|0,1\rangle$ cannot couple to the field, but then the second sum in (6.73) becomes important.

From Eq. (6.77c) we find that we need the elements already calculated in (6.81). Adding all $q$-modes together we find

$$\dot{\rho}_{11} = \Gamma\rho_{22}, \tag{6.90}$$

as we expect for spontaneous emission. Thus the results (6.82), (6.88), and

(6.90) give the spontaneous emission terms as introduced in Eqs. (1.110)–(1.112). This is our only case where the coherences decay more slowly than the populations. The factor $\frac{1}{2}$ in (6.88) derives from the fact that only the state $|2\rangle$ in $\rho_{21}$ is capable of decaying spontaneously.

The quantum field degrees of freedom thus lead to a decay of the atomic degrees of freedom, if the field is treated as an incoherent bath lacking the capability to sustain coherence between its states.

## 6.6. RESONANCE FLUORESCENCE IN A STRONG FIELD

In the preceding section we calculated the spontaneous decay effects observable in the atomic system only. If we look at the spectrum of the emitted light we can obtain new information; in the foregoing discussion this was allowed to get lost.

To discover what is observed by a spectrally selective photon counter we must consider the possibility of determining the probability that after the decay we have a photon present in the state $|q\rangle$. This is, however, given by $\langle q,1|\rho|q,1\rangle$. The counting rate must be proportional to the rate at which this element increases with time, or we propose as the observable

$$W_q = \frac{d}{dt}[\langle q,1|\rho|q,1|\rangle]_{\text{spont}}, \tag{6.91}$$

where only the growth due to spontaneous emission can be seen in the detector. From Eqs. (6.77c) we obtain

$$W_q = -2\lambda(q)\text{Re}\langle q,1|\rho|0,2\rangle. \tag{6.92}$$

When we assume that no strong fields act on the atomic system, but the only influence is from spontaneous decay, we can use the result (6.81) calculated in Sec. 6.5. We obtain

$$W_q = \frac{2\pi}{\hbar}|\hbar\lambda(q)|^2\delta(\hbar\omega_{21} - \hbar\Omega_q)\rho_{22}. \tag{6.93}$$

This is the result expected from time-dependent perturbation theory with the coupling matrix element given by (6.75). It describes elastic emission at the resonance frequency $\Omega_q = \omega_{21}$, once there is an excited state population given by $\rho_{22}$. This result shows why one can use spontaneous emission to measure the transfer of population to a level, assuming that the initial and final levels of spontaneous emission are not coupled by any external fields.

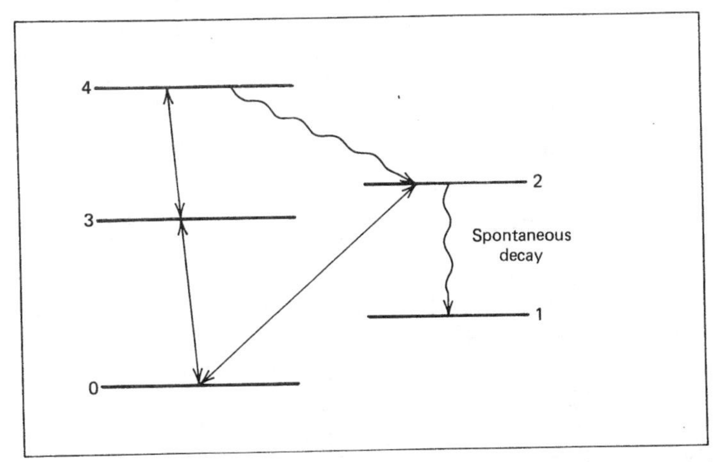


**Fig. 6.1** The situation where the spontaneous decay between levels $|2\rangle$ and $|1\rangle$ can be taken to be independent of other fields. If $|1\rangle$ is not coupled to any level, the decay can be added as if the other fields were not present. Only if the field on the transition $|0\rangle$ to $|2\rangle$ becomes very strong will the rate of emission out of $|2\rangle$ to $|1\rangle$ be affected.

The situation is illustrated in Fig. 6.1. When strong field effects enter, spontaneous decay to an uncoupled level also becomes more complex.

It is usual to create a population $\rho_{22}$ by letting a light field connect levels 1 and 2. If this is not too strong we can assume that the spontaneous emission proceeds unperturbed and the derivation of the preceding section still holds true. Then Eq. (6.93) remains valid, and $\rho_{22}$ is just taken from the treatment of a two-level system in a classical field. The spontaneous decay is not affected by the addition of the independent classical coupling. This assumption has been inherent in our treatment so far.

When the field is highly coherent we can no longer assume that the spontaneous emission remains unaffected. The spectrum of spontaneous emission changes and depends on the strong field. It becomes an interesting question how this modifies the spontaneous emission of energy.

The experimental setup is shown schematically in Fig. 6.2. A strong laser beam is made to cross an atomic beam at right angle, and the detector observes emission into the third orthogonal direction. In this configuration there are no Doppler effects involved in any radiation. The experimental difficulty is to achieve enough intensity for the recording with essentially single atoms interacting independently with the field.

For a strong field, the upper level is occupied with the probability $\frac{1}{2}$, and the rate of spontaneous emission events is given by $\frac{1}{2}\Gamma$. Those are not

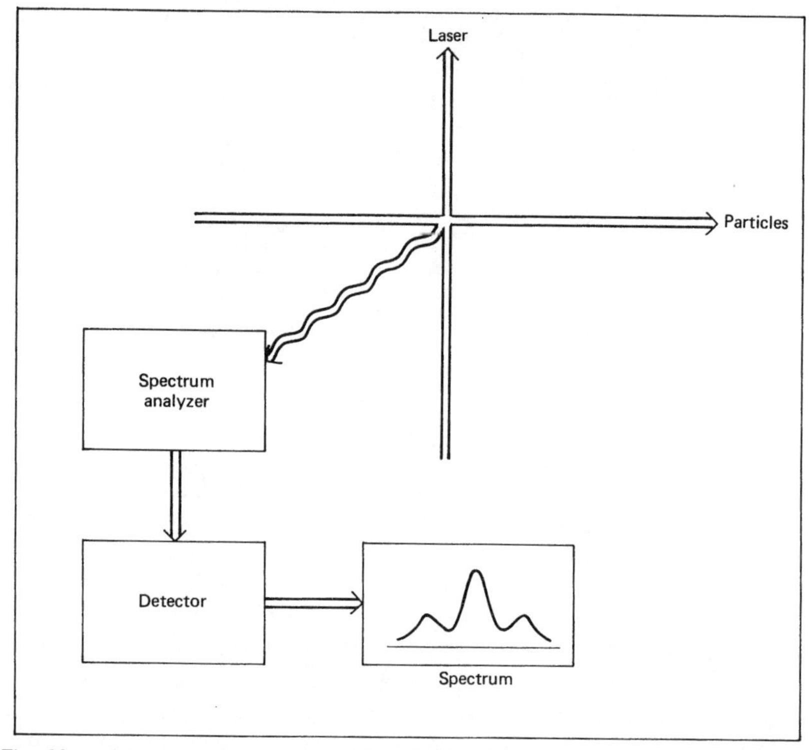


**Fig. 6.2** A beam of particles crosses a laser beam. The fluorescence is monitored in the orthogonal direction, and the spectral distribution of emitted photons is observed.

uncorrelated, and for $n$ photons emitted we must absorb $n$ quanta from the laser field of frequency $\Omega$. Overall energy conservation requires only that

$$\sum_{\nu=1}^{n} \Omega_{q\nu} = n\Omega. \tag{6.94}$$

The individual frequencies $\Omega_q$ need no longer coincide with $\omega_{21}$ or $\Omega$. The spectral measurement (6.92) will record the statistical distribution of the frequencies $\Omega_q$. The ensuing situation is highly nonperturbative in the laser field, and a nonlinear problem must be solved.

From Eq. (6.92) we see that we need to calculate the quantity

$$\langle 0, 2|\rho|q, 1\rangle = \langle 2|\rho a_q^\dagger|1\rangle. \tag{6.95}$$

Instead of calculating the equation of motion for $\rho$ we must now calculate that of $\rho a_q^\dagger$. As only $\rho$ depends on time in the Schrödinger picture, we have

$$i\hbar\frac{d}{dt}\left(\rho a_q^\dagger\right) = [H, \rho]a_q^\dagger = \left[H, \rho a_q^\dagger\right] - \rho\left[H, a_q^\dagger\right]. \tag{6.96}$$

The first term on the right will give the same terms as for $\rho$ alone, see Eq. (6.77), but the presence of the commutator $[H, a_q^\dagger]$ will serve to modify the result. Our Hamiltonian is (6.74) with the coherent field added

$$\begin{aligned} H = {} & \hbar\omega_{21}|2\rangle\langle 2| + \hbar\sum_q \Omega_q a_q^\dagger a_q \\ & + i\hbar\sum_q \lambda(q)\left(|2\rangle a_q\langle 1| - |1\rangle a_q^\dagger\langle 2|\right) \\ & - \tfrac{1}{2}\mu_{12}E\left[|2\rangle\langle 1|e^{-i\Omega t} + |1\rangle\langle 2|e^{i\Omega t}\right]. \end{aligned} \tag{6.97}$$

From Eq. (6.96) we find

$$i\hbar\frac{d}{dt}\left(\rho a_q^\dagger\right) = \left[H, \rho a_q^\dagger\right] - \hbar\Omega_q\rho a_q^\dagger + i\hbar\lambda(q)\rho|2\rangle\langle 1|. \tag{6.98}$$

From the commutator we have the earlier well-known terms due to the strong field $E$ and the terms due to spontaneous emission just as in Eqs. (6.77). In this part of the equation of motion the spontaneous emission is treated as previously; we must calculate density matrix elements and must neglect field coherences. Then we can define field-averaged density matrix elements by a trick similar to that of Eq. (6.73), namely,

$$\rho_{ijq} \equiv \langle 0, i|\rho|q, j\rangle + \sum_{q'}\langle q', i|\rho|q', q, j\rangle. \tag{6.99}$$

With this definition we obtain the decay rates of spontaneous emission again.

The term of (6.98) with $\hbar\Omega_q$ adds this frequency to that of the density matrix elements $\langle i|\rho a_q^\dagger|j\rangle = \langle 0, i|\rho|q, j\rangle$, as seen, for example, in Eq. (6.77d). The term $\rho|2\rangle\langle 1|$ is new; we have not encountered its kind before. It contains no photon variables, but couples, for example, the density matrix element $\langle 2|\rho|1\rangle$ to $\langle 2|\rho|2\rangle$ and $\langle 1|\rho|1\rangle$ to $\langle 1|\rho|2\rangle$. In this way the one-photon density matrix $\rho_{ijq}$ of (6.92) will be driven by the values of the ordinary density matrix $\rho_{22}$ and $\rho_{12}$ as defined by Eq. (6.73). These are, of course, to be calculated in the presence of the strong field $E$. When all this is done we obtain the equations

$$i\frac{d}{dt}\rho_{21q} = \left(\omega_{21} - \Omega_q - i\frac{1}{2}\Gamma\right)\rho_{21q} - \frac{1}{2\hbar}\mu_{12}Ee^{-i\Omega t}\left(\rho_{11q} - \rho_{22q}\right) - i\lambda(q)\rho_{22} \tag{6.100a}$$

$$i\frac{d}{dt}\rho_{11q} = -\Omega_q\rho_{11q} + i\Gamma\rho_{22q} - \frac{1}{2\hbar}\mu_{12}E\left(e^{i\Omega t}\rho_{21q} - e^{-i\Omega t}\rho_{12q}\right) - i\lambda(q)\rho_{12} \tag{6.100b}$$

$$i\frac{d}{dt}\rho_{22q} = -\left(\Omega_q + i\Gamma\right)\rho_{22q} - \frac{1}{2\hbar}\mu_{12}E\left(e^{-i\Omega t}\rho_{12q} - e^{i\Omega t}\rho_{21q}\right) \tag{6.100c}$$

$$i\frac{d}{dt}\rho_{12q} = -\left(\Omega_q + \omega_{21} + i\frac{1}{2}\Gamma\right)\rho_{12q} - \frac{1}{2\hbar}\mu_{12}Ee^{i\Omega t}\left(\rho_{22q} - \rho_{11q}\right). \tag{6.100d}$$

Note here that $\rho_{12q} \neq (\rho_{21q})^*$. This is due to the position of $a_q^\dagger$ in Eq. (6.98). If we now introduce a rotating wave approximation by setting

$$\begin{aligned} \rho_{11q} &= e^{i\Omega t}\tilde{\rho}_{11q} \qquad & \rho_{22q} &= e^{i\Omega t}\tilde{\rho}_{22q} \\ \rho_{12q} &= e^{2i\Omega t}\tilde{\rho}_{12q} \qquad & \rho_{21q} &= \tilde{\rho}_{21q} \\ \rho_{12} &= \tilde{\rho}_{12}e^{i\Omega t} \end{aligned} \tag{6.101}$$

and require steady state, we find the equation

$$\begin{bmatrix} \nu + \Delta - i\frac{1}{2}\Gamma & -\alpha & \alpha & 0 \\ -\alpha & \nu & i\Gamma & \alpha \\ \alpha & 0 & \nu - i\Gamma & -\alpha \\ 0 & \alpha & -\alpha & \nu - \Delta - i\frac{1}{2}\Gamma \end{bmatrix} \begin{bmatrix} \tilde{\rho}_{21q} \\ \tilde{\rho}_{11q} \\ \tilde{\rho}_{22q} \\ \tilde{\rho}_{12q} \end{bmatrix} = i\lambda(q) \begin{bmatrix} \rho_{22} \\ \tilde{\rho}_{12} \\ 0 \\ 0 \end{bmatrix} \tag{6.102}$$

where for simplicity we have set

$$\Delta = \omega_{21} - \Omega, \qquad \nu = \Omega - \Omega_q, \qquad \alpha = \frac{1}{2\hbar}\mu_{12}E. \tag{6.103}$$

$\Delta$ is the laser detuning from atomic resonance, $\nu$ is the spontaneous emission frequency referred to the laser frequency $\Omega$, and $\alpha$ is the coupling constant frequency of the strong field.

It is easy to solve the element $\tilde{\rho}_{21q}$ from here by inverting the matrix to obtain the observed quantity (6.92). We must also take $\rho_{22}$ and $\tilde{\rho}_{12}$ from the solution of the two-level system in the strong field. These equations are

$$i\frac{d}{dt}\rho_{22} = -i\Gamma\rho_{22} - \alpha(\tilde{\rho}_{12} - \tilde{\rho}_{21}) \tag{6.104a}$$

$$i\frac{d}{dt}\tilde{\rho}_{21} = (\Delta - i\tfrac{1}{2}\Gamma)\tilde{\rho}_{21} - \alpha(\rho_{11} - \rho_{22}) \tag{6.104b}$$

$$\rho_{22} + \rho_{11} = 1, \tag{6.105}$$

where the rotating wave approximation, the notation of this section, and the spontaneous decays of the preceding section have been introduced. The steady state solution becomes

$$\rho_{22} = \frac{\alpha^2}{\Delta^2 + (\Gamma/2)^2 + 2\alpha^2} \tag{6.106a}$$

$$\tilde{\rho}_{21} = \alpha\frac{\Delta + i\frac{1}{2}\Gamma}{\Delta^2 + (\Gamma/2)^2 + 2\alpha^2} = \frac{\Delta + i\frac{1}{2}\Gamma}{\alpha}\rho_{22}\,. \tag{6.106b}$$

These occur on the right-hand side of Eq. (6.102).

The matrix of Eq. (6.102) is singular at $\nu = 0$, which is most easily seen from the fact that the second and third row are the same with opposite signs when $\nu = 0$. Thus the determinant becomes zero. This zero signifies the elastic scattering of the incoming photon at frequency $\Omega$ into an outgoing one at $\Omega_q = \Omega$. This totally elastic scattering, Rayleigh scattering, thus prevails independently of the saturation behavior of the two-level system.

The existence of the elastic peak follows directly from a classical consideration. The atomic dipole is driven by the external field, and it oscillates at the frequency of the external driving force just like a classical oscillator. The induced dipole moment radiates, and this radiation has just the frequency of the incoming light.

When the quantity $\tilde{\rho}_{21q}$ is solved from (6.102) at $\nu \sim 0$, we see that the dependence on $\rho_{22}$ disappears, and solving straightforwardly we obtain

$$\begin{aligned}\tilde{\rho}_{21q} &= \frac{i\alpha\lambda(q)(\Delta + i\Gamma/2)}{\nu\left[\Delta^2 + (\Gamma/2)^2 + 2\alpha^2\right]}\tilde{\rho}_{12} \\ &= \frac{i}{\nu}\lambda(q)|\tilde{\rho}_{21}|^2,\end{aligned} \tag{6.107}$$

where (6.106b) has been introduced. The observable (6.92) now becomes

$$W_q = -2\lambda(q)^2|\tilde{\rho}_{21}|^2 \mathrm{Re}\left|\frac{i}{\nu - i\eta}\right|_{\eta\to 0}$$

$$= \frac{2\pi}{\hbar}|\hbar, \lambda(q)\tilde{\rho}_{21}|^2\delta(\hbar\nu). \tag{6.108}$$

This is the quantum mechanical emission rate due to an oscillating dipole $\hbar\lambda(q)\tilde{\rho}_{21}$, just as we argued before. The delta function ensures strict energy conservation in the scattering.

In the limit of low intensity $\alpha \to 0$, we obtain

$$W_q = \frac{2\pi}{\hbar}\frac{|\hbar\lambda\alpha|^2}{\Delta^2 + (\Gamma/2)^2}\delta(\hbar\nu). \tag{6.109}$$

With (6.106a) this agrees with our earlier result (6.93) except that in the present case the outgoing frequency agrees exactly with the incoming one. *This is the correct result*, following from strict energy conservation between the incoming and outgoing photons. The atom is both initially and finally in the ground state. Because it cannot decay, its energy value is sharp, and the burden of satisfying energy conservation falls entirely on the fields. When the lower level has a finite lifetime, the elastic peak becomes broadened because of the width of the lower level.

In large intensities, $\alpha \gg |\Delta|$ and $\Gamma$, the induced dipole saturates and the upper-level population becomes the main source of the radiation. The atom spends about half of its time on each level, see Sec. 2.2, and we have from Eqs. (6.106)

$$\rho_{22} \simeq \frac{1}{2}, \qquad \tilde{\rho}_{21} \simeq O\left(\frac{1}{\alpha}\right). \tag{6.110}$$

For simplicity we choose to look at the centrally tuned case $\Delta = 0$; the strong driving field is in resonance with the two-level system.

The determinant of the matrix becomes

$$D \equiv \nu\left(\nu - \frac{i\Gamma}{2}\right)\left[(\nu - i\Gamma)\left(\nu - i\frac{1}{2}\Gamma\right) - 4\alpha^2\right]$$

$$\simeq \nu\left(\nu - \frac{i\Gamma}{2}\right)\left[\nu^2 - 4\alpha^2 - i\frac{3}{2}\Gamma\nu\right]. \tag{6.111}$$

The required density matrix element $\tilde{\rho}_{21q}$ can easily be obtained with the assumption (6.110) to be

$$\tilde{\rho}_{21q} = i\lambda(q)\frac{\nu\left[\nu^2 - i\frac{3}{2}\Gamma\nu - 2\alpha^2\right]}{D}\rho_{22}. \tag{6.112}$$

Here as in Eq. (6.111) we have neglected terms of order $\Gamma^2$ as compared with $\alpha^2$. The factor $\nu$ cancels exactly, and we write

$$\frac{\nu\left(\nu^2 - i\frac{3}{2}\Gamma\nu - 2\alpha^2\right)}{D} \simeq \frac{1}{2}\left[\left(\frac{1}{\nu - i\frac{1}{2}\Gamma}\right) + \frac{1}{2}\left(\frac{1}{\nu - 2\alpha - i\frac{3}{4}\Gamma}\right) + \frac{1}{2}\left(\frac{1}{\nu + 2\alpha - i\frac{3}{4}\Gamma}\right)\right], \tag{6.113}$$

where we have retained only the terms to leading order in $\alpha$. Inserting this back into (6.112) and then back into (6.92) we obtain

$$W_q = \frac{\lambda(q)^2}{2}\left[\frac{\frac{1}{2}\Gamma}{\nu^2 + (\Gamma/2)^2} + \frac{1}{2}\left(\frac{3\Gamma/4}{(\nu - 2\alpha)^2 + (3\Gamma/4)^2}\right) + \frac{1}{2}\left(\frac{3\Gamma/4}{(\nu + 2\alpha)^2 + (3\Gamma/4)^2}\right)\right]. \tag{6.114}$$

The observed spectrum thus has the shape shown in Fig. 6.3 for large values of $\alpha$. The two side peaks split by $4\alpha$ are 50% broader than the central peak; at resonance their peak value reaches only one-third that of the central peak. This result has been verified experimentally and derived theoretically in many different ways.

In this section we have shown how a strong field can modify the spectrum of the emitted light. The Rabi flipping at frequency $2\alpha$ generates sidebands on each side. Is it also possible to observe the effects of the strong field on the pure atomic variables? Is it possible to see a modification of the spontaneous emission rate due to the field amplitude?

From Eq. (6.77a) we have

$$\frac{d}{dt}\rho_{22} = 2\sum_q \lambda(q)\mathrm{Re}\langle 0,2|\rho|q,1\rangle. \tag{6.115}$$

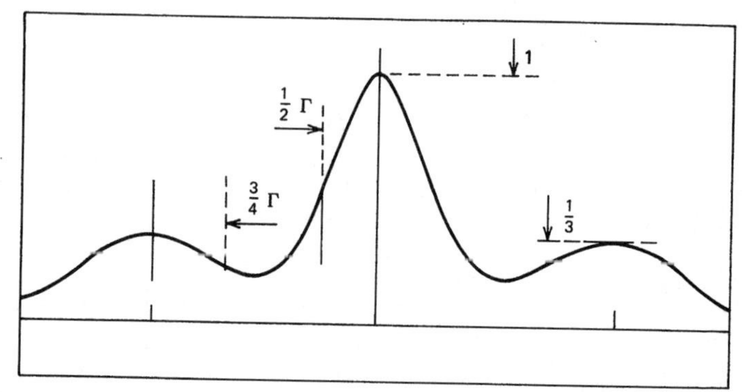


**Fig. 6.3** The spontaneous emission spectrum observed in Fig. 6.2. The central peak is three times higher than the side bands, which in turn are 50% broader than the central peak. This spectrum thus differs in an essential way from what one would expect in low-order perturbation theory.

Inserting the large amplitude result (6.112) into this we find again

$$\dot{\rho}_{22} = -\Gamma\rho_{22} \tag{6.116}$$

with

$$\Gamma = \sum_q \lambda(q)^2 \left[ \frac{\frac{1}{2}\Gamma}{\nu^2 + (\Gamma/4)^2} + \frac{1}{2}\frac{3\Gamma/4}{(\nu - 2\alpha)^2 + (3\Gamma/4)^2} + \frac{1}{2}\frac{3\Gamma/4}{(\nu + 2\alpha)^2 + (3\Gamma/4)^2} \right]$$

$$\simeq \pi \sum_q \lambda(q)^2 \left[ \delta(\nu) + \frac{1}{2}\delta(\nu - 2\alpha) + \frac{1}{2}\delta(\nu + 2\alpha) \right], \tag{6.117}$$

where the last form results if $\Gamma$ is small compared with the range over which $\lambda(q)$ varies near the frequency $\Omega$, which determines the center of the spectrum. Assuming all other factors to be slowly varying, we can compare (6.117) with (6.83) and find that both agree to a lowest approximation.

Thus even if the spectrum is split up into three peaks when $\alpha$ becomes large, the integrated intensity remains the same, and this will determine the rate of decay $\Gamma$. It is not easy to observe any intensity effect even if the spectrum of the emitted photons is modified in this essential way. This conclusion holds also for the exact result from (6.102) even without our approximations. Further complications derive from the detuning $\Delta$; if this is not zero, the spectrum becomes displaced from the atomic resonance, but it retains its general shape. It is easily plotted numerically from (6.102).

## 6.7. COMMENTS AND REFERENCES

Here we only outline the foundations of the quantization of fields. For our treatment the approach by Fermi (1932) is adequate. A good presentation is found in the introductory chapter of Sakurai (1967). A presentation that is adapted to the needs of quantum electronics is given by Loudon (1974). Many useful techniques and methods are presented by Louisell (1973). A review of quantum field aspects of quantum electronics is given in Stenholm (1973). For laser spectroscopy these considerations are continued in Stenholm (1978a).

The spontaneous decay was first derived by Weisskopf and Wigner (1930). In our discussion we follow the presentation given by Cohen–Tannoudji (1977). We do, however, simplify the treatment somewhat.

The first calculation of the Lamb shift was published by Bethe (1947). The calculation of the resonance fluorescence spectrum is discussed in the textbook by Heitler (1954) in Sec. 20. He obtains the narrow elastic scattering peak at the exciting frequency. It has recently been observed in strong laser fields by Eisenberger et al. (1976) and Gibbs and Venkatesan (1976). The calculation of the resonance fluorescence spectrum in a strong exciting laser light was first carried out by Burshtein (1966a), Newstein (1968), and Mollow (1969). The first experimental observation was by Shuda et al. (1974) and further observation by Grove et al. (1977) and Hartig et al. (1976). Detailed theoretical considerations are found in Cohen–Tannoudji (1977) and from a different point of view in Kimble and Mandel (1976).

Our present approach is developed from one originally introduced by Baklanov (1974); some further details can be found in Stenholm (1978a).

The quantum mechanical description of laser operation was formulated and developed by Haken (1970b) and his collaborators. Here quantum effects were treated as noise sources, an approach also pursued extensively by Lax (1968). A quantum theory following the spirit of the present work was presented by Scully and Lamb (1967) and is discussed extensively in Sargent et al. (1974). A different approach to the quantum theory of the laser is presented by Kazantsev and Surdutovich (1969) and in greater detail in Kazantsev and Surdutovich (1975).

# References

Section numbers given at the end of each reference are those in which the particular reference is cited.

Abragam, A. (1961), *The Principles of Nuclear Magnetism*, Clarendon Press, Oxford (Secs. 1.11, 5.5).

Abramowitz, M. and I. A. Stegun (1970), *Handbook of Mathematical Functions*, Dover, New York (Secs. 2.4, 5.2.a).

Agarwal, G. S. (1976), *Phys. Rev. Lett.* **37**, 1383 (Sec. 5.5).

Allen, L. and J. H. Eberly (1975), *Optical Resonances and Two-Level Atoms*, Wiley, New York (Secs. 1.10, 1.11, 3.2).

Allen, L. and C. R. Stroud Jr. (1982), *Phys. Rep.* **91**, 1 (Sec. 4.5.e).

Aminoff, C. -G. and S. Stenholm (1976), *J. Phys. B: At. Mol. Phys.* **9**, 1039 (Secs. 2.8, 5.5).

Anderson, P. W. (1949), *Phys. Rev.* **76**, 647 (Sec. 1.11).

Anderson, P. W. (1954), *J. Phys. Soc. Jap.* **9**, 316 (Sec. 5.5).

Arimondo, E., A. Bambini, and S. Stenholm (1981), *Phys. Rev.* **A24**, 898 (Sec. 4.5.e).

Autler, S. H. and C. H. Townes (1955), *Phys. Rev.* **100**, 703 (Secs. 2.8, 4.3.f).

Avan, P. and C. Cohen–Tannoudji (1977), *J. Phys. B: At. Mol. Phys.* **10**, 155 (Sec. 5.5).

Bagaev, S. N., Yu. D. Kolomnikov, V. N. Lisitsyn, and V. P. Chebotaev (1968), *IEEE J.Q.E.* **QE-4**, 868 (Sec. 4.2.c).

Bagaev, S. N., L. S. Vasilenko, V. G. Goldort, A. K. Dmitriyev, A. S. Dychkov, and V. P. Chebotaev (1977), *Appl. Phys.* **13**, 291 (Sec. 2.8).

Baklanov, E. V. (1974), *Sov. Phys. JETP* **38**, 1100 (Sec. 6.7).

Baklanov, E. V. and V. P. Chebotaev (1971), *Sov. Phys. JETP* **33**, 300 (Secs. 2.8, 4.2.c, 4.3.f).

Baklanov, E. V. and V. P. Chebotaev (1972), *Sov. Phys. JETP* **34**, 490 (Sec. 4.2.c).

Bennett Jr., W. R. (1962), *Phys. Rev.* **126**, 580 (Sec. 2.8).

Berman, P. R. (1975), *Appl. Phys.* **6**, 283 (Sec. 1.11).

Berman, P. R. (1978), *Phys. Rep.* **43**, 101 (Secs. 1.11, 4.3.f).

Bernhardt, A. F. and B. W. Shore (1981), *Phys. Rev.* **A23**, 1290 (Sec. 4.5.e).

Berry, M. V. (1966), *The Diffraction of Light by Ultrasound*, Academic, New York (Sec. 4.5.e).

Beterov, I. M., V. N. Lisitsyn, and V. P. Chebotaev (1971a), *Opt. Spectrosc.* **30**, 497 (Sec. 3.7).

Beterov, I. M., V. N. Lisitsyn, and V. P. Chebotaev (1971b), *Opt. Spectrosc.* **30**, 592 (Sec. 3.7).

Bethe, H. (1947), *Phys. Rev.* **72**, 339 (Sec. 6.7).

Bjorkholm, J. E. and P. F. Liao (1976), *Phys. Rev.* **A14**, 751 (Sec. 4.3.f).

Blombergen, N. (1965), *Nonlinear Optics*, W. A. Benjamin, New York (Sec. 2.8).

Bordé, C. (1976), *C.R. Acad. Sci. Paris* **B283**, 181 (Sec. 2.8).

Bowden, C. M., M. Ciftan, and H. R. Robl (1981), Eds., *Optical Bistability*, Plenum, New York (Secs. 3.7, 4.1).

Breit, G. (1933), *Rev. Mod. Phys.* **5**, 91 (Sec. 4.4.f).

Burshtein, A. I. (1965), *Sov. Phys. JETP* **21**, 567 (Sec. 5.5).

Burshtein, A. I. (1966a), *Sov. Phys. JETP* **22**, 939 (Sec. 6.7).

Burshtein, A. I. (1966b), *Sov. Phys.-Dokl.* **11**, 65 (Sec. 5.5).

Burshtein, A. I. and Yu. S. Oseledchik (1967), *Sov. Phys. JETP* **24**, 716 (Sec. 5.5).

Cohen–Tannoudji, C. (1968), in M. Levý, Ed., *Chargése Lectures in Physics*, Vol. 2, Gordon and Breach, New York (Secs. 2.8, 4.3.f).

Cohen–Tannoudji, C. (1977), in R. Balian, S. Haroche, and S. Liberman, Eds., *Frontiers in Laser Spectroscopy*, Les Houches Summer School 1975, North-Holland, Amsterdam (Sec. 6.7).

Close, D. H. (1967), *Phys. Rev.* **153**, 360 (Sec. 3.7).

Corbalan, R., G. Orriols, L. Roso, R. Vilaseca, and E. Arimondo (1981), *Opt. Comm.* **40**, 29 (Sec. 2.8).

Corney, A. (1977), *Atomic and Laser Spectroscopy*, Clarendon Press, Oxford (Secs. 4.2.c, 4.4.f).

Dabkiewicz, Ph., T. W. Hänsch, D. R. Lyons, A. L. Schawlow, A. Siegel, Z. -Y. Wang, and G. -Y. Yan (1981) in A. R. W. McKellar, T. Oka and B. P. Stoicheff, Eds., *Laser Spectroscopy*, Springer-Verlag, Heidelberg (Sec. 4.2.c).

Decomps, B., M. Dumont, and M. Ducloy (1976), in H. Walther, Ed., *Laser Spectroscopy of Atoms and Molecules*, Springer-Verlag, Heidelberg (Sec. 4.4.f).

DeWitt, C., A. Blandin, and C. Cohen–Tannoudji (1964), Eds., *Quantum Optics and Electronics*, Les Houches Summer School of Theoretical Physics 1964, Benjamin, New York (Sec. 3.7).

Dixit, S. N., P. Zoller, and P. Lambropoulos (1980), *Phys. Rev.* **A21**, 1289 (Secs. 4.3.f, 5.5).

Druet, S. A. J. and J. -P. E. Taran (1981), *Prog. Quantum Electron.* **7**, 1 (Sec. 2.8).

Eberly, J. H. (1976), *Phys. Rev. Lett.* **37**, 1387 (Sec. 5.5).

Eberly, J. H. (1980), in D. F. Walls and J. D. Harvey, Eds., *Laser Physics*, Proceedings of the Second New Zealand Summer School in Laser Physics, Academic, New York (Sec. 4.5.e).

Eberly, J. H. and P. Lambropoulos (1978), Eds., *Multiphoton Processes*, Proceedings of the International Conference at the University of Rochester 1977, Wiley, New York (Sec. 4.5.e).

Eisenberger, P., P. M. Platzman, and H. Winick (1976), *Phys. Rev. Lett.* **36**, 623 (Sec. 6.7).

Fano, U. (1957), *Rev. Mod. Phys.* **29**, 74 (Sec. 1.11).

Feld, M. S. (1973), in A. Javan, N. A. Kurnit and M. S. Feld, Eds., *Fundamental and Applied Laser Physics*, Proceedings of the Esfahan Symposium 1971, Wiley, New York (Sec. 4.4.f).

Feld, M. S. (1977) in *Frontiers in Laser Spectroscopy*, R. Balian, S. Haroche, and S. Liberman, Eds., Les Houches Summer School 1975, North-Holland, Amsterdam (Sec. 4.5.e).

Feld, M. S. and A. Javan (1969), *Phys. Rev.* **177**, 540 (Sec. 4.3.f).

Feldman, B. J. and M. S. Feld (1970), *Phys. Rev.* **A1**, 1375 (Sec. 2.8).

Feldman, B. J. and M. S. Feld (1972), *Phys. Rev.* **A5**, 899 (Sec. 4.3.f).

Fermi, E. (1932), *Rev. Mod. Phys.* **4**, 87 (Sec. 6.7).

Freund, S. M., M. Römheld, and T. Oka (1975), *Phys. Rev. Lett.* **35**, 1497 (Sec. 2.8).

Gibbs, H. M. and T. N. C. Venkatesan (1976) *Opt. Comm.* **17**, 87 (Sec. 6.7).

Grove, R. E., F. Y. Wu, and S. Ezekiel (1977), *Phys. Rev.* **A15**, 227 (Sec. 6.7).

ter Haar, D. (1961), *Rep. Prog. Phys.* **24**, 304 (Sec. 1.11).

Haken, H. (1970a), in S. M. Kay and A. Maitland, Eds., *Quantum Optics*, Scottish Universities Summer School in Physics 1969, Academic, New York (Sec. 3.7).

Haken, H. (1970b), in L. Genzel, Ed., *Light and Matter*, Handbuch der Physik, Band XXV/2c, Springer-Verlag, Heidelberg (Secs. 2.8, 3.7, 6.7).

Hall, J. L., C. J. Brodé, and K. Uehara (1976), *Phys. Rev. Lett.* **37**, 1339 (Sec. 2.8).

Hanle, W. (1924), *Z. Phys.* **30**, 93 (Sec. 4.4.f).

Hänsch, Th. W. (1973), in F. P. Schäfer, Ed., *Dye Lasers*, Springer-Verlag, Heidelberg (Sec. 4.2.c).

Hänsch, Th. W. and P. Toschek (1970), *Z. Phys.* **236**, 213 (Sec. 4.3.f).

Haroche, S. and F. Hartmann (1972), *Phys. Rev.* **A6**, 1280 (Secs. 2.8, 4.2.c).

Hartig, W., W. Rasmussen, R. Schieder, and H. Walther (1976), *Z. Phys.* **A278**, 205 (Sec. 6.7).

Hellwarth, R. W. (1977), *Prog. Quantum Electron.* **5**, 1 (Sec. 2.8).

Heitler, W. (1954), *The Quantum Theory of Radiation*, 3rd ed., Clarendon Press, Oxford (Sec. 6.7).

Holt, H. K. (1970), *Phys. Rev.* **A2**, 233 (Sec. 2.8).

Jackson, J. D. (1975), *Classical Electrodynamics*, Wiley, New York (Sec. 1.11).

Jancel, R. (1969), *Foundations of Classical and Quantum Statistical Mechanics*, Pergamon, Oxford (Sec. 1.11).

Janossy, M. and S. Varro (1980), *Invited Papers*, 2nd International Conference on Multiphoton Processes, Budapest, April 1980 (Sec. 4.5.e).

Javan, A. (1977), in R. Balian, S. Haroche and S. Liberman, Eds., *Frontiers in Laser Spectroscopy*, Les Houches Summer School 1975, North-Holland, Amsterdam (Sec. 4.2.c).

Jones, W. B. and W. J. Thron (1980), *Continued Fractions*, *Encyclopedia of Mathematics and Its Applications*, Vol. 11, Addison-Wesley, New York (Sec. 2.8).

van Kampen, N. G. (1976), *Phys. Rep.* **24**, 171 (Sec. 5.5).

van Kampen, N. G. (1982), *Stochastic Processes in Physics and Chemistry*, North-Holland, Amsterdam (Sec. 5.5).

Kazantsev, A. P. and G. I. Surdutovich (1969), *Sov. Phys. JETP* **29**, 1075 (Sec. 6.7).

Kazantsev, A. P. and G. I. Surdutovich (1975) in J. H. Sanders and S. Stenholm, Eds., *Progress in Quantum Electronics*, Pergamon, Oxford, (Sec. 6.7).

Kimble, H. J. and L. Mandel (1976), *Phys. Rev.* **A13**, 2123 (Sec. 6.7).

Kol'chenko, A. B., S. G. Rautian, and R. I. Sokolovskii (1969), *Sov. Phys. JETP* **28**, 986 (Sec. 2.8).

Kyrölä, E. and R. Salomaa (1981), *Phys. Rev.* **A23**, 1874 (Sec. 4.3.f).

Kyrölä, E. and S. Stenholm (1977), *Opt. Comm.* **22**, 123 (Sec. 2.8).

Kyrölä, E. and S. Stenholm (1979), *Opt. Comm.* **30**, 37 (Sec. 2.8).

Lamb, Jr., W. E. (1963) in P. A. Miles, Ed., *Quantum Electronics and Coherent Light*, International School of Physics, "Enrico Fermi," p. 78, Academic, New York (Sec. 2.8).

Lamb, Jr., W. E. (1964), *Phys. Rev.* **A134**, 1429 (Secs. 1.11, 2.8, 3.7).

Lamb, Jr., W. E. and T. M. Sanders (1960), *Phys. Rev.* **119**, 1901 (Sec. 1.11).

Landau, L. D. (1927), *Z. Phys.* **45**, 430 (Sec. 1.11).

Lax, M. (1968), in M. Chretien, E. P. Gross, and S. Deser, Eds., *Brandeis Summer Institute Lectures* 1966, Vol. 2, Gordon and Breach, New York (Secs. 3.7, 6.7).

Lee, P. H., P. B. Schoefer, and W. B. Barker (1968), *Appl. Phys. Lett.* **13**, 373 (Sec. 3.7).

Letokhov, V. S. and V. P. Chebotaev (1969), *JETP Lett.* **9**, 215 (Sec. 4.2.c).

Letokhov, V. S. and V. P. Chebotaev (1977), *Nonlinear Laser Spectroscopy*, Springer-Verlag, Heidelberg (Secs. 1.11, 2.8, 3.7, 4.2.c).

Letokhov, V. S. and A. A. Makarov (1981), *Sov. Phys. Usp.* **24**, 366 (Sec. 4.5.e).

Levenson, M. D. (1982) *Introduction to Nonlinear Laser Spectroscopy*, Academic, New York (Secs. 1.11, 2.8, 4.2.c).

Lisitsyn, V. N. and V. P. Chebotaev (1968a), *Sov. Phys. JETP* **27**, 227 (Sec. 3.7).

Lisitsyn, V. N. and V. P. Chebotaev (1968b), *JETP Lett.* **7**, 1 (Sec. 3.7).

Loudon, R. (1974), *The Quantum Theory of Light*, Oxford University Press, Oxford (Sec. 6.7).

Louisell, W. H. (1973), *Quantum Statistical Properties of Radiation*, Wiley, New York (Secs. 5.5, 6.7).

Mattick, A. T., A. Sanchez, N. A. Kurnit and A. Javan (1973), *Appl. Phys. Lett.* **23**, 675 (Sec. 4.2.c).

McFarlane, R. A., W. R. Bennett, and W. E. Lamb Jr. (1963), *Appl. Phys. Lett.* **2**, 189 (Sec. 2.8).

Mollow, B. R. (1969), *Phys. Rev.* **188**, 1969 (Sec. 6.7).

Mostowski, J. and S. Stenholm (1982), *Phys. Scr.* **26**, 221 (Sec. 5.5).

von Neumann, J. (1927), *Göttinger Nachr.* **1**, 245, 273 (Sec. 1.11).

Newstein, M. (1968), *Phys. Rev.* **167**, 89 (Sec. 6.7).

Pauli, W. (1928), in A. Sommerfelds, Ed., *Probleme der modernen Physik*, Festschrift zum 50. Geburtstag, P. Debye, Hirzel, Leipzig (Sec. 1.11).

Popova, T. Ya., A. K. Popov, S. G. Rautian, and A. A. Feoktistov (1970a), *Sov. Phys. JETP* **30**, 243 (Sec. 4.3.f).

Popova, T. Ya., A. K. Popov, S. G. Rautian, and R. I. Sokolovski (1970b), *Sov. Phys. JETP* **30**, 466 (Sec. 4.3.f).

Poulsen, O. and N. I. Winstrup (1981), *Phys. Rev. Lett.* **47**, 1522 (Sec. 4.3.f).

Read, J. and T. Oka (1977), *Phys. Rev. Lett.* **38**, 67 (Sec. 2.8).

Sakurai, J. J. (1967), *Advanced Quantum Mechanics*, Addison-Wesley, New York (Sec. 6.7).

Salomaa, R. (1977), *J. Phys. B: At. Mol. Phys.* **10**, 3005 (Sec. 4.3.f).

Salomaa, R. and S. Stenholm (1973a), *Phys. Rev.* **A8**, 2695 (Sec. 3.7).

Salomaa, R. and S. Stenholm (1973b), *Phys. Rev.* **A8**, 2711 (Sec. 3.7).

Salomaa, R. and S. Stenholm (1975), *J. Phys. B: At. Mol. Phys.* **8**, 1795 (Sec. 4.3.f).

Salomaa, R. and S. Stenholm (1976a), *J. Phys. B: At. Mol. Phys.* **9**, 1221 (Sec. 4.3.f).

Salomaa, R. and S. Stenholm (1976b), *Opt. Comm.* **16**, 292 (Sec. 4.3.f).

Salomaa, R. and S. Stenholm (1978), *Appl. Phys.* **17**, 309 (Sec. 4.3.f).

Sargent III, M. (1978), *Phys. Rep.* **43**, 223 (Secs. 4.2.c, 4.3.f, 4.4.f).

Sargent III, M., M. O. Scully, and W. E. Lamb Jr. (1974), *Laser Physics*, Addison-Wesley, New York (Secs. 1.11, 2.8, 3.7, 6.7).

Schawlow, A. L. and C. H. Townes (1958), *Phys. Rev.* **112**, 1940 (Secs. 3.1, 3.7).

Schenzle, A. and R. G. Brewer (1978), *Phys. Rep.* **43**, 456 (Sec. 4.3.f).

Schuda, F., C. R. Stroud, Jr., and M. Hercher (1974), *J. Phys. B: At. Mol. Phys.* **7**, L198 (Sec. 6.7).

Schrödinger, E. (1935), *Naturwissenschaften* **23**, 807, 823, 844 (Sec. 1.11).

Scully, M. O. and W. E. Lamb, Jr. (1967), *Phys. Rev.* **A2**, 2529 (Sec. 6.7).

Shimoda, K. (1976), Ed., *High-Resolution Spectroscopy*, Springer-Verlag, Heidelberg (Secs. 2.8, 4.2.c, 4.3.f, 4.4.f).

Shirley, J. H. (1965), *Phys. Rev.* **138B**, 979 (Sec. 2.8).

Shirley, J. H. (1980), *J. Phys. B: Mol. Phys.* **13**, 1537 (Sec. 2.8).

Shirley, J. H. and S. Stenholm (1977), *J. Phys. A: Math. Gen.* **10**, 613 (Sec. 2.8).

Slichter, C. P. (1978), *Principles of Magnetic Resonance*, Springer-Verlag, Heidelberg (Secs. 1.11, 5.5).

Smith, P. W. (1966), *IEEE J.Q.E.* **QR-2**, 62 (Sec. 3.7).

Stenholm, S. (1973), *Phys. Rep.* **6**, 2 (Sec. 6.7).

Stenholm, S. (1974), *J. Phys. B: At. Mol. Phys.* **7**, 1235 (Sec. 2.8).

Stenholm, S. (1978a), *Phys. Rep.* **43**, 151 (Secs. 2.8, 6.7).

Stenholm, S. (1978b), in W. Hanle and H. Kleinpoppen, Eds., *Progress in Atomic Spectroscopy*, Plenum, New York (Sec. 2.8).

Stenholm, S. and A. Bambini (1981), *IEEE J. Quantum Electron.* **QE-17**, 1363 (Sec. 4.5.e).

Stenholm, S. and W. E. Lamb Jr. (1969), *Phys. Rev.* **181**, 618 (Sec. 2.8).

Svelto, O. (1976), *Principles of Lasers*, Heyden & Son, London (Sec. 3.7).

Szöke, A. and A. Javan (1963), *Phys. Rev. Lett.* **10**, 521 (Sec. 2.8).

Tolman, R. C. (1938), *The Principles of Statistical Mechanics*, Oxford University Press, Oxford (Sec. 1.11).

Vasilenko, L. S., V. P. Chebotaev, A. V. Shishaev (1970), *JETP Lett.* **12**, 113 (Sec. 4.3.f).

Wax, N. (1954), Ed., *Selected Papers on Noise and Stochastic Processes*, Dover, New York (Secs. 1.7, 1.11, 5.5).

Weisskopf, V. (1932), *Z. Phys.* **75**, 287 (Sec. 1.11).

Weisskopf, V. (1933), *Phys. Z.* **34**, 1 (Sec. 1.11).

Weisskopf, V. and E. Wigner (1930), *Z. Phys.* **63**, 54 (Sec. 6.7).

Wieman, C. E. and T. W. Hänsch (1976), *Phys. Rev. Lett.* **36**, 1170 (Sec. 4.2.c).

Wilcox, L. R. and W. E. Lamb Jr. (1960), *Phys. Rev.* **119**, 1915 (Sec. 1.11).

Wodkiewicz, K. (1979a), *J. Math. Phys.* **20**, 45 (Sec. 5.5).

Wodkiewicz, K. (1979b), *Phys. Rev.* **A19**, 1686 (Sec. 5.5).

Yariv, A. (1976), *Introduction to Optical Electronics*, 2nd ed., Holt, Rinehart and Winston, New York, (Sec. 1.11).

Zoller, P. (1977), *J: Phys. B: At. Mol. Phys.* **10**, L321 (Sec. 5.5).

Zoller, P. (1979a), *Phys. Rev.* **A19**, 1151 (Sec. 5.5).

Zoller, P. (1979b), *Phys. Rev.* **A20**, 2420 (Sec. 5.5).

Zoller, P., G. Alber, and R. Salvador (1981), *Phys. Rev.* **A24**, 398 (Sec. 5.5).

Zusman, L. D. and A. I. Burshtein (1972), *Sov. Phys. JETP* **34**, 520 (Sec. 5.5).

# Index

A CATALOG OF SELECTED

# DOVER BOOKS

IN SCIENCE AND MATHEMATICS

# Astronomy

BURNHAM'S CELESTIAL HANDBOOK, Robert Burnham, Jr. Thorough guide to the stars beyond our solar system. Exhaustive treatment. Alphabetical by constellation: Andromeda to Cetus in Vol. 1; Chamaeleon to Orion in Vol. 2; and Pavo to Vulpecula in Vol. 3. Hundreds of illustrations. Index in Vol. 3. 2,000pp. 6⅛ x 9¼.
Vol. I: 23567-X
Vol. II: 23568-8
Vol. III: 23673-0

EXPLORING THE MOON THROUGH BINOCULARS AND SMALL TELESCOPES, Ernest H. Cherrington, Jr. Informative, profusely illustrated guide to locating and identifying craters, rills, seas, mountains, other lunar features. Newly revised and updated with special section of new photos. Over 100 photos and diagrams. 240pp. 8¼ x 11. 24491-1

THE EXTRATERRESTRIAL LIFE DEBATE, 1750–1900, Michael J. Crowe. First detailed, scholarly study in English of the many ideas that developed from 1750 to 1900 regarding the existence of intelligent extraterrestrial life. Examines ideas of Kant, Herschel, Voltaire, Percival Lowell, many other scientists and thinkers. 16 illustrations. 704pp. 5⅜ x 8½. 40675-X

THEORIES OF THE WORLD FROM ANTIQUITY TO THE COPERNICAN REVOLUTION, Michael J. Crowe. Newly revised edition of an accessible, enlightening book recreates the change from an earth-centered to a sun-centered conception of the solar system. 242pp. 5⅜ x 8½. 41444-2

A HISTORY OF ASTRONOMY, A. Pannekoek. Well-balanced, carefully reasoned study covers such topics as Ptolemaic theory, work of Copernicus, Kepler, Newton, Eddington's work on stars, much more. Illustrated. References. 521pp. 5⅜ x 8½.
65994-1

A COMPLETE MANUAL OF AMATEUR ASTRONOMY: Tools and Techniques for Astronomical Observations, P. Clay Sherrod with Thomas L. Koed. Concise, highly readable book discusses: selecting, setting up and maintaining a telescope; amateur studies of the sun; lunar topography and occultations; observations of Mars, Jupiter, Saturn, the minor planets and the stars; an introduction to photoelectric photometry; more. 1981 ed. 124 figures. 26 halftones. 37 tables. 335pp. 6½ x 9¼.
42820-6

AMATEUR ASTRONOMER'S HANDBOOK, J. B. Sidgwick. Timeless, comprehensive coverage of telescopes, mirrors, lenses, mountings, telescope drives, micrometers, spectroscopes, more. 189 illustrations. 576pp. 5⅝ x 8¼. (Available in U.S. only.)
24034-7

STARS AND RELATIVITY, Ya. B. Zel'dovich and I. D. Novikov. Vol. 1 of *Relativistic Astrophysics* by famed Russian scientists. General relativity, properties of matter under astrophysical conditions, stars, and stellar systems. Deep physical insights, clear presentation. 1971 edition. References. 544pp. 5⅝ x 8¼. 69424-0

# Chemistry

THE SCEPTICAL CHYMIST: The Classic 1661 Text, Robert Boyle. Boyle defines the term "element," asserting that all natural phenomena can be explained by the motion and organization of primary particles. 1911 ed. viii+232pp. 5⅜ x 8½. 42825-7

RADIOACTIVE SUBSTANCES, Marie Curie. Here is the celebrated scientist's doctoral thesis, the prelude to her receipt of the 1903 Nobel Prize. Curie discusses establishing atomic character of radioactivity found in compounds of uranium and thorium; extraction from pitchblende of polonium and radium; isolation of pure radium chloride; determination of atomic weight of radium; plus electric, photographic, luminous, heat, color effects of radioactivity. ii+94pp. 5⅜ x 8½. 42550-9

CHEMICAL MAGIC, Leonard A. Ford. Second Edition, Revised by E. Winston Grundmeier. Over 100 unusual stunts demonstrating cold fire, dust explosions, much more. Text explains scientific principles and stresses safety precautions. 128pp. 5⅜ x 8½. 67628-5

THE DEVELOPMENT OF MODERN CHEMISTRY, Aaron J. Ihde. Authoritative history of chemistry from ancient Greek theory to 20th-century innovation. Covers major chemists and their discoveries. 209 illustrations. 14 tables. Bibliographies. Indices. Appendices. 851pp. 5⅜ x 8½. 64235-6

CATALYSIS IN CHEMISTRY AND ENZYMOLOGY, William P. Jencks. Exceptionally clear coverage of mechanisms for catalysis, forces in aqueous solution, carbonyl- and acyl-group reactions, practical kinetics, more. 864pp. 5⅜ x 8½. 65460-5

ELEMENTS OF CHEMISTRY, Antoine Lavoisier. Monumental classic by founder of modern chemistry in remarkable reprint of rare 1790 Kerr translation. A must for every student of chemistry or the history of science. 539pp. 5⅜ x 8½. 64624-6

THE HISTORICAL BACKGROUND OF CHEMISTRY, Henry M. Leicester. Evolution of ideas, not individual biography. Concentrates on formulation of a coherent set of chemical laws. 260pp. 5⅜ x 8½. 61053-5

A SHORT HISTORY OF CHEMISTRY, J. R. Partington. Classic exposition explores origins of chemistry, alchemy, early medical chemistry, nature of atmosphere, theory of valency, laws and structure of atomic theory, much more. 428pp. 5⅜ x 8½. (Available in U.S. only.) 65977-1

GENERAL CHEMISTRY, Linus Pauling. Revised 3rd edition of classic first-year text by Nobel laureate. Atomic and molecular structure, quantum mechanics, statistical mechanics, thermodynamics correlated with descriptive chemistry. Problems. 992pp. 5⅜ x 8½. 65622-5

FROM ALCHEMY TO CHEMISTRY, John Read. Broad, humanistic treatment focuses on great figures of chemistry and ideas that revolutionized the science. 50 illustrations. 240pp. 5⅜ x 8½. 28690-8

# Engineering

DE RE METALLICA, Georgius Agricola. The famous Hoover translation of greatest treatise on technological chemistry, engineering, geology, mining of early modern times (1556). All 289 original woodcuts. 638pp. 6¾ x 11. 60006-8

FUNDAMENTALS OF ASTRODYNAMICS, Roger Bate et al. Modern approach developed by U.S. Air Force Academy. Designed as a first course. Problems, exercises. Numerous illustrations. 455pp. 5⅜ x 8½. 60061-0

DYNAMICS OF FLUIDS IN POROUS MEDIA, Jacob Bear. For advanced students of ground water hydrology, soil mechanics and physics, drainage and irrigation engineering, and more. 335 illustrations. Exercises, with answers. 784pp. 6⅛ x 9¼. 65675-6

THEORY OF VISCOELASTICITY (Second Edition), Richard M. Christensen. Complete, consistent description of the linear theory of the viscoelastic behavior of materials. Problem-solving techniques discussed. 1982 edition. 29 figures. xiv+364pp. 6⅛ x 9¼. 42880-X

MECHANICS, J. P. Den Hartog. A classic introductory text or refresher. Hundreds of applications and design problems illuminate fundamentals of trusses, loaded beams and cables, etc. 334 answered problems. 462pp. 5⅜ x 8½. 60754-2

MECHANICAL VIBRATIONS, J. P. Den Hartog. Classic textbook offers lucid explanations and illustrative models, applying theories of vibrations to a variety of practical industrial engineering problems. Numerous figures. 233 problems, solutions. Appendix. Index. Preface. 436pp. 5⅜ x 8½. 64785-4

STRENGTH OF MATERIALS, J. P. Den Hartog. Full, clear treatment of basic material (tension, torsion, bending, etc.) plus advanced material on engineering methods, applications. 350 answered problems. 323pp. 5⅜ x 8½. 60755-0

A HISTORY OF MECHANICS, René Dugas. Monumental study of mechanical principles from antiquity to quantum mechanics. Contributions of ancient Greeks, Galileo, Leonardo, Kepler, Lagrange, many others. 671pp. 5⅜ x 8½. 65632-2

STABILITY THEORY AND ITS APPLICATIONS TO STRUCTURAL MECHANICS, Clive L. Dym. Self-contained text focuses on Koiter postbuckling analyses, with mathematical notions of stability of motion. Basing minimum energy principles for static stability upon dynamic concepts of stability of motion, it develops asymptotic buckling and postbuckling analyses from potential energy considerations, with applications to columns, plates, and arches. 1974 ed. 208pp. 5⅜ x 8½. 42541-X

METAL FATIGUE, N. E. Frost, K. J. Marsh, and L. P. Pook. Definitive, clearly written, and well-illustrated volume addresses all aspects of the subject, from the historical development of understanding metal fatigue to vital concepts of the cyclic stress that causes a crack to grow. Includes 7 appendixes. 544pp. 5⅜ x 8½. 40927-9

ROCKETS, Robert Goddard. Two of the most significant publications in the history of rocketry and jet propulsion: "A Method of Reaching Extreme Altitudes" (1919) and "Liquid Propellant Rocket Development" (1936). 128pp. 5⅜ x 8½. 42537-1

STATISTICAL MECHANICS: Principles and Applications, Terrell L. Hill. Standard text covers fundamentals of statistical mechanics, applications to fluctuation theory, imperfect gases, distribution functions, more. 448pp. 5⅜ x 8½. 65390-0

ENGINEERING AND TECHNOLOGY 1650–1750: Illustrations and Texts from Original Sources, Martin Jensen. Highly readable text with more than 200 contemporary drawings and detailed engravings of engineering projects dealing with surveying, leveling, materials, hand tools, lifting equipment, transport and erection, piling, bailing, water supply, hydraulic engineering, and more. Among the specific projects outlined–transporting a 50-ton stone to the Louvre, erecting an obelisk, building timber locks, and dredging canals. 207pp. 8⅜ x 11¼. 42232-1

THE VARIATIONAL PRINCIPLES OF MECHANICS, Cornelius Lanczos. Graduate level coverage of calculus of variations, equations of motion, relativistic mechanics, more. First inexpensive paperbound edition of classic treatise. Index. Bibliography. 418pp. 5⅜ x 8½. 65067-7

PROTECTION OF ELECTRONIC CIRCUITS FROM OVERVOLTAGES, Ronald B. Standler. Five-part treatment presents practical rules and strategies for circuits designed to protect electronic systems from damage by transient overvoltages. 1989 ed. xxiv+434pp. 6⅛ x 9¼. 42552-5

ROTARY WING AERODYNAMICS, W. Z. Stepniewski. Clear, concise text covers aerodynamic phenomena of the rotor and offers guidelines for helicopter performance evaluation. Originally prepared for NASA. 537 figures. 640pp. 6⅛ x 9¼. 64647-5

INTRODUCTION TO SPACE DYNAMICS, William Tyrrell Thomson. Comprehensive, classic introduction to space-flight engineering for advanced undergraduate and graduate students. Includes vector algebra, kinematics, transformation of coordinates. Bibliography. Index. 352pp. 5⅜ x 8½. 65113-4

HISTORY OF STRENGTH OF MATERIALS, Stephen P. Timoshenko. Excellent historical survey of the strength of materials with many references to the theories of elasticity and structure. 245 figures. 452pp. 5⅜ x 8½. 61187-6

ANALYTICAL FRACTURE MECHANICS, David J. Unger. Self-contained text supplements standard fracture mechanics texts by focusing on analytical methods for determining crack-tip stress and strain fields. 336pp. 6⅛ x 9¼. 41737-9

STATISTICAL MECHANICS OF ELASTICITY, J. H. Weiner. Advanced, self-contained treatment illustrates general principles and elastic behavior of solids. Part 1, based on classical mechanics, studies thermoelastic behavior of crystalline and polymeric solids. Part 2, based on quantum mechanics, focuses on interatomic force laws, behavior of solids, and thermally activated processes. For students of physics and chemistry and for polymer physicists. 1983 ed. 96 figures. 496pp. 5⅜ x 8½. 42260-7

# Mathematics

FUNCTIONAL ANALYSIS (Second Corrected Edition), George Bachman and Lawrence Narici. Excellent treatment of subject geared toward students with background in linear algebra, advanced calculus, physics, and engineering. Text covers introduction to inner-product spaces, normed, metric spaces, and topological spaces; complete orthonormal sets, the Hahn-Banach Theorem and its consequences, and many other related subjects. 1966 ed. 544pp. 6⅛ x 9¼. 40251-7

ASYMPTOTIC EXPANSIONS OF INTEGRALS, Norman Bleistein & Richard A. Handelsman. Best introduction to important field with applications in a variety of scientific disciplines. New preface. Problems. Diagrams. Tables. Bibliography. Index. 448pp. 5⅜ x 8½. 65082-0

VECTOR AND TENSOR ANALYSIS WITH APPLICATIONS, A. I. Borisenko and I. E. Tarapov. Concise introduction. Worked-out problems, solutions, exercises. 257pp. 5⅝ x 8¼. 63833-2

THE ABSOLUTE DIFFERENTIAL CALCULUS (CALCULUS OF TENSORS), Tullio Levi-Civita. Great 20th-century mathematician's classic work on material necessary for mathematical grasp of theory of relativity. 452pp. 5⅝ x 8¼. 63401-9

AN INTRODUCTION TO ORDINARY DIFFERENTIAL EQUATIONS, Earl A. Coddington. A thorough and systematic first course in elementary differential equations for undergraduates in mathematics and science, with many exercises and problems (with answers). Index. 304pp. 5⅜ x 8½. 65942-9

FOURIER SERIES AND ORTHOGONAL FUNCTIONS, Harry F. Davis. An incisive text combining theory and practical example to introduce Fourier series, orthogonal functions and applications of the Fourier method to boundary-value problems. 570 exercises. Answers and notes. 416pp. 5⅜ x 8½. 65973-9

COMPUTABILITY AND UNSOLVABILITY, Martin Davis. Classic graduate-level introduction to theory of computability, usually referred to as theory of recurrent functions. New preface and appendix. 288pp. 5⅜ x 8½. 61471-9

ASYMPTOTIC METHODS IN ANALYSIS, N. G. de Bruijn. An inexpensive, comprehensive guide to asymptotic methods–the pioneering work that teaches by explaining worked examples in detail. Index. 224pp. 5⅜ x 8½ 64221-6

APPLIED COMPLEX VARIABLES, John W. Dettman. Step-by-step coverage of fundamentals of analytic function theory–plus lucid exposition of five important applications: Potential Theory; Ordinary Differential Equations; Fourier Transforms; Laplace Transforms; Asymptotic Expansions. 66 figures. Exercises at chapter ends. 512pp. 5⅜ x 8½. 64670-X

INTRODUCTION TO LINEAR ALGEBRA AND DIFFERENTIAL EQUATIONS, John W. Dettman. Excellent text covers complex numbers, determinants, orthonormal bases, Laplace transforms, much more. Exercises with solutions. Undergraduate level. 416pp. 5⅜ x 8½. 65191-6

# Math–Decision Theory, Statistics, Probability

ELEMENTARY DECISION THEORY, Herman Chernoff and Lincoln E. Moses. Clear introduction to statistics and statistical theory covers data processing, probability and random variables, testing hypotheses, much more. Exercises. 364pp. 5⅜ x 8½. 65218-1

STATISTICS MANUAL, Edwin L. Crow et al. Comprehensive, practical collection of classical and modern methods prepared by U.S. Naval Ordnance Test Station. Stress on use. Basics of statistics assumed. 288pp. 5⅜ x 8½. 60599-X

SOME THEORY OF SAMPLING, William Edwards Deming. Analysis of the problems, theory, and design of sampling techniques for social scientists, industrial managers, and others who find statistics important at work. 61 tables. 90 figures. xvii +602pp. 5⅜ x 8½. 64684-X

LINEAR PROGRAMMING AND ECONOMIC ANALYSIS, Robert Dorfman, Paul A. Samuelson and Robert M. Solow. First comprehensive treatment of linear programming in standard economic analysis. Game theory, modern welfare economics, Leontief input-output, more. 525pp. 5⅜ x 8½. 65491-5

PROBABILITY: An Introduction, Samuel Goldberg. Excellent basic text covers set theory, probability theory for finite sample spaces, binomial theorem, much more. 360 problems. Bibliographies. 322pp. 5⅜ x 8½. 65252-1

GAMES AND DECISIONS: Introduction and Critical Survey, R. Duncan Luce and Howard Raiffa. Superb nontechnical introduction to game theory, primarily applied to social sciences. Utility theory, zero-sum games, n-person games, decision-making, much more. Bibliography. 509pp. 5⅜ x 8½. 65943-7

INTRODUCTION TO THE THEORY OF GAMES, J. C. C. McKinsey. This comprehensive overview of the mathematical theory of games illustrates applications to situations involving conflicts of interest, including economic, social, political, and military contexts. Appropriate for advanced undergraduate and graduate courses; advanced calculus a prerequisite. 1952 ed. x+372pp. 5⅜ x 8½. 42811-7

FIFTY CHALLENGING PROBLEMS IN PROBABILITY WITH SOLUTIONS, Frederick Mosteller. Remarkable puzzlers, graded in difficulty, illustrate elementary and advanced aspects of probability. Detailed solutions. 88pp. 5⅜ x 8½. 65355-2

PROBABILITY THEORY: A Concise Course, Y. A. Rozanov. Highly readable, self-contained introduction covers combination of events, dependent events, Bernoulli trials, etc. 148pp. 5⅜ x 8¼. 63544-9

STATISTICAL METHOD FROM THE VIEWPOINT OF QUALITY CONTROL, Walter A. Shewhart. Important text explains regulation of variables, uses of statistical control to achieve quality control in industry, agriculture, other areas. 192pp. 5⅜ x 8½. 65232-7

# Math–Geometry and Topology

ELEMENTARY CONCEPTS OF TOPOLOGY, Paul Alexandroff. Elegant, intuitive approach to topology from set-theoretic topology to Betti groups; how concepts of topology are useful in math and physics. 25 figures. 57pp. 5⅜ x 8½. 60747-X

COMBINATORIAL TOPOLOGY, P. S. Alexandrov. Clearly written, well-organized, three-part text begins by dealing with certain classic problems without using the formal techniques of homology theory and advances to the central concept, the Betti groups. Numerous detailed examples. 654pp. 5⅜ x 8½. 40179-0

EXPERIMENTS IN TOPOLOGY, Stephen Barr. Classic, lively explanation of one of the byways of mathematics. Klein bottles, Moebius strips, projective planes, map coloring, problem of the Koenigsberg bridges, much more, described with clarity and wit. 43 figures. 210pp. 5⅜ x 8½. 25933-1

CONFORMAL MAPPING ON RIEMANN SURFACES, Harvey Cohn. Lucid, insightful book presents ideal coverage of subject. 334 exercises make book perfect for self-study. 55 figures. 352pp. 5⅝ x 8¼. 64025-6

THE GEOMETRY OF RENÉ DESCARTES, René Descartes. The great work founded analytical geometry. Original French text, Descartes's own diagrams, together with definitive Smith-Latham translation. 244pp. 5⅜ x 8½. 60068-8

PRACTICAL CONIC SECTIONS: The Geometric Properties of Ellipses, Parabolas and Hyperbolas, J. W. Downs. This text shows how to create ellipses, parabolas, and hyperbolas. It also presents historical background on their ancient origins and describes the reflective properties and roles of curves in design applications. 1993 ed. 98 figures. xii+100pp. 6½ x 9¼. 42876-1

THE THIRTEEN BOOKS OF EUCLID'S ELEMENTS, translated with introduction and commentary by Thomas L. Heath. Definitive edition. Textual and linguistic notes, mathematical analysis. 2,500 years of critical commentary. Unabridged. 1,414pp. 5⅜ x 8½. Three-vol. set. Vol. I: 60088-2 Vol. II: 60089-0 Vol. III: 60090-4

GEOMETRY OF COMPLEX NUMBERS, Hans Schwerdtfeger. Illuminating, widely praised book on analytic geometry of circles, the Moebius transformation, and two-dimensional non-Euclidean geometries. 200pp. 5⅝ x 8¼. 63830-8

DIFFERENTIAL GEOMETRY, Heinrich W. Guggenheimer. Local differential geometry as an application of advanced calculus and linear algebra. Curvature, transformation groups, surfaces, more. Exercises. 62 figures. 378pp. 5⅜ x 8½. 63433-7

CURVATURE AND HOMOLOGY: Enlarged Edition, Samuel I. Goldberg. Revised edition examines topology of differentiable manifolds; curvature, homology of Riemannian manifolds; compact Lie groups; complex manifolds; curvature, homology of Kaehler manifolds. New Preface. Four new appendixes. 416pp. 5⅜ x 8½. 40207-X

# History of Math

THE WORKS OF ARCHIMEDES, Archimedes (T. L. Heath, ed.). Topics include the famous problems of the ratio of the areas of a cylinder and an inscribed sphere; the measurement of a circle; the properties of conoids, spheroids, and spirals; and the quadrature of the parabola. Informative introduction. clxxxvi+326pp; supplement, 52pp. 5⅜ x 8½. 42084-1

A SHORT ACCOUNT OF THE HISTORY OF MATHEMATICS, W. W. Rouse Ball. One of clearest, most authoritative surveys from the Egyptians and Phoenicians through 19th-century figures such as Grassman, Galois, Riemann. Fourth edition. 522pp. 5⅜ x 8½. 20630-0

THE HISTORY OF THE CALCULUS AND ITS CONCEPTUAL DEVELOPMENT, Carl B. Boyer. Origins in antiquity, medieval contributions, work of Newton, Leibniz, rigorous formulation. Treatment is verbal. 346pp. 5⅜ x 8½. 60509-4

THE HISTORICAL ROOTS OF ELEMENTARY MATHEMATICS, Lucas N. H. Bunt, Phillip S. Jones, and Jack D. Bedient. Fundamental underpinnings of modern arithmetic, algebra, geometry, and number systems derived from ancient civilizations. 320pp. 5⅜ x 8½. 25563-8

A HISTORY OF MATHEMATICAL NOTATIONS, Florian Cajori. This classic study notes the first appearance of a mathematical symbol and its origin, the competition it encountered, its spread among writers in different countries, its rise to popularity, its eventual decline or ultimate survival. Original 1929 two-volume edition presented here in one volume. xxviii+820pp. 5⅜ x 8½. 67766-4

GAMES, GODS & GAMBLING: A History of Probability and Statistical Ideas, F. N. David. Episodes from the lives of Galileo, Fermat, Pascal, and others illustrate this fascinating account of the roots of mathematics. Features thought-provoking references to classics, archaeology, biography, poetry. 1962 edition. 304pp. 5⅜ x 8½. (Available in U.S. only.) 40023-9

OF MEN AND NUMBERS: The Story of the Great Mathematicians, Jane Muir. Fascinating accounts of the lives and accomplishments of history's greatest mathematical minds–Pythagoras, Descartes, Euler, Pascal, Cantor, many more. Anecdotal, illuminating. 30 diagrams. Bibliography. 256pp. 5⅜ x 8½. 28973-7

HISTORY OF MATHEMATICS, David E. Smith. Nontechnical survey from ancient Greece and Orient to late 19th century; evolution of arithmetic, geometry, trigonometry, calculating devices, algebra, the calculus. 362 illustrations. 1,355pp. 5⅜ x 8½. Two-vol. set. Vol. I: 20429-4 Vol. II: 20430-8

A CONCISE HISTORY OF MATHEMATICS, Dirk J. Struik. The best brief history of mathematics. Stresses origins and covers every major figure from ancient Near East to 19th century. 41 illustrations. 195pp. 5⅜ x 8½. 60255-9

# Physics

OPTICAL RESONANCE AND TWO-LEVEL ATOMS, L. Allen and J. H. Eberly. Clear, comprehensive introduction to basic principles behind all quantum optical resonance phenomena. 53 illustrations. Preface. Index. 256pp. 5⅜ x 8½. 65533-4

QUANTUM THEORY, David Bohm. This advanced undergraduate-level text presents the quantum theory in terms of qualitative and imaginative concepts, followed by specific applications worked out in mathematical detail. Preface. Index. 655pp. 5⅜ x 8½. 65969-0

ATOMIC PHYSICS: 8th edition, Max Born. Nobel laureate's lucid treatment of kinetic theory of gases, elementary particles, nuclear atom, wave-corpuscles, atomic structure and spectral lines, much more. Over 40 appendices, bibliography. 495pp. 5⅜ x 8½. 65984-4

A SOPHISTICATE'S PRIMER OF RELATIVITY, P. W. Bridgman. Geared toward readers already acquainted with special relativity, this book transcends the view of theory as a working tool to answer natural questions: What is a frame of reference? What is a "law of nature"? What is the role of the "observer"? Extensive treatment, written in terms accessible to those without a scientific background. 1983 ed. xlviii+172pp. 5⅜ x 8½. 42549-5

AN INTRODUCTION TO HAMILTONIAN OPTICS, H. A. Buchdahl. Detailed account of the Hamiltonian treatment of aberration theory in geometrical optics. Many classes of optical systems defined in terms of the symmetries they possess. Problems with detailed solutions. 1970 edition. xv+360pp. 5⅜ x 8½. 67597-1

PRIMER OF QUANTUM MECHANICS, Marvin Chester. Introductory text examines the classical quantum bead on a track: its state and representations; operator eigenvalues; harmonic oscillator and bound bead in a symmetric force field; and bead in a spherical shell. Other topics include spin, matrices, and the structure of quantum mechanics; the simplest atom; indistinguishable particles; and stationary-state perturbation theory. 1992 ed. xiv+314pp. 6⅛ x 9¼. 42878-8

LECTURES ON QUANTUM MECHANICS, Paul A. M. Dirac. Four concise, brilliant lectures on mathematical methods in quantum mechanics from Nobel Prize–winning quantum pioneer build on idea of visualizing quantum theory through the use of classical mechanics. 96pp. 5⅜ x 8½. 41713-1

THIRTY YEARS THAT SHOOK PHYSICS: The Story of Quantum Theory, George Gamow. Lucid, accessible introduction to influential theory of energy and matter. Careful explanations of Dirac's anti-particles, Bohr's model of the atom, much more. 12 plates. Numerous drawings. 240pp. 5⅜ x 8½. 24895-X

ELECTRONIC STRUCTURE AND THE PROPERTIES OF SOLIDS: The Physics of the Chemical Bond, Walter A. Harrison. Innovative text offers basic understanding of the electronic structure of covalent and ionic solids, simple metals, transition metals and their compounds. Problems. 1980 edition. 582pp. 6⅛ x 9¼. 66021-4